An Introduction to Mathematical Methods of Physics

An Introduction to Mathematical Methods of Physics

Lorella M. Jones
University of Illinois at Urbana-Champaign

Portions of this text have been adapted from
Mathematical Methods of Physics
by Jon Mathews and R. L. Walker,
California Institute of Technology

The Benjamin/Cummings Publishing Company, Inc.
Menlo Park, California / Reading, Massachusetts
London / Amsterdam / Don Mills, Ontario / Sydney

Library of Congress Cataloging in Publication Data

Jones, Lorella M. 1943–
An introduction to Mathematical methods of physics.

"Portions of this text have been adapted from Mathematical methods of physics by Jon Mathews and R. L. Walker."
Bibliography: p.
Includes index.
1. Mathematical physics. I. Mathews, Jon. Mathematical methods of physics. II. Title.
QA401.J66 530.1'5 78-57377
ISBN 0-8053-5130-2

 Printed in the United States of America.
Published simultaneously in Canada.
Library of Congress Catalog Card No. 78-57377

ISBN 0-8053-5130-2
CDEFGHIJKL-MA-89876543210

The Benjamin/Cummings Publishing Company, Inc.
2727 Sand Hill Road
Menlo Park, California 94025

Preface

This book covers the material for a course in mathematical physics at the advanced undergraduate level. This course has evolved over a period of years at the University of Illinois in an attempt to provide a good transition from the students' mathematical training (which typically contains one semester of differential equations) to the relative sophistication expected in graduate level quantum mechanics and electromagnetism courses. The general philosophy adopted has been to maintain the emphasis on problem solving of the Mathews and Walker book (*Mathematical Methods of Physics* by J. Mathews and R. Walker, W. A. Benjamin, Inc.) while restricting the topics covered and supplying a large amount of background material that is needed at the undergraduate level.

The first half of the course focuses on the solutions of ordinary and partial differential equations encountered in physics and their applications to boundary value problems (Chapters 1, 2, 3, and 4 and Sections 12-1 through 12-4). We begin by reviewing the basic ideas of vector spaces and matrices to provide a background for the theory of expansion in orthogonal sets; then we discuss basic techniques of dealing with ordinary differential equations. A great deal of stress is placed on the relation between Sturm-Liouville operators and Hermitian matrices because we have found (in teaching graduate quantum mechanics courses) that most students have not been exposed to this in their mathematics course. The bra-ket notation commonly used in advanced quantum courses is used throughout this section of the book. Green's functions and perturbation theory are developed first with matrix equations. We conclude this section (in Chapter 12) with the technique of separation of variables in partial differential equations.

The second half has generally been billed as "Applications of Complex Variables in Physics"; it consists of Chapters 7, 8, 9, 10, and the remainder of Chapter 12. No prior knowledge of complex variables is assumed; in fact, students who have completed an undergraduate course in complex variable theory prior to this course do not stay ahead of the others for very long. Basic calculational aspects such as evaluation of integrals, transform techniques, and conformal mapping are discussed in Chapters 8 and 9, while other more sophisticated applications are separated in Chapter 10. The material on complex variables (Chapters 7–10) is, to a great extent, independent of the material in the previous chapters, and it could form the basis for a separate course, if desired.

Chapters 5 and 6 contain material on infinite series and integrals which may be new to some students; this material may be interjected into the course as needed. Chapter 11 ("Integral Equations") is a nice way to wrap up either the differential equations section or the whole course.

We should stress that the course is not a mathematics course in the classical sense; rather it teaches methods of thinking about mathematics which are useful to physicists. The student is frequently referred to other works for more rigorous treatments and, in some cases, for proofs. Also we do not attempt to cover many of the modern topics, such as statistics, group theory, and non-linear mathematics, in the graduate curriculum. Instead, our course is directed students will study these topics, as well as more complicated applications of linear mathematics in the graduate curriculum. Instead, our course is directed to students who are wondering about the general framework behind methods used in their electromagnetism or quantum mechanics class; we hope to satisfy their curiosity and give them a basis for further explorations in the field.

ACKNOWLEDGMENT

When this book was first outlined nine years ago, it was perceived as an adaptation for undergraduates of the well-known graduate text by Mathews and Walker. Because of this, a substantial number of the elementary problems and examples from the original book are included. Despite the shift in emphasis, this material still plays an important part in the text. I would like to thank Jon Mathews and Bob Walker for consenting to this arrangement and for their helpful comments and criticisms over the intervening years.

Lorella M. Jones

Contents

CHAPTER 1 Vectors and Matrices

Many of the practical calculations in physics involve large amounts of algebra. These seemingly tedious computations can usually be made more transparent and interesting by grouping quantities according to their role in the manipulations. The language of vectors and linear operators provides a convenient method for such grouping.

In this chapter we explain the terminology of linear vector spaces and linear operators. These abstract quantities are often treated in matrix form; a given operator will be represented by a different matrix in each coordinate system. Hence the algebraic properties of matrices and their behavior under coordinate transformations are extremely important in practical work, and we discuss this behavior in some detail (Sections 1-2 and 1-3).

It frequently happens that a complicated algebra problem may be reduced to that of finding the eigenvalues of some matrix. We treat these problems in Section 1-4, with emphasis on Hermitian matrices because of their importance in physics.

1-1 LINEAR VECTOR SPACES

A linear vector space is a set of objects (vectors) **a**, **b**, **c**, ... on which two operations are defined:

1. Addition, which is commutative and associative

$$\begin{aligned} \mathbf{a} + \mathbf{b} = \mathbf{c} = \mathbf{b} + \mathbf{a} \\ (\mathbf{a} + \mathbf{b}) + \mathbf{c} = \mathbf{a} + (\mathbf{b} + \mathbf{c}) \end{aligned} \tag{1-1}$$

2. Multiplication by a scalar (any complex number), which is distributive and associative; that is,

$$\lambda(\mathbf{a} + \mathbf{b}) = \lambda\mathbf{a} + \lambda\mathbf{b}$$
$$\lambda(\mu\mathbf{a}) = (\lambda\mu)\mathbf{a} \tag{1-2}$$
$$(\lambda + \mu)\mathbf{a} = \lambda\mathbf{a} + \mu\mathbf{a}$$

In addition, we assume a null vector **0** exists such that for all **a**

$$\mathbf{a} + \mathbf{0} = \mathbf{a} \tag{1-3}$$

that multiplication by the scalar 1 leaves every vector unchanged,

$$1\mathbf{a} = \mathbf{a}$$

and finally, that for every **a**, a vector $-\mathbf{a}$ exists such that

$$\mathbf{a} + (-\mathbf{a}) = \mathbf{0} \tag{1-4}$$

Example

A simple physical example of a vector space is ordinary three-dimensional space. With each point (x, y, z) we associate a vector, drawn to that point from the origin. The reader should already be familiar with the addition of vectors in this space, and multiplication of a vector by a scalar number; these operations produce new vectors in the same space.

One generalization of this case that often occurs in physics is the simultaneous description of two points in three-dimensional space. For example, we might wish to describe the positions of two projectiles located at (x_1, y_1, z_1) and (x_2, y_2, z_2). For this problem we would use a six-dimensional space with coordinates $(x_1, y_1, z_1, x_2, y_2, z_2)$. This space is called the direct product of the two independent three-dimensional spaces.

As a more abstract example, the set (C^∞) of all infinitely differentiable functions on the interval $(-1, 1)$ also constitutes a vector space: the sum $f_1(x) + f_2(x)$ and the product $cf(x)$ are also infinitely differentiable functions.

A set of vectors **a**, **b**, ..., **u** is said to be *linearly independent* provided no equation

$$\lambda\mathbf{a} + \mu\mathbf{b} + \cdots + \sigma\mathbf{u} = 0 \tag{1-5}$$

holds except the trivial one with $\lambda = \mu = \cdots = \sigma = 0$.

If in a particular vector space there exist n linearly independent vectors but no set of $n + 1$ linearly independent ones, the space is said to be *n-dimensional.* For example (x, y, z) space is three-dimensional, whereas the space of functions on $(-1, 1)$ is infinite-dimensional. (This will become clearer after you have read Chapter 4.)

Let $\mathbf{e}_1, \mathbf{e}_2, \ldots, \mathbf{e}_n$ be a set of n linearly independent vectors in an n-dimensional vector space. Then if $\mathbf{x}$ is an arbitrary vector in the space, there exists a relation

$$\lambda\mathbf{e}_1 + \mu\mathbf{e}_2 + \cdots + \sigma\mathbf{e}_n + \tau\mathbf{x} = 0$$

with not all the constants equal to zero, and in particular $\tau \neq 0$. Thus, $\mathbf{x}$ can be written as a linear combination of the $\mathbf{e}_i$:

$$\mathbf{x} = \sum_{i=1}^{n} x_i \mathbf{e}_i \tag{1-6}$$

The vectors $\mathbf{e}_i$ are said to form a *basis*, or *coordinate system*, and the numbers x_i are the *components* of x in this system. The $\mathbf{e}_i$ are called *base vectors*. The fact that an arbitrary vector $\mathbf{x}$ can be written as a linear combination of the $\mathbf{e}_i$ is often expressed thus: The set of base vectors $\mathbf{e}_i$ is *complete*. The idea of completeness is very important, and will occur several times later in this text.

Generally it is convenient to choose base vectors that correspond to some physical symmetry of the problem at hand. Consider a simple mechanical system composed of two beads sliding on a rod, and connected with a spring (Figure 1-1). We could set the origin of the coordinate system at 0, and describe

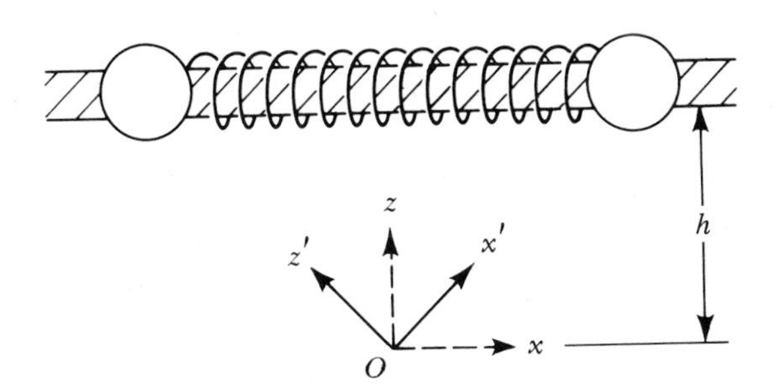

FIGURE 1-1 System with motion in only one dimension

the positions of the beads by the coordinates in the (tilted) system $(x_1', y_1', z_1', x_2', y_2', z_2')$. However, it is clear that the description of the motion will be simplest if we choose the unprimed system, where $(x_1, y_1, z_1, x_2, y_2, z_2) = (x_1, 0, h, x_2, 0, h)$. Then the only coordinates that change with time under the vibrations of the spring are x_1 and x_2 and we can neglect the other four coordinates. (Life becomes even simpler if we move the origin up so that $h = 0$.)

Note that the constraint forcing the beads to remain on the rod tells us immediately that only the change with time of two coordinates (the position of the beads along the rods) is important. We can save a lot of time by realizing at the beginning that a two-dimensional space will give a complete description of the motion.

PROBLEM

1-1 Throughout this chapter we allude to the set (C^∞) of all infinitely differentiable functions on the interval $(-1, 1)$. Convince yourself that this is indeed a vector space and that it has an infinite number of dimensions. (*Hint*: think about expanding the functions in terms of sin $n\pi x$ and cos $n\pi x$, or in a Taylor's series about the origin.)

1-2 LINEAR OPERATIONS

Next we consider a *linear vector function* of a vector, that is, a rule that associates with every vector $\mathbf{x}$ a vector $\boldsymbol{\phi}(\mathbf{x})$, in a linear way,

$$\boldsymbol{\phi}(\lambda\mathbf{a} + \mu\mathbf{b}) = \lambda\boldsymbol{\phi}(\mathbf{a}) + \mu\boldsymbol{\phi}(\mathbf{b}) \tag{1-7}$$

Because of this linearity, it is sufficient to know the n vectors $\boldsymbol{\phi}(\mathbf{e}_i)$. These may be conveniently described in terms of the basis $\mathbf{e}_i$; that is,

$$\boldsymbol{\phi}(\mathbf{e}_i) = \sum_{j=1}^{n} A_{ji}\mathbf{e}_j \tag{1-8}$$

where A_{ji} is thus the jth component of the vector $\boldsymbol{\phi}(\mathbf{e}_i)$.

If we now consider an arbitrary vector $\mathbf{x}$, and call $\boldsymbol{\phi}(\mathbf{x}) = \mathbf{y}$, we have

$$\mathbf{y} = \boldsymbol{\phi}\left(\sum_i x_i\mathbf{e}_i\right) = \sum_i x_i \sum_j A_{ji}\mathbf{e}_j$$

Thus the components of $\mathbf{x}$ and $\mathbf{y}$ are related by

$$y_j = \sum_i A_{ji}x_i \tag{1-9}$$

We may also describe the above relations in another way, saying that the association of $\mathbf{y}$ with $\mathbf{x}$ is accomplished by a *linear operator* $\mathscr{A}$ operating on $\mathbf{x}$. Symbolically,

$$\mathbf{y} = \mathscr{A}\mathbf{x} \tag{1-10}$$

Then the number A_{ji} are the *components* of the linear operator $\mathscr{A}$ (or the vector function $\boldsymbol{\phi}$) in the coordinate system $\mathbf{e}_i$. In particular, from (1-8), A_{ji} is the jth component of the vector $\mathscr{A}\mathbf{e}_i = \boldsymbol{\phi}(\mathbf{e}_i)$.

Just as with vectors, a linear operator often has a physical meaning that is independent of a specific coordinate system, and may be described without reference to a specific system.

Example

We may describe a simple rotation operator $\mathscr{R}$ acting on three dimensional space by the rule: $\mathscr{R}\mathbf{x}$ is the vector obtained by rotating $\mathbf{x}$ by 30° about a vertical axis in the positive (right-hand) sense. This operation is illustrated in Figure 1-2. The coordinate system in the figure is not necessary for the

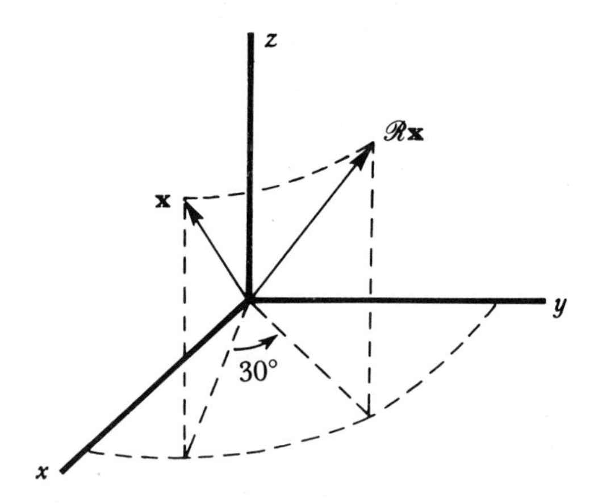

FIGURE 1-2 The result of a simple rotation operator

definition of the operator, since "vertical" has a meaning independent of coordinate systems. However, we could also define $\mathscr{R}$ by giving its components in the system shown.

The sum and product of linear operators and the product of an operator and a scalar number may be defined by the relations

$$(\mathscr{A} + \mathscr{B})\mathbf{x} = \mathscr{A}\mathbf{x} + \mathscr{B}\mathbf{x} \tag{1-11}$$

$$(\mathscr{A}\mathscr{B})\mathbf{x} = \mathscr{A}(\mathscr{B}\mathbf{x}) \tag{1-12}$$

$$(\lambda\mathscr{A})\mathbf{x} = \lambda(\mathscr{A}\mathbf{x}) \tag{1-13}$$

In general $\mathscr{A}\mathscr{B} \neq \mathscr{B}\mathscr{A}$, but if $\mathscr{A}\mathscr{B}\mathbf{x} = \mathscr{B}\mathscr{A}\mathbf{x}$ for every $\mathbf{x}$, $\mathscr{A}$ and $\mathscr{B}$ are said to *commute*.

The null and identity operators 0 and 1 have obvious meanings, namely,

$$0\mathbf{x} = \mathbf{0} \qquad \text{and} \qquad 1\mathbf{x} = \mathbf{x} \tag{1-14}$$

for every vector $\mathbf{x}$ in our space. Two operators $\mathscr{A}$ and $\mathscr{B}$ are equal if $\mathscr{A}\mathbf{x} = \mathscr{B}\mathbf{x}$ for every vector $\mathbf{x}$. Finally, if an operator $\mathscr{A}^{-1}$ exists with the properties

$$\mathscr{A}\mathscr{A}^{-1} = \mathscr{A}^{-1}\mathscr{A} = 1 \tag{1-15}$$

$\mathscr{A}^{-1}$ is called the *inverse* of $\mathscr{A}$. Operators that have an inverse are said to be *nonsingular*.

In the above discussion, specifically in Equation 1-8, we have assumed that the vector $\phi(\mathbf{x}) = \mathscr{A}\mathbf{x}$ is in the same space as the vector $\mathbf{x}$. This need not be so. The only change required in the above analysis if $\phi(\mathbf{x})$ is in a different space from

$\mathbf{x}$ is to express $\phi(\mathbf{e}_i)$ in terms of a basis $\mathbf{f}_i$ in the ϕ space, so that (1-8) becomes

$$\phi(\mathbf{e}_i) = \sum_j A_{ji}\mathbf{f}_j \tag{1-16}$$

Then the components A_{ji} of the operator $\mathscr{A}$ refer to the two bases $\mathbf{e}_i$ and $\mathbf{f}_i$. It is clear, furthermore, that the two spaces may have different numbers of dimensions; if so, the operator $\mathscr{A}$ cannot have an inverse.

Example

Consider the space of three-dimensional position vectors $\mathbf{x}$ and a conventional xyz Cartesian coordinate system such as is shown in Figure 1-2. We define a projection operator $\mathscr{P}$ such that $\mathscr{P}\mathbf{x}$ is the projection of $\mathbf{x}$ on the xy-plane, that is, it has the same x- and y-coordinates as $\mathbf{x}$ but zero z-component. In this case, the two-dimensional space of vectors $\mathscr{P}\mathbf{x}$ is a subspace of the original space of vectors. Notice that this operator $\mathscr{P}$ does not have an inverse, because we cannot get back to a particular point in the third dimension from each point in the plane.

Example

Let us return to the system of two beads connected by a spring in Figure 1-1. We write the equations for the forces on the beads due to extension of the spring from its normal length a:

$$k(x_2 - x_1 - a) = -m\ddot{x}_2$$

$$k(x_2 - x_1 - a) = m\ddot{x}_1$$

Physically only deviations from the equilibrium position are interesting, so we use the new coordinate $x_2' = x_2 - a$. The equations simplify to

$$-k(x_2' - x_1) = m\ddot{x}_2'$$

$$k(x_2' - x_1) = m\ddot{x}_1$$

These equations define a linear map relating the position vector (x_1, x_2') to the force vector $m(\ddot{x}_1, \ddot{x}_2')$. The two spaces—position and acceleration—are quite different, although they happen to have the same dimension.

MATRICES

The numbers A_{ij} introduced above can be displayed as a *matrix*

$$A = \begin{pmatrix} A_{11} & A_{12} & \dots & A_{1n} \\ A_{21} & A_{22} & \dots & A_{2n} \\ \vdots & & & \\ A_{m1} & A_{m2} & \dots & A_{mn} \end{pmatrix} \tag{1-17}$$

The numbers A_{ij} are called the elements of the matrix. Matrices are not the only possible way to represent the action of a linear operator. For example, operators on function spaces may be represented by differentiation or integration. [The operators d/dx and $(x^2 - 1)d^2/dx^2 + 2x\,d/dx$ act linearly on C^∞.] The action of such differential operators will be considered in more detail in Chapters 3 and 4; this section will concentrate on the properties of matrix representations.

In general, a matrix is a rectangular array of numbers that obeys certain rules of algebra. If the linear operators $\mathscr{A}$ and $\mathscr{B}$ are represented by the matrices A and B, respectively, we would like the operator $\lambda\mathscr{A} + \mu\mathscr{B}$ to be represented by the matrix $\lambda A + \mu B$. This allows us to find the rules that matrices must follow if they are to represent linear operators.

For example, the j-components in a given coordinate system of the operator equations (1-11), (1-12), and (1-13) are

for equation (1-11): $\sum_i (A + B)_{ji} x_i = \sum_i A_{ji} x_i + \sum_i B_{ji} x_i$

for equation (1-12): $\sum_i (AB)_{ji} x_i = \sum_l A_{jl}(Bx)_l = \sum_{li} A_{jl} B_{li} x_i$

for equation (1-13): $\sum_i (\lambda A)_{ji} x_i = \lambda \sum_i A_{ji} x_i$

Since $\mathbf{x}$ is arbitrary, these give immediately the rules for adding and multiplying matrices, and for multiplication by a constant:

$$(A + B)_{ji} = A_{ji} + B_{ji} \tag{1-18}$$

$$(AB)_{ji} = \sum_l A_{jl} B_{li} \tag{1-19}$$

$$(\lambda A)_{ji} = \lambda A_{ji} \tag{1-20}$$

The least trivial of these is (1-19), about which we note the following:

1. Matrix multiplication is not, in general, commutative: AB need not equal BA.

2. The ji element of AB is the sum of products of elements from the jth row of A and the ith column of B.

3. The number of columns in A must equal the number of rows in B if the definition of the product (1-19) is to make sense.

The elements of the null matrix 0 are all zero, and the elements of the identity 1 are δ_{ij}, the Kronecker delta, defined by

$$\delta_{ij} = 1 \quad \text{if} \quad i = j; \qquad \delta_{ij} = 0 \quad \text{if} \quad i \neq j \tag{1-21}$$

That is, 1 is *diagonal* with diagonal elements unity.

For an inverse matrix A^{-1} we have the conditions

$$A^{-1}A = AA^{-1} = 1 \tag{1-22}$$

corresponding to the operator equations of the same form. If A is an $n \times m$ matrix (one with n rows and m columns), then A^{-1} must be an $m \times n$ matrix so that $A^{-1}A$ and AA^{-1} are defined (rule 3 above). The mn entries in A^{-1} must satisfy m^2 conditions in order that $A^{-1}A = 1$, and n^2 conditions in order that $AA^{-1} = 1$. This is possible only if $mn \geq n^2$ and $mn \geq m^2$; that is, if $n = m$. Thus only square matrices can have inverses.

However, not all square matrices have inverses. The projection operator described in the example above can be represented by a square matrix,

$$\begin{pmatrix} 1 & 0 & 0 \\ 0 & 1 & 0 \\ 0 & 0 & 0 \end{pmatrix}$$

but we saw by physical reasoning that the inverse of this operator does not exist. Abstracting from this example, we see that the dimension of the *actual* image space must be the same as that of the original space if an inverse operator is to be possible; if the actual image is a small subspace of the original space, then we cannot form an inverse. We need some criterion to tell when the image of a linear transformation has the same dimension as the original space. This is provided by the determinant of the square matrix representing the transformation. If $\varepsilon_{ijk\ldots p}$ is the completely antisymmetric tensor in n dimensions, then the determinant of an $n \times n$ matrix Q is given by

$$\det Q = \sum_{ijk\ldots p} \varepsilon_{ijk\ldots p} Q_{i1} Q_{j2} \ldots Q_{pn} \tag{1-23}$$

Example

In a two-dimensional space, consider the number ε_{ij} such that $\varepsilon_{ij} = -\varepsilon_{ji}$ and $\varepsilon_{12} = +1$. Clearly these four numbers must be $\varepsilon_{11} = 0 = \varepsilon_{22}$, $\varepsilon_{12} = +1 = -\varepsilon_{21}$. For a 2×2 matrix A_{ij}, the sum

$$\det A = \sum_{ij} \varepsilon_{ij} A_{i1} A_{j2} = A_{11}A_{22} - A_{21}A_{12}$$

is called the determinant of A.

For a three-dimensional space, the completely antisymmetric tensor ε_{ijk} is defined by $\varepsilon_{ijk} = -\varepsilon_{jik} = -\varepsilon_{ikj}$ and $\varepsilon_{123} = +1$. The determinant of a 3×3 matrix is defined by

$$\sum_{ijk} \varepsilon_{ijk} B_{i1} B_{j2} B_{k3} = \det B$$

Notice that the determinant will vanish if one of the columns is a linear combination of the others, e.g., $B_{i1} = \alpha B_{i2} + \beta B_{i3}$, for

$$\sum_{ijk} \varepsilon_{ijk} B_{i2} B_{j2} B_{k3} = \sum_{ijk} \varepsilon_{jik} B_{j2} B_{i2} B_{k3} \qquad \textit{(relabeling)}$$

$$= -\sum_{ijk} \varepsilon_{ijk} B_{j2} B_{i2} B_{k3} \qquad \textit{(antisymmetry)}$$

$$= -\sum_{ijk} \varepsilon_{ijk} B_{i2} B_{j2} B_{k3}$$

Hence $\sum_{ijk} \varepsilon_{ijk} B_{i2} B_{j2} B_{k3} = 0$, and similarly $\sum_{ijk} \varepsilon_{ijk} B_{i3} B_{j2} B_{k3} = 0$.

In general, if the image space has fewer dimensions than the original space, the determinant of the transformation matrix is zero. To prove this, we note that the column vectors given by columns of the matrix, Q_{i1}, Q_{i2}, etc., are the image vectors of the basis vectors in the original space. These n vectors will therefore span the image space (i.e., every vector in the image space can be represented as a linear combination of these). If the actual dimension of the final space is smaller than n, then some of these vectors will be linearly dependent on the others. Hence one or more of the columns Q_{i1}, etc., will be a linear combination of the other columns. It is easy to demonstrate that the complete antisymmetry of the tensor $\varepsilon_{ijk\ldots p}$ will then force the determinant to vanish.

If n is a very large number, computation of the determinant by direct application of (1-23) can be cumbersome. Some simplification is achieved by a procedure called *expansion in cofactors*. The *cofactor* of a matrix entry X_{ij} is defined to be $(-1)^{i+j}$ times the determinant of the matrix obtained by eliminating the ith row and the jth column from X. In general (see Problem 1-4 for a specific case)

$$\det X = \sum_i X_{ij} (\text{cofactor of } X_{ij}) \tag{1-24}$$

In terms of the cofactors, an explicit equation for the inverse of a square matrix can be given:

$$(A^{-1})_{ij} = \frac{\text{cofactor of } A_{ji}}{\det A} \tag{1-25}$$

provided that the determinant of the matrix A is non-zero.

A number of matrices closely related to a given matrix A are defined in Table 1-1.

Some further definitions are the following:

A is *real*	if A^*	$= A$
symmetric	*if* $\tilde{A}$	$= A$
antisymmetric	if $\tilde{A}$	$= -A$

$$
\begin{array}{lll}
A \text{ is } \textit{Hermitian} & \text{if } A^\dagger = A & \\
\qquad \textit{orthogonal} & \text{if } A^{-1} = \tilde{A} & \\
\qquad \textit{unitary} & \text{if } A^{-1} = A^\dagger & \\
\qquad \textit{diagonal} & \text{if } A_{ij} = 0 & \text{for } i \neq j \\
\qquad \textit{idempotent} & \text{if } A^2 = A &
\end{array}
$$

These terms will be used throughout the rest of the book.

The *trace* or *spur* of a square matrix A is the sum of its diagonal elements:

$$\text{Tr } A = \sum_i A_{ii}$$

TABLE 1-1

Matrices Related to a Matrix A

Matrix	*Components*	*Example*
A	A_{ij}	$\begin{pmatrix} 1 & i \\ 1+i & 2 \end{pmatrix}$
Transpose $\tilde{A}$	$(\tilde{A})_{ij} = A_{ji}$	$\begin{pmatrix} 1 & 1+i \\ i & 2 \end{pmatrix}$
Complex conjugate A^*	$(A^*)_{ij} = (A_{ij})^*$	$\begin{pmatrix} 1 & -i \\ 1-i & 2 \end{pmatrix}$
Hermitian conjugate or adjoint $A^\dagger = (\tilde{A})^*$	$(A^\dagger)_{ij} = (A_{ji})^*$	$\begin{pmatrix} 1 & 1-i \\ -i & 2 \end{pmatrix}$
Inverse A^{-1} (square matrices with determinant $\neq 0$ only)	$(A^{-1})_{ij} = \dfrac{\text{cofactor of } A_{ji}}{\det A}$	$\dfrac{1}{(3-i)}\begin{pmatrix} 2 & -i \\ -1-i & 1 \end{pmatrix}$

We list below a few relations involving products of square matrices, which are frequently useful and may be easily verified by the student:

$$(ABC)^{-1} = C^{-1}B^{-1}A^{-1} \qquad \text{(provided all inverses exist)}$$

$$(\widetilde{ABC}) = \tilde{C}\tilde{B}\tilde{A}$$

$$\text{Tr } (AB) = \text{Tr } (BA)$$

$$\det (AB) = (\det A)(\det B) = \det (BA)$$

Finally, we point out that matrices may sometimes be subdivided into submatrices or *blocks* in such a way as to simplify certain algebraic relations (and work).

Example

If we subdivide A and B as follows,

$$A = \left(\begin{array}{cc|ccc} a_{11} & a_{12} & a_{13} & a_{14} & a_{15} \\ a_{21} & a_{22} & a_{23} & a_{24} & a_{25} \\ \hline a_{31} & a_{32} & a_{33} & a_{34} & a_{35} \\ a_{41} & a_{42} & a_{43} & a_{44} & a_{45} \\ a_{51} & a_{52} & a_{53} & a_{54} & a_{55} \end{array}\right) = \begin{pmatrix} A_{11} & A_{12} \\ A_{21} & A_{22} \end{pmatrix}$$

$$B = \left(\begin{array}{ccc|cc} b_{11} & b_{12} & b_{13} & b_{14} & b_{15} \\ b_{21} & b_{22} & b_{23} & b_{24} & b_{25} \\ \hline b_{31} & b_{32} & b_{33} & b_{34} & b_{35} \\ b_{41} & b_{42} & b_{43} & b_{44} & b_{45} \\ b_{51} & b_{52} & b_{53} & b_{54} & b_{55} \end{array}\right) = \begin{pmatrix} B_{11} & B_{12} \\ B_{21} & B_{22} \end{pmatrix}$$

then A and B have the form of 2×2 *block matrices* whose elements A_{ij}, B_{ij} are themselves matrices. We can easily see that the correct product AB results if the block matrices are multiplied according to the usual rule,

$$AB = \begin{pmatrix} A_{11} & A_{12} \\ A_{21} & A_{22} \end{pmatrix} \begin{pmatrix} B_{11} & B_{12} \\ B_{21} & B_{22} \end{pmatrix} = \begin{pmatrix} A_{11}B_{11} + A_{12}B_{21} & A_{11}B_{12} + A_{12}B_{22} \\ A_{21}B_{11} + A_{22}B_{21} & A_{21}B_{12} + A_{22}B_{22} \end{pmatrix}$$

provided all the matrix products in the last matrix make sense. This will be true provided the original division of the columns in the first matrix is the same as the division of the rows in the second. Thus, the above division is not suitable for working out the product BA in terms of the 2×2 block matrices. (Is there a different subdivision of B that will work for both products AB and BA?)

PROBLEMS

1-2 Show that

(a) $(\widetilde{AB}) = \tilde{B}\tilde{A}$

(b) $(AB)^\dagger = B^\dagger A^\dagger$

(c) $A(BC) = (AB)C$

(d) $\text{Tr } ABC = \text{Tr } BCA$

1-3 Prove that if $B_{i1} = p_1^i$, $B_{j2} = p_2^j$, and $B_{k3} = p_3^k$ for simple vectors $\mathbf{p}_1$, $\mathbf{p}_2$, and $\mathbf{p}_3$ in three-dimensional space, then

$$\det B = \mathbf{p}_1 \cdot (\mathbf{p}_2 \times \mathbf{p}_3)$$

1-4 Consider

$$\det B = \sum_{ijk} \varepsilon_{ijk} B_{i1} B_{j2} B_{k3}$$

Prove by explicit computation that this is

$$\det B = B_{13} \det \begin{pmatrix} B_{21} & B_{22} \\ B_{31} & B_{32} \end{pmatrix} - B_{23} \det \begin{pmatrix} B_{11} & B_{12} \\ B_{31} & B_{32} \end{pmatrix}$$
$$+ B_{33} \det \begin{pmatrix} B_{11} & B_{12} \\ B_{21} & B_{22} \end{pmatrix}$$

1-5 Use expansion in cofactors to compute

$$\det \begin{pmatrix} H_{11} & H_{12} & H_{13} & H_{14} \\ H_{21} & H_{22} & H_{23} & H_{24} \\ H_{31} & H_{32} & H_{33} & H_{34} \\ H_{41} & H_{42} & H_{43} & H_{44} \end{pmatrix}$$

1-6 Given a matrix Q, consider the matrix X where

$$X_{ij} = \frac{\text{cofactor } Q_{ji}}{\det Q}$$

For the matrix product XQ, prove that

$$(XQ)_{lm} = (QX)_{lm} = \delta_{lm}$$

Hence $X = Q^{-1}$.

1-7 In the text we found a system of equations to describe the motion of the two beads displayed in Figure 1-1:

$$-k(x_2' - x_1) = m\ddot{x}_2'$$
$$k(x_2' - x_1) = m\ddot{x}_1$$

(a) Write these in the form of a matrix equation

$$\begin{pmatrix} M_{11} & M_{12} \\ M_{21} & M_{22} \end{pmatrix} \begin{pmatrix} x_1 \\ x_2' \end{pmatrix} = \begin{pmatrix} \ddot{x}_1 \\ \ddot{x}_2' \end{pmatrix}$$

What properties does the matrix M have? (i.e., is it orthogonal, antisymmetric, etc?).

(b) Show that a change of coordinates to the set

$$y_1 = x_2' - x_1 \qquad y_2 = x_2' + x_1$$

simplifies the equations, and give the solution of these equations for arbitrary initial positions x_1^0, $x_2'^0$ and velocities $\dot{x}_1^0$, $\dot{x}_2'^0$. Is there any

physical reason why y_1 and y_2 should have simple time dependence?
(c) Define a matrix U such that $Uy = x$. What properties from the list on page 9 does U have?

1-8 A theorem, which is commonly used in problem solving, states: If one has n linear *inhomogeneous* equations in n unknowns, then there is a unique solution only if the determinant of the coefficients of the unknowns is *nonzero*. Explain why this theorem is true.

1-9 Another important theorem states: If one has n linear *homogeneous* equations in n unknowns, then there is a non-trivial solution for the unknowns only if the determinant of the coefficients of the unknowns is *zero*. Why must this be so?

1-3 COORDINATE TRANSFORMATIONS

As we have seen, a linear operator is represented by a matrix in a given coordinate system. The right side of equation (1-9), $y_j = \sum_i A_{ji} x_i$, is a special case of a matrix product Ax, in which the right-hand matrix x (and thus the product y) has a single column. Such matrices are called *column vectors.*

We now ask how the components of vectors and linear operators transform when we change the coordinate system. Consider a new basis $\mathbf{e}'_j$ such that

$$\mathbf{e}_i = \sum_j u_{ji} \mathbf{e}'_j \tag{1-26}$$

Then the n^2 coefficients u_{ij} form the elements of a *transformation matrix* u, which effects the transformation from one system to the other.

Consider an arbitrary vector $\mathbf{x}$ with components x_i and x'_j in the two systems. Then

$$\mathbf{x} = \sum_i x_i \mathbf{e}_i = \sum_i x_i \sum_j u_{ji} \mathbf{e}'_j = \sum_j \left(\sum_i x_i u_{ji} \right) \mathbf{e}'_j = \sum_j x'_j \mathbf{e}'_j \tag{1-27}$$

This transformation of coordinates $x'_j = \sum_i x_i u_{ji}$ is equivalent to the matrix equation

$$x' = ux \tag{1-28}$$

Next we find the transformation law for the components of linear operators by writing the operator equation

$$\mathbf{y} = A\mathbf{x}$$

as matrix equations in the two coordinate systems

$$y = Ax \qquad y' = A'x'$$

Using $y' = uy$ and $x' = ux$, we arrive at

$$uy = A'ux; \qquad y = u^{-1} A'ux$$

Thus the desired transformation is

$$A = u^{-1} A' u; \qquad u A u^{-1} = A' \tag{1-29}$$

This is an example of a *similarity transformation*, which is defined to be a transformation of square matrices of the form

$$A' = S^{-1} A S$$

Any algebraic matrix equation remains unchanged under a similarity transformation; for example, the equation

$$ABC + \lambda D = 0$$

implies

$$S^{-1} A (SS^{-1}) B (SS^{-1}) CS + S^{-1} \lambda D S = 0$$

or

$$A'B'C' + \lambda D' = 0$$

SCALAR PRODUCTS AND UNITARY TRANSFORMATIONS

Coordinate transformations in which the transformation matrix u is orthogonal or unitary play an especially important role in physics, because these transformations preserve the physically important length of a vector. In order to demonstrate this, we must introduce a mathematical definition of length together with some useful related ideas.

First, we consider a *linear scalar function* of a vector. This is a concept related to that of a linear vector function, which was used above to introduce matrices. A linear scalar function is a rule which, for every vector $\mathbf{x}$, defines a scalar $\phi(\mathbf{x})$ in a linear way:

$$\phi(\lambda \mathbf{x} + \mu \mathbf{y}) = \lambda \phi(\mathbf{x}) + \mu \phi(\mathbf{y}) \tag{1-30}$$

If we have a basis $\mathbf{e}_i$, the linear scalar function may be conveniently specified by the n numbers $\alpha_i = \phi(\mathbf{e}_i)$, which we shall call the *components* of the scalar function ϕ. Then,

$$\phi(\mathbf{x}) = \phi\left(\sum_i x_i \mathbf{e}_i\right) = \sum_i \alpha_i x_i \tag{1-31}$$

The expression fits the rule for matrix multiplication if the components α_i form a *row vector*, that is, a matrix having a single row. Then (1-31) is the matrix equation

$$\phi(\mathbf{x}) = \alpha x$$

The function ϕ is a map from vectors to numbers which can be defined completely independently of any coordinate system. Hence the image of a given vector under this map should not change if we change the coordinate system; instead the row vector α_i must transform in such a way as to keep the result the same:

$$\phi(\mathbf{x}) = \alpha x = \alpha' x' = \alpha' u x$$

Thus the requirement of invariance determines the law of transformation of the α_i

$$\alpha = \alpha' u \qquad \alpha' = \alpha u^{-1}$$

or

$$\alpha_i = \sum_j \alpha'_j u_{ji} \equiv \alpha'_j u_{ji} \tag{1-32}$$

[We shall often use the *summation convention*; that is, if an index is repeated on the same side of an equation, it is to be summed over. In equation (1-32) the index j is summed.]

Notice that the components α_i transform by a different rule than the $x_i(x'_j = u_{ji}x_i)$, but by the same rule as the basis vectors $(\mathbf{e}_i = u_{ji}\mathbf{e}'_j)$. One says that α_i and x_i transform *contragrediently* to each other, and that α_i and $\mathbf{e}_i$ transform *cogrediently* to each other. The linear scalar functions may be thought of as forming a vector space that is distinct from the space of the original vectors, but which has the same dimensionality. It is sometimes called the *dual space*.

Example

Consider the dot product between two vectors in three-dimensional space, $\mathbf{x} \cdot \mathbf{y} = |\mathbf{x}|\,|\mathbf{y}| \cos\theta$ where θ is the angle between the two vectors and $|\mathbf{x}|$ is the usual length. If we hold the vector $\mathbf{x}$ fixed, and consider its dot products with all the other vectors in the space, we see that this defines a linear scalar function on all the other vectors

$$\begin{aligned}\phi_{\mathbf{x}}(\lambda\mathbf{y}_1 + \mu\mathbf{y}_2) &= \mathbf{x} \cdot (\lambda\mathbf{y}_1 + \mu\mathbf{y}_2) = \lambda\mathbf{x} \cdot \mathbf{y}_1 + \mu\mathbf{x} \cdot \mathbf{y}_2 \\ &= \lambda\phi_{\mathbf{x}}(\mathbf{y}_1) + \mu\phi_{\mathbf{x}}(\mathbf{y}_2)\end{aligned}$$

Since we have such a function for each vector $\mathbf{x}$, the space of all such functions is clearly the same size as the space of all vectors $\mathbf{x}$.

In the above example, we used our knowledge of the physical dot product to assign a member of the dual space $\phi_{\mathbf{x}}(\ )$ to every vector $\mathbf{x}$. The length squared of the vector in this case is $\phi_{\mathbf{x}}(\mathbf{x})$, just a special case of the dot product. If we can generalize the idea of a dot product (i.e., assignment of a scalar to any pair of vectors), we will therefore be able both to formalize the notion of length, and to obtain a more general rule for relating dual space vectors to the original vectors. This leads us to the general definition of scalar product.

For every pair of vectors $\mathbf{a}$ and $\mathbf{b}$, we define a *scalar product*

$$\mathbf{a} \cdot \mathbf{b} \qquad \text{or} \qquad (\mathbf{a}, \mathbf{b})$$

as a scalar function of the two vectors **a**, **b** with the following properties:

$$1.\ \mathbf{a}\cdot\mathbf{b} = (\mathbf{b}\cdot\mathbf{a})^* \tag{1-33}$$

$$2.\ \mathbf{a}\cdot(\lambda\mathbf{b} + \mu\mathbf{c}) = \lambda\mathbf{a}\cdot\mathbf{b} + \mu\mathbf{a}\cdot\mathbf{c} \tag{1-34}$$

$$3.\ \mathbf{a}\cdot\mathbf{a} \geq 0, \text{ and } \mathbf{a}\cdot\mathbf{a} = 0 \text{ implies } \mathbf{a} = 0 \tag{1-35}$$

Two vectors whose scalar product is zero are said to be orthogonal, and the *length* of a vector **a** is defined to be $|\mathbf{a}\cdot\mathbf{a}|^{1/2}$.

Notice that properties (1-33) and (1-34) imply that the scalar product $\mathbf{a}\cdot\mathbf{b}$ is *antilinear* (or conjugate linear) in its first argument, that is,

$$(\lambda\mathbf{a} + \mu\mathbf{b})\cdot\mathbf{c} = \lambda^*\mathbf{a}\cdot\mathbf{c} + \mu^*\mathbf{b}\cdot\mathbf{c}$$

It is often convenient to choose basis vectors that are orthonormal, that is, orthogonal and normalized to unit length.

$$\mathbf{e}_i\cdot\mathbf{e}_j = \delta_{ij} \tag{1-36}$$

In this case the scalar product of two vectors takes a very simple form when expressed in components:

$$\mathbf{x}\cdot\mathbf{y} = \left(\sum_i x_i\mathbf{e}_i\right)\cdot\left(\sum_j y_i\mathbf{e}_j\right) = \sum_{ij} x_i^* y_j \mathbf{e}_i\cdot\mathbf{e}_j = \sum_i x_i^* y_i \tag{1-37}$$

Also, the row vector (dual space vector, or linear scalar function) corresponding to a given vector in the original space may be obtained by just taking the complex conjugate transpose of the original column vector. [Why is this obvious from equation (1-37)?]. Hence, no matter what form the scalar product takes, it is convenient to have an orthonormal basis set.

Suppose that we have a vector space in which a metric (i.e., scalar product) is defined, and a coordinate system with orthonormal base vectors $\mathbf{e}_i$. If we wish to introduce a new basis $\mathbf{e}_i'$, what condition must we impose on the transformation matrix u in order that the new base vectors be also orthonormal?

$$\mathbf{e}_i\cdot\mathbf{e}_j = \delta_{ij} = \left(\sum_k u_{ki}\mathbf{e}_k'\right)\cdot\left(\sum_l u_{lj}\mathbf{e}_l'\right) = \sum_{kl} u_{ki}^* u_{lj}\mathbf{e}_k'\cdot\mathbf{e}_l' \tag{1-38}$$

If the new base vectors are to be orthonormal, we must have $\mathbf{e}_k'\cdot\mathbf{e}_l' = \delta_{kl}$. Thus equation (1-38) becomes

$$\begin{aligned} \delta_{ij} &= \sum_k u_{ki}^* u_{kj} = \sum_k (u^+)_{ik} u_{kj} \\ 1 &= u^+ u;\ u^+ = u^{-1} \end{aligned} \tag{1-39}$$

Thus *u must be unitary in order that the form of our scalar product* (and hence the lengths of vectors) *remain unchanged.* In classical physics, u is usually real as well as unitary; that is, it is orthogonal.

Other important properties of unitary transformations have to do with transformations of matrices rather than vectors. A similarity transformation by a unitary matrix leaves a unitary matrix unitary and a Hermitian matrix Hermitian. These properties may be easily demonstrated by the student (see Problem 1-10).

PROBLEMS

1-10 Suppose the matrices A and B are Hermitian and the matrices C and D are unitary. Prove that

(a) $C^{-1}AC$ is Hermitian
(b) $C^{-1}DC$ is unitary
(c) $i(AB - BA)$ is Hermitian

1-11 Using a particular coordinate system, a linear transformation in an abstract vector space is represented by the matrix

$$\begin{pmatrix} 2 & 1 & 0 \\ 1 & 2 & 0 \\ 0 & 0 & 5 \end{pmatrix}$$

and a particular (abstract) vector by the column vector

$$\begin{pmatrix} 1 \\ 2 \\ 3 \end{pmatrix}$$

Give the matrix and column vector in a new coordinate system, in terms of which the old base vectors are represented by

$$\mathbf{e}_1 = \begin{pmatrix} 1 \\ 1 \\ 0 \end{pmatrix} \qquad \mathbf{e}_2 = \begin{pmatrix} 1 \\ -1 \\ 0 \end{pmatrix} \qquad \mathbf{e}_3 = \begin{pmatrix} 0 \\ 0 \\ 1 \end{pmatrix}$$

1-12 Show that the set of all linear functions of one vector space into itself (or into some other space) form a vector space.

1-13 In mechanics one commonly uses Euler angles to specify the orientation of a rigid body. Imagine a set of axes $x'y'z'$ fixed within the body. We wish to specify their orientation relative to the xyz axes fixed in the room. It has been discovered that one can always go from the xyz axes to the $x'y'z'$ axes by a series of three steps:

(a) Rotate by an angle ϕ counterclockwise about the z axis. This takes the xyz axes into a new set, the $\xi\eta\zeta$ axes.
(b) Now rotate the $\xi\eta\zeta$ axes counterclockwise about the ξ axis by an angle θ. This will give the $\xi'\eta'\zeta'$ axes.
(c) Finally, rotate the $\xi'\eta'\zeta'$ axes counterclockwise by an angle ψ about the ζ' axis. This gives the $x'\,y'\,z'$ axes.

The angles ϕ, θ, ψ are called the Euler angles.

(i) Write matrices for the linear rotation operators of (a), (b) and (c). Specify the set of basis vectors used in each matrix.

(ii) Write the matrix giving the overall mapping from xyz to $x'y'z'$. Also, give its inverse matrix. What properties does the matrix have (i.e., is it orthogonal, antisymmetric, Hermitian, etc.)?

1-14 Despite the inherent simplicity of orthonormal basis systems, it is sometimes more convenient for a particular physical problem to use a basis that is not orthogonal. One important case arises in the description of noncubic crystal lattices. Here the greatest symmetry in the problem is provided by the periodic repetition of the lattice structure; we must adjust our thinking to mesh with this symmetry.

Suppose that (for a two-dimensional crystal) the sides of the unit cell can be specified by the vectors $\mathbf{a} = \mathbf{e}_x$, and $\mathbf{b} = \mathbf{e}_x \cos\theta + \mathbf{e}_y \sin\theta$. Here $\mathbf{e}_x$ and $\mathbf{e}_y$ are our usual orthonormal basis vectors. We wish to preserve our standard notion of length in this case, and use the usual dot product with $\mathbf{e}_x \cdot \mathbf{e}_y = 0$ as our scalar product; however, the convenient basis vectors $\mathbf{a}$ and $\mathbf{b}$ are not orthogonal in this case.

(a) What is the dot product of $\alpha\mathbf{a} + \beta\mathbf{b}$ with $\gamma\mathbf{a} + \delta\mathbf{b}$?

(b) Using $\mathbf{a}$ and $\mathbf{b}$ as basis vectors, any point in the crystal can be specified by $\mathbf{v} = v_1\mathbf{a} + v_2\mathbf{b}$; in matrix form we would then write

$$\mathbf{v} \leftrightarrow \begin{pmatrix} v_1 \\ v_2 \end{pmatrix}$$

The action of any linear scalar function ϕ on this vector would be given by $\phi(\mathbf{v}) = v_1\phi(\mathbf{a}) + v_2\phi(\mathbf{b})$. We see that computation of the coordinates v_1 and v_2 can be greatly simplified if there are special functions ϕ_1 and ϕ_2 such that $\phi_1(\mathbf{a}) = 1, \phi_1(\mathbf{b}) = 0; \phi_2(\mathbf{a}) = 0, \phi_2(\mathbf{b}) = 1$. If these functions exist, we have $v_1 = \phi_1(\mathbf{v})$, $v_2 = \phi_2(\mathbf{v})$. The row vectors corresponding to these special linear scalar functions are clearly $\phi_1 \leftrightarrow (1\ 0)$, $\phi_2 \leftrightarrow (0\ 1)$.

We know that our scalar product defines a relationship between column vectors $\mathbf{w}$ in the original space and linear scalar functions $\phi_{\mathbf{w}}(\ )$ by

$$\mathbf{w} \cdot \mathbf{v} = \phi_{\mathbf{w}}(\mathbf{v}).$$

Hence there must be a vector $\mathbf{A}$ in the original space such that $\mathbf{A} \cdot \mathbf{v} = \phi_1(\mathbf{v})$, and a vector $\mathbf{B}$ such that $\mathbf{B} \cdot \mathbf{v} = \phi_2(\mathbf{v})$. Find $\mathbf{A}$ and $\mathbf{B}$. The vectors $\mathbf{A}$ and $\mathbf{B}$ are called reciprocal lattice vectors.

1-15 (a) Let $\mathbf{a}$ and $\mathbf{b}$ be any two vectors in a linear vector space, and define $\mathbf{c} = \mathbf{a} + \lambda\mathbf{b}$, where λ is a scalar. By requiring that $\mathbf{c} \cdot \mathbf{c} \geq 0$ for all λ, derive the *Cauchy-Schwartz inequality*

$$(\mathbf{a} \cdot \mathbf{a})(\mathbf{b} \cdot \mathbf{b}) \geq (\mathbf{a} \cdot \mathbf{b})^2$$

(When does equality hold?)

(b) In an infinite-dimensional space, questions of convergence arise and it may be difficult to properly define the expansion $\mathbf{x} = \Sigma x_i \mathbf{e}_i$ of an arbitary vector $\mathbf{x}$ in terms of the base vectors $\mathbf{e}_1, \mathbf{e}_2, \ldots$. A useful result can, however, be derived: Assume $\mathbf{e}_i \cdot \mathbf{e}_j = \delta_{ij}$, define $x_k = \mathbf{e}_k \cdot \mathbf{x}$, and define

$$\mathbf{x}^{(n)} = \sum_{i=1}^{n} x_i \mathbf{e}_i$$

Apply the Cauchy-Schwartz inequality (Part a) and derive the inequality

$$\sum_{i=1}^{n} |x_n|^2 \leq \mathbf{x} \cdot \mathbf{x}$$

This result, which is valid for any n, no matter how large, is known as *Bessel's inequality.*

1-4 EIGENVALUE PROBLEMS

When an operator $\mathscr{A}$ acts on a vector $\mathbf{x}$, the resulting vector $\mathscr{A}\mathbf{x}$ is, in general, distinct from $\mathbf{x}$. However, there may exist certain (nonzero) vectors for which $\mathscr{A}\mathbf{x}$ is just $\mathbf{x}$ multiplied by a constant λ. That is,

$$\mathscr{A}\mathbf{x} = \lambda \mathbf{x} \tag{1-40}$$

Such a vector is called an *eigenvector* of the operator $\mathscr{A}$ and the constant λ is called an *eigenvalue.* The eigenvector is said to "belong" to the eigenvalue.

Example

Consider the rotation operator $\mathscr{R}$ illustrated in Figure 1-1. Any vector lying along the axis of rotation (z-axis) is an eigenvector of $\mathscr{R}$ belonging to the eigenvalue 1. (Are there any other eigenvectors?)

In a given coordinate system, the i-component of equation (1-40) is

$$\sum_j A_{ij} x_j = \lambda x_i \qquad i = 1, 2, \ldots, n \tag{1-41}$$

or, in matrix notation,

$$Ax = \lambda x \tag{1-42}$$

It is important to be able to find the eigenvalues λ for which the system of linear equations (1-41) has a nontrivial solution.

Example

If $\mathscr{A}$ is the matrix

$$A = \begin{pmatrix} 1 & 2 & 3 \\ 4 & 5 & 6 \\ 7 & 8 & 9 \end{pmatrix}$$

the system of equations (1-41) is

$$(1 - \lambda)x_1 + 2x_2 + 3x_3 = 0$$

$$4x_1 + (5 - \lambda)x_2 + 6x_3 = 0$$

$$7x_1 + 8x_2 + (9 - \lambda)x_3 = 0$$

To make this more transparent, let us write it in the form

$$Yx = 0$$

where Y is the matrix $A - \lambda 1$. Y is a square matrix; hence it might have an inverse, Y^{-1}. If this were the case we could multiply both sides of our equation by Y^{-1} and obtain

$$x = Y^{-1}0 = 0$$

This is not a very interesting solution. We conclude that if the system of equations is to have a non-trivial solution, Y cannot have an inverse. Hence from equation (1-25), Y must have determinant zero, that is,

$$\det Y \equiv \begin{vmatrix} 1 - \lambda & 2 & 3 \\ 4 & 5 - \lambda & 6 \\ 7 & 8 & 9 - \lambda \end{vmatrix} = 0$$

This gives a third-order polynomial in λ whose three roots are the eigenvalues λ_i.

In general, the eigenvalues of a matrix A are determined by the equation

$$\det (A - \lambda 1) = 0 \qquad (\textit{secular equation}) \tag{1-43}$$

If A is a $n \times n$ matrix, there will be n roots λ, not necessarily all different.

Clearly the eigenvalues and eigenvectors of a particular matrix contain a lot of information about the operator represented by the matrix. For certain types of matrices, knowing the eigenvalues and eigenvectors is equivalent to knowing the action of the operator on the entire vector space. Matrices of this kind are particularly well suited to physical applications, because their eigenvalues and eigenvectors can easily be given physical interpretations.

Hermitian matrices belong to this physically important class of "eigenvector-determined" operators. In the following paragraphs we shall prove some important theorems concerning their eigenvectors and eigenvalues.

PROPERTIES OF HERMITIAN MATRICES

Let H be a Hermitian matrix, λ_1 and λ_2 two of its eigenvalues, and x_1 and x_2 two eigenvectors, belonging to λ_1 and λ_2, respectively. That is

$$Hx_1 = \lambda_1 x_1 \qquad Hx_2 = \lambda_2 x_2 \tag{1-44}$$

Take the scalar product of x_2 with the first equation and x_1 with the second. Then

$$x_2^\dagger H x_1 = \lambda_1 x_2^\dagger x_1 \qquad x_1^\dagger H x_2 = \lambda_2 x_1^\dagger x_2 \tag{1-45}$$

where we have written $u^\dagger v = \sum_i u_i^* v_i$ for the scalar product of two column vectors u and v.

The left-hand sides of equation (1-45) are complex conjugates, since

$$\begin{aligned}(x_2^\dagger H x_1)^* &= \sum_{ij} (x_{i2}^* H_{ij} x_{j1})^* \\ &= \sum_{ij} x_{i2} H_{ij}^* x_{j1}^* = \sum_{ij} x_{j1}^* H_{ji} x_{i2}\end{aligned}$$

that is,

$$(x_2^\dagger H x_1)^* = x_1^\dagger H x_2 \tag{1-46}$$

Therefore, from (1-45),

$$(\lambda_1 - \lambda_2^*) x_2^\dagger x_1 = 0$$

Suppose first that $\lambda_1 = \lambda_2$, $x_1 = x_2 \neq 0$. Then $x_2^\dagger x_1 = x_1^\dagger x_1 > 0$, so that $\lambda_1 = \lambda_1^*$. *The eigenvalues of a Hermitian matrix are real.*

Alternatively, suppose $\lambda_1 \neq \lambda_2$. Then $x_2^\dagger x_1 = 0$. *Eigenvectors of a Hermitian matrix belonging to different eigenvalues are orthogonal.*

Several eigenvectors may belong to the same eigenvalue; such an eigenvalue is said to be *degenerate*. The eigenvectors in question are also often called degenerate. We have seen above that eigenvectors of a Hermitian matrix belonging to different eigenvalues are orthogonal. What about eigenvectors belonging to the same (degenerate) eigenvalue? Can these be made orthogonal?

Suppose the three independent eigenvectors $\mathbf{x}_1$, $\mathbf{x}_2$, and $\mathbf{x}_3$ belong to the eigenvalue λ. Clearly, any linear combination of these is also an eigenvector. Let

$$\begin{aligned}\mathbf{y}_1 &= \mathbf{x}_1 \\ \mathbf{y}_2 &= \mathbf{x}_2 + \alpha \mathbf{y}_1\end{aligned}$$

We wish to choose α so that $\mathbf{y}_1$ and $\mathbf{y}_2$ are orthogonal. Therefore

$$\mathbf{y}_1 \cdot \mathbf{y}_2 = 0 = \mathbf{y}_1 \cdot \mathbf{x}_2 + \alpha \mathbf{y}_1 \cdot \mathbf{y}_1$$

$$\alpha = -\frac{\mathbf{y}_1 \cdot \mathbf{x}_2}{\mathbf{y}_1 \cdot \mathbf{y}_1}$$

Now set

$$\mathbf{y}_3 = \mathbf{x}_3 + \beta\mathbf{y}_1 + \gamma\mathbf{y}_2$$

We wish to choose β and γ so that $\mathbf{y}_3$ is orthogonal to both $\mathbf{y}_1$ and $\mathbf{y}_2$. Thus

$$0 = \mathbf{y}_1 \cdot \mathbf{x}_3 + \beta\mathbf{y}_1 \cdot \mathbf{y}_1 \qquad \beta = -\frac{\mathbf{y}_1 \cdot \mathbf{x}_3}{\mathbf{y}_1 \cdot \mathbf{y}_1}$$

$$0 = \mathbf{y}_2 \cdot \mathbf{x}_3 + \gamma\mathbf{y}_2 \cdot \mathbf{y}_2 \qquad \gamma = -\frac{\mathbf{y}_2 \cdot \mathbf{x}_3}{\mathbf{y}_2 \cdot \mathbf{y}_2}$$

Thus we have constructed three mutually orthogonal eigenvectors $\mathbf{y}_i$. This procedure, known as the *Gram-Schmidt orthogonalization procedure*, clearly can be extended to an arbitrary number of degenerate eigenvectors. Thus, we can arrange things so that all n eigenvectors of a Hermitian matrix are mutually orthogonal.

COORDINATE TRANSFORMATIONS

The eigenvalues and eigenvectors of a linear operator are independent of the particular coordinate system used in finding them, as is clear from equation (1-40). It is also clear from the matrix equation (1-42)

$$Ax = \lambda x$$

which implies

$$uAu^{-1}ux = \lambda ux$$

or

$$A'x' = \lambda x'$$

Thus, if x is an eigenvector of A, its transform $x' = ux$ is an eigenvector of the transformed matrix A', and the eigenvalues are the same.

Suppose we make a transformation to a coordinate system in which the base vectors $\mathbf{e}_i'$ are eigenvectors of $\mathscr{A}$ (assuming these form a linearly independent set),

$$\mathscr{A}\mathbf{e}_i' = \lambda_i \mathbf{e}_i' \tag{1-47}$$

In this system, the matrix element A_{ij}' is the ith component of the vector $\mathscr{A}\mathbf{e}_j'$, which, from (1-47), is zero or λ_j according as $i \neq j$ or $i = j$, respectively. That is,

$$A_{ij}' = \lambda_j \delta_{ij} \tag{1-48}$$

Thus the matrix A' is diagonal, and the diagonal elements are the eigenvalues!

Because of its importance, we write the specific form of the transformation matrix u in this case. From equation (1-26), we find $\mathbf{e}_j' = \sum_i (u^{-1})_{ij}\mathbf{e}_i$. Thus the jth column of u^{-1} consists of the components of the eigenvector $\mathbf{e}_j'$ in the unprimed coordinate system. If the transformation is to be unitary, the eigenvectors must be mutually orthogonal (made so by the Gram-Schmidt procedure if necessary) and they must be normalized.

We have seen above that the eigenvalues of a matrix remain unchanged when the matrix undergoes a similarity transformation. They are invariants of the matrix. Other invariants are the trace and the determinant:

$$\text{Tr } A' = \text{Tr } S^{-1}AS = \text{Tr } ASS^{-1} = \text{Tr } A$$

$$\det A' = \det (S^{-1}AS) = (\det S^{-1})(\det A)(\det S) = \det A$$

These invariants are not independent of the eigenvalues. In fact, we may evaluate them in the system in which A is diagonal, with the result

$$\text{Tr } A = \sum_i \lambda_i \tag{1-49}$$

$$\det A = \prod_i \lambda_i \tag{1-50}$$

The eigenvalues are the only independent invariants.

One place where matrix eigenvalue problems arise is in the treatment of normal modes of vibration in mechanical or electrical systems.

Example

Consider the problem of molecular vibrations, or, in general, "small vibrations" of a (classical) mechanical system. For example, in Figure 1-3

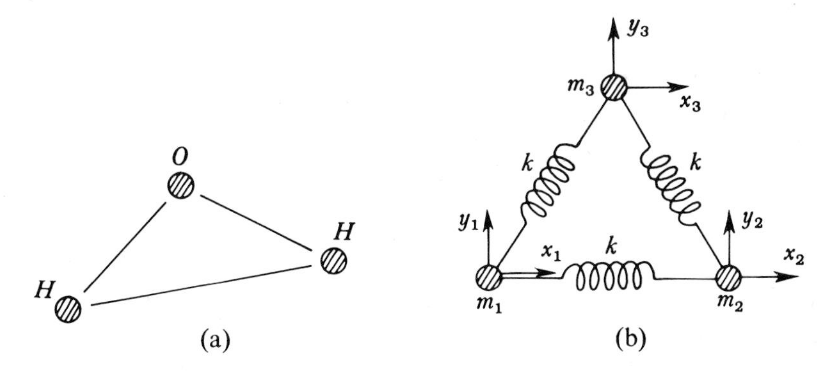

FIGURE 1-3 Simple vibrating systems: (a) a water molecule; (b) three masses connected by springs

are shown schematically (a) a water molecule and (b) a simple vibrating system consisting of three masses connected by springs.

The configuration or state of such a system may be conveniently described by an n-dimensional vector $\mathbf{x}$ where n is the number of degrees of freedom, that is, three times the number of masses. A simple coordinate system consists of the base vectors:

$\mathbf{e}_1$: mass 1 displaced a unit distance in the x-direction
$\mathbf{e}_2$: mass 1 displaced a unit distance in the y-direction
$\mathbf{e}_3$: mass 1 displaced a unit distance in the z-direction
$\mathbf{e}_4$: mass 2 displaced a unit distance in the x-direction
etc.

In terms of components in this coordinate system, the kinetic energy is

$$T = \frac{1}{2}\sum_i m_i \dot{x}_i^2 = \frac{1}{2}\tilde{\dot{x}} M \dot{x}$$

where m_i is the mass associated with x_i (that is, $m_1 = m_2 = m_3 =$ mass of particle 1; $m_4 = m_5 = m_6 =$ mass of particle 2, etc.), $\dot{x}_i = dx_i/dt$, and M is the diagonal matrix $M_{ij} = m_i \delta_{ij}$.

For vibrations of small amplitude, we expand the potential energy in a Taylor's series

$$V = V_0 + \sum_i \left(\frac{\partial V}{\partial x_i}\right)_0 x_i + \sum_{ij} \frac{1}{2}\left(\frac{\partial^2 V}{\partial x_i\, \partial x_j}\right)_0 x_i x_j + \cdots$$

The force $(\partial V/\partial x_i)_0$ at equilibrium must be zero, and we can choose $V_0 = 0$, so neglecting third-order terms the potential energy has the form

$$V = \frac{1}{2}\sum_{ij} V_{ij} x_i x_j = \frac{1}{2}\tilde{x} V x$$

where V is a real symmetric matrix; $V_{ij} = [\partial^2 V/(\partial x_i\, \partial x_j)]_0$.

From the Lagrangian

$$L = T - V = \frac{1}{2}\sum_i m_i \dot{x}_i^2 - \frac{1}{2}\sum_{ij} V_{ij} x_i x_j$$

Lagrange's equations,

$$\frac{d}{dt}\left(\frac{\partial L}{\partial \dot{x}_i}\right) - \frac{\partial L}{\partial x_i} = 0$$

yield

$$m_i \ddot{x}_i + \sum_j V_{ij} x_j = 0$$

By definition, the normal modes will move with simple harmonic motion at particular frequencies. To find these frequencies, we set $\ddot{x}_i = -\omega^2 x_i$, and solve the resulting set of equations

$$-\omega^2 m_i x_i + \sum_j V_{ij} x_j = 0$$

In matrix notation these read $(-\omega^2 M + V)x = 0$. This closely resembles an eigenvalue equation. We can put it into the standard form (1-40) by introducing a matrix S such that $\tilde{S}MS = 1 (S_{ij} = \delta_{ij} 1/\sqrt{m_i})$. Then the equation can be written as $\tilde{S}(-\omega^2 M + V)SS^{-1}x = 0 = (-\omega^2 + \tilde{S}VS)S^{-1}x$. $\tilde{S}VS$ is a real symmetric matrix; hence its eigenvalues ω_l^2 are real and their corresponding eigenvectors v_l are orthogonal. The numbers ω_l are the normal frequencies of the system, and the corresponding vectors $\mathbf{x}_l = \mathbf{S}\mathbf{v}_l$ are the normal modes. Sometimes zero frequencies appear as eigenvalues. These usually correspond to translations and rotations of the system as a whole.

Note: the Lagrangian techniques used here are elegant, but certainly not essential for solution of the problem. We obtain the same matrix equations by writing out Newton's laws in the form $\vec{F} = m\vec{a}$. See Problem 1-20.

We have seen above that a square matrix can be diagonalized by a unitary matrix formed from its eigenvectors, provided the eigenvectors form a complete orthonormal set. We now show that this is true for a Hermitian matrix H.

There is at least one eigenvalue and eigenvector

$$Hu_1 = \lambda_1 u_1 \tag{1-51}$$

because the secular equation (1-43) has at least one solution. Choose an orthonormal coordinate system $\mathbf{e}'_j$ in which the first member $\mathbf{e}'_1 = \mathbf{u}_1$. (This can be done because there exist sets of n linearly independent vectors that include $\mathbf{u}_1$, and these can be rearranged in orthogonal combinations by the Gram-Schmidt procedure if necessary.) Upon transforming H to this system, the elements in the first column are

$$H'_{i1} = \mathbf{e}'_i \cdot H\mathbf{u}_1 = \lambda_1 \delta_{i1} \tag{1-52}$$

Then, the fact that H' is Hermitian determines the first row, and we have the following form for H':

$$H' = \left(\begin{array}{c|cccc} \lambda_1 & 0 & 0 & \dots & 0 \\ \hline 0 & & & & \\ 0 & & & & \\ \vdots & & & \text{G} & \\ 0 & & & & \end{array}\right) \tag{1-53}$$

where G is a Hermitian matrix of dimensionality $n - 1$, in the vector subspace normal to $\mathbf{u}_1$. We now repeat the same process with G and continue in this way until H is completely diagonalized. We have found in the process n independent eigenvectors of H, from which we may construct a unitary transformation matrix which diagonalizes H.

Similar considerations apply to Hermitian differential operators on function spaces. This is particularly important for physics because in quantum mechanics the observables, such as linear momentum, energy, and angular momentum, are all represented by Hermitian operators. The values allowed these observables by quantum mechanics are the eigenvalues of the operators. The eigenvectors of the operators then span the space of wave functions on which the observable is defined.

MATRICES WITH THE SAME EIGENVECTORS

Now suppose we have two Hermitian matrices H_1 and H_2. Can they be diagonalized "simultaneously" by the same unitary transformation? That is, can we find a unitary matrix U such that both

$$D_1 = U^{-1}H_1U \qquad \text{and} \qquad D_2 = U^{-1}H_2U \tag{1-54}$$

are diagonal?

Since diagonal matrices clearly commute with each other,

$$0 = D_1D_2 - D_2D_1 = U^{-1}(H_1H_2 - H_2H_1)U$$

$$H_1H_2 - H_2H_1 = 0$$

Therefore, a *necessary* condition is that H_1 and H_2 commute.

This condition is also *sufficient*. For suppose that H_1 and H_2 commute. Let

$$U^{-1}H_1U = D \qquad \text{(diagonal)}$$

$$U^{-1}H_2U = M \qquad \text{(maybe not diagonal)}$$

Now D and M commute

$$DM = MD$$

The ij element of this equation is

$$D_{ii}M_{ij} = M_{ij}D_{jj}$$

(here the indices i and j are *not* summed).

Thus, if $D_{ii} \neq D_{jj}$, $M_{ij} = 0$. This does not mean that M is diagonal, however, because H_1 may have some degenerate eigenvalues; that is, several of the elements of D may be equal. Suppose, for example, the first three are equal. Then

$$D = \left(\begin{array}{ccc|ccc} \lambda & 0 & 0 & & & \\ 0 & \lambda & 0 & & 0 & \\ 0 & 0 & \lambda & & & \\ \hline & & & \lambda_4 & & 0 \\ & 0 & & & \cdot & \\ & & & & & \cdot \\ & & & 0 & & \cdot \end{array}\right) \qquad M = \left(\begin{array}{ccc|c} a & b & c & \\ d & e & f & 0 \\ g & h & i & \\ \hline & 0 & & N \end{array}\right)$$

That is, M is in block diagonal form. The submatrix in the upper left-hand corner of M is Hermitian and can be diagonalized by a unitary transformation that involves only the first three rows and columns. This corner of D is just a multiple of the unit matrix and is therefore unaffected. Repeating this operation, clearly both M and D can be simultaneously diagonalized.

This leads to the general condition that a matrix be diagonalizable by means of a unitary transformation. Consider an arbitrary matrix M. We can write

$$M = A + iB \tag{1-55}$$

where A and B are both Hermitian, by choosing

$$A = \frac{M + M^\dagger}{2} \qquad B = \frac{M - M^\dagger}{2i} \tag{1-56}$$

(This is just like splitting a complex number into its real and imaginary parts.) Now A and B can be diagonalized separately, but in order that M may be diagonalized, we must be able to diagonalize A and B *simultaneously*. The requirement for this is that A and B commute, which is equivalent to requiring that M and $M^\dagger$ commute. A matrix that commutes with its Hermitian conjugate is said to be *normal*; a matrix M can be diagonalized by a unitary transformation if, and only if, M is normal. Note that unitary matrices as well as Hermitian matrices are normal.

PROBLEMS

1-16 Find the eigenvalues and normalized eigenvectors of the matrix

$$\begin{pmatrix} 1 & 2 & 3 \\ 4 & 5 & 6 \\ 7 & 8 & 9 \end{pmatrix}$$

Express your answers numerically (3 significant figures).

1-17 Let U be a unitary matrix and let x_1, x_2 be two eigenvectors of U belonging to the eigenvalues λ_1, λ_2, respectively. Show that

(a) $|\lambda_1| = |\lambda_2| = 1$

(b) If $\lambda_1 \neq \lambda_2$, $x_1^\dagger x_2 = 0$

1-18 Transform the matrix A and vector x given below to a coordinate system in which A is diagonal.

$$A = \begin{pmatrix} 0 & -i & 0 & 0 & 0 \\ i & 0 & 0 & 0 & 0 \\ 0 & 0 & 3 & 0 & 0 \\ 0 & 0 & 0 & 1 & -i \\ 0 & 0 & 0 & i & 1 \end{pmatrix} \qquad x = \begin{pmatrix} 1 \\ a \\ i \\ b \\ -1 \end{pmatrix}$$

1-19 Refer to your solution to Problem 1-13.

(a) Calculate the eigenvectors and eigenvalues of the matrices found in Part (i) of that problem. Are there any obvious symmetries in your answer?

(b) From the answer to (a), predict some properties of the eigenvalues and eigenvectors of the overall transformation matrix calculated in Part (ii) of 1-13. Calculate the eigenvectors and eigenvalues and see whether your predictions are verified.

1-20 In the example related to Figure 1-3, the equations of motion were derived using Lagrangian techniques. These provide a compact summary of the dynamics; however the same results can be obtained from Newton's laws. Use the usual relationship $\mathbf{F} = m\ddot{\mathbf{x}}$, and the potential energy for a compressed spring, to write down equations of motion for a "balls-and-springs" model of the water molecule. Do these have the form discussed in the text?

1-21 Find the normal modes and normal frequencies for the linear vibrations of the CO_2 molecule (that is, vibrations in the line of the molecule). Use the model of Figure 1-4.

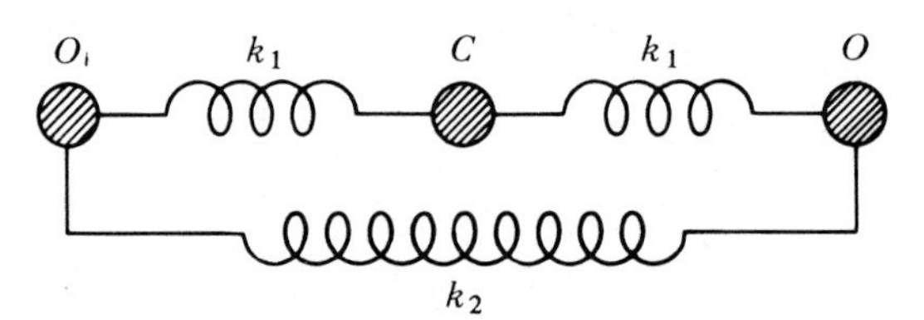

FIGURE 1-4 A mechanical model for the CO_2 molecule

1-22 The angular momentum of a rigid body

$$\mathbf{L} = \sum_i m_i(\mathbf{r}_i \times \mathbf{v}_i)$$

(where we have summed over all masses in the body) takes on a simple form if the body is rotating about some fixed point. In that case, $\mathbf{v}_i = \omega x \mathbf{r}_i$, and

$$\mathbf{L} = \sum_i m_i[\omega r_i^2 - \mathbf{r}_i(\mathbf{r}_i \cdot \omega)]$$

so the components of angular momentum $\mathbf{L}$ and angular velocity ω are related by a linear transformation

$$L_x = \omega_x \sum_i m_i(r_i^2 - x_i^2) - \omega_y \sum_i m_i x_i y_i - \omega_z \sum_i m_i x_i z_i$$

or

$$L_x = I_{xx}\omega_x + I_{xy}\omega_y + I_{xz}\omega_z$$

with similar equations for the other components.

The numbers I_{ij} are referred to as the inertia tensor of the body, and the relation between $\mathbf{L}$ and $\boldsymbol{\omega}$ can then be written in matrix form as

$$L = I\omega$$

The kinetic energy of rotation is a scalar quantity formed from the two vectors $\boldsymbol{\omega}$ and $\mathbf{L}$. We see it must be proportional to $\boldsymbol{\omega} \cdot \mathbf{L}$, or $\omega^\dagger L$ in matrix form. Comparison with a simple case gives us the proportionality constant

$$T = \frac{1}{2}\omega^+ L = \frac{1}{2}\omega^+ I\omega$$

The general relation between kinetic energy and the angular velocity is therefore rather complicated:

$$T = \frac{1}{2}[I_{xx}\omega_x^2 + I_{xy}\omega_x\omega_y + I_{xz}\omega_x\omega_z + I_{yx}\omega_y\omega_x + I_{yy}\omega_y^2 + I_{yz}\omega_y\omega_z + I_{zx}\omega_z\omega_x + I_{zy}\omega_z\omega_y + I_{zz}\omega_z^2]$$

(a) Show that the inertia tensor has a symmetric matrix.

(b) Use the result of (a) to demonstrate that there are axes 1, 2, 3 in the body (the so-called principal axes) such that transformation to these axes simplifies the form of T:

$$T = \frac{1}{2}(I_1\omega_1^2 + I_2\omega_2^2 + I_3\omega_3^2)$$

The numbers $I_1, I_2,$ and I_3 are called the principal moments of inertia.

(c) Given the inertia tensor

$$\begin{pmatrix} \frac{5}{2} & \sqrt{\frac{3}{2}} & \sqrt{\frac{3}{4}} \\ \sqrt{\frac{3}{2}} & \frac{7}{3} & \sqrt{\frac{1}{18}} \\ \sqrt{\frac{3}{4}} & \sqrt{\frac{1}{18}} & \frac{13}{6} \end{pmatrix}$$

in the x, y, z coordinate system, find the components (in the x, y, z basis) of the principal axes. What are the principal moments of inertia for this case?

SUMMARY

Many physical quantities can be cast into the mathematical form of vector spaces. In an n-dimensional vector space, every vector can be put in the form

$$\mathbf{v} = \sum_{m=1}^{n} v_m \mathbf{e}_m$$

where the $\{\mathbf{e}_m\}$ are n linearly independent vectors. The vector $\mathbf{v}$ can also be written as a column matrix,

$$\begin{pmatrix} v_1 \\ v_2 \\ \vdots \\ v_n \end{pmatrix}$$

Operators which act on elements of one vector space and take them into elements of another (or the same) vector space can be represented by matrices

$$\begin{pmatrix} A_{11} & A_{12} & \dots & A_{1n} \\ \vdots & & & \\ A_{p1} & A_{p2} & \dots & A_{pn} \end{pmatrix}$$

When we change the basis (that is, the coordinate system) in a vector space, the column matrix representing a particular vector changes by

$$v \to uv$$

The matrix representing a linear operator on this vector space changes by a similarity transformation

$$A \to uAu^{-1}$$

Elements of the dual space have matrices that change according to

$$v^\dagger \to v^\dagger u^\dagger$$

If the coordinate transformation is to preserve lengths of vectors, u must be a unitary matrix (that is, $u^\dagger = u^{-1}$).

An eigenvector $\mathbf{v}^n$ of an operator A is one that obeys the equation

$$A\mathbf{v}^n = \lambda_n \mathbf{v}^n$$

for some number λ_n (called the eigenvalue). For Hermitian matrices (ones such that $A^+ = A$), all the eigenvalues are real and eigenvectors belonging to different eigenvalues are orthogonal. It is possible to choose eigenvectors such that they span the space. Two Hermitian matrices can be diagonalized simultaneously if and only if they commute.

CHAPTER 2

Abstract Formalism of Vector Spaces

Not all vector spaces encountered in physical applications are finite dimensional; in quantum mechanics one often meets spaces with an infinite number of dimensions. Matrix representations are of little use here, but the basic operator ideas are unchanged. In Section 2-1 we discuss an abstract notation for linear vector spaces that is useful regardless of their dimensionality. The notation, together with the powerful eigenvalue concepts, is applied in the following two sections to the derivation of general formulae for Green's functions and perturbation theory.

2-1 BRA-KET NOTATION

Often in practical applications of vector space techniques, it is helpful to phrase the problem in an abstract way first, and then to translate it into particular vectors and matrices (or functions and differential operators). In this section we describe a concise method for such abstract phrasing. This bra-ket notation, as it is called, was first introduced by P. A. M. Dirac (12). In this section we sketch the basic ideas and illustrate their application to finite dimensional vector spaces; applications to function spaces are discussed in Section 4-2.

Each vector is written as a symbol (*ket*).

$$|v_1\rangle, \qquad |v_2\rangle, \qquad \text{etc.} \tag{2-1}$$

Suppose we have a set of base vectors $|n\rangle$. We know that every vector in the space can be expanded in terms of these base vectors as

$$|v\rangle = \sum_n c_n |n\rangle \tag{2-2}$$

We also know that c_n is the scalar product of $|v\rangle$ with $|n\rangle$, provided the base vectors are orthonormal.

As discussed in Section 1-3, the scalar product allows us to associate another vector space (the dual, or adjoint space) with our original one. The adjoint space has the same dimension as the initial space. For two vectors in the initial space $|v_1\rangle$ and $|v_2\rangle$, we define a new symbol $\langle v_2 | v_1\rangle$ to represent the scalar product of $|v_1\rangle$ with $|v_2\rangle$. (Clearly the orthonormality of the base vectors gives us $\langle n \mid m\rangle = \delta_{nm}$). In terms of the base vectors $|n\rangle$,

$$|v_1\rangle = \sum c_n^1 |n\rangle$$

$$|v_2\rangle = \sum c_n^2 |n\rangle$$

and

$$\langle v_2 \mid v_1\rangle = \sum_n c_n^{2*} c_n^1 = \sum_{n,m} c_m^{2*} c_n^1 \langle m \mid n\rangle \tag{2-3}$$

We can formally separate the symbol

$$\langle v_2 \mid v_1\rangle = \{\langle v_2|\} \{|v_1\rangle\}$$

to obtain

$$\langle v_2| = \sum_n c_n^{2*} \langle n| \tag{2-4}$$

as the symbol for a vector in the dual space corresponding to $|v_2\rangle$. (This symbol $\langle v_2|$ is called a *bra*.)

Some simple properties of the bra-ket notation are worth noting:

1. From our definition of the scalar product,

$$\langle v_1 \mid v_2\rangle = \langle v_2 \mid v_1\rangle^* \tag{2-5}$$

2. From the decomposition $|v\rangle = \sum_n c_n |n\rangle$, we find $c_n = \langle n \mid v\rangle$, so that

$$|v\rangle = \sum_n |n\rangle \langle n \mid v\rangle \tag{2-6}$$

Again, if we formally separate this as $|v\rangle = \left\{\sum_n |n\rangle \langle n|\right\} |v\rangle$, we find

$$\sum_n |n\rangle \langle n| = 1 \tag{2-7}$$

and we have found an expression for the identity operator in terms of the base vectors of the space.

Similar expressions can be found for other operators. Consider a general operator H, with coordinates H_{mn} in the basis $|n\rangle$ defined by

$$H |n\rangle = \sum_m H_{mn} |m\rangle \tag{2-8}$$

that is,

$$\langle m| H |n\rangle = H_{mn} \tag{2-9}$$

We then have

$$H\,|v\rangle = \sum_n c_n H\,|n\rangle = \sum_{n,m} c_n H_{mn}\,|m\rangle$$

$$\langle v|\,H\,|v\rangle = \sum_{p,n} c_p^* c_n H_{pn}$$

or

$$\langle v|\,H\,|v\rangle = \sum_{p,n} \langle v \mid p\rangle \langle p|\,H\,|n\rangle \langle n \mid v\rangle \tag{2-10}$$

This allows us to write

$$H = \sum_{p,n} |p\rangle \langle p|\,H\,|n\rangle \langle n| \tag{2-11}$$

If the base states $|n\rangle$ happen to be eigenvectors of H, then

$$H\,|n\rangle = f_n\,|n\rangle$$

and (2-11) reduces to a particularly simple form

$$H = \sum_n f_n\,|n\rangle \langle n| \tag{2-12}$$

These abstract expressions can easily be translated into matrix notation. The kets are represented by column vectors and the bras by row vectors. If we were working in a three-dimensional space, for example, we might have

$$|1\rangle \leftrightarrow \begin{pmatrix} 1 \\ 0 \\ 0 \end{pmatrix}; \qquad |2\rangle \leftrightarrow \begin{pmatrix} 0 \\ 1 \\ 0 \end{pmatrix}; \qquad |3\rangle \leftrightarrow \begin{pmatrix} 0 \\ 0 \\ 1 \end{pmatrix} \tag{2-13}$$

An operator of the form $|m\rangle \langle n|$ is represented by a matrix. Consider for example $|1\rangle \langle 2|$. This acts on a vector

$$|v\rangle \leftrightarrow \begin{pmatrix} v_1 \\ v_2 \\ v_3 \end{pmatrix}$$

by first taking the scalar product of $|v\rangle$ with $|2\rangle (\langle 2 \mid v\rangle = v_2)$ and then multiplying the unit vector $|1\rangle$ by this number. Hence

$$|1\rangle \langle 2 \mid v\rangle \leftrightarrow \begin{pmatrix} v_2 \\ 0 \\ 0 \end{pmatrix}$$

This is the same as multiplying $|v\rangle$ by a matrix θ^{12}:

$$\begin{pmatrix} 0 & 1 & 0 \\ 0 & 0 & 0 \\ 0 & 0 & 0 \end{pmatrix} \begin{pmatrix} v_1 \\ v_2 \\ v_3 \end{pmatrix} = \theta^{12}\,v$$

The ith row of the matrix is obtained by multiplying the elements of the row vector representing $\langle 2|$ by the ith element of the column vector representing $|1\rangle$.

Thus we may write

$$|1\rangle\langle 2| \leftrightarrow \begin{pmatrix} 1 \\ 0 \\ 0 \end{pmatrix}^* (0 \quad 1 \quad 0) = \begin{pmatrix} 0 & 1 & 0 \\ 0 & 0 & 0 \\ 0 & 0 & 0 \end{pmatrix}$$

$$|2\rangle\langle 2| \leftrightarrow \begin{pmatrix} 0 \\ 1 \\ 0 \end{pmatrix}^* (0 \quad 1 \quad 0) = \begin{pmatrix} 0 & 0 & 0 \\ 0 & 1 & 0 \\ 0 & 0 & 0 \end{pmatrix}$$

The operator $|n\rangle\langle n|$ is a projection operator; it projects vectors onto the subspace spanned by the base vector $|n\rangle$. In general any Hermitian operator $\mathscr{P}$ such that $\mathscr{P}^2 = \mathscr{P}$ is called a projection operator. For example, in the projection of three-dimensional vectors onto the xy-plane (discussed in Section 1-2), one would use $\mathscr{P} = |e_x\rangle\langle e_x| + |e_y\rangle\langle e_y|$.

A Hermitian operator may be written in terms of projection operators which project onto the subspaces spanned by its eigenvalues. For example, an operator H' such that

$$H'\,|1\rangle = 2\,|1\rangle \qquad H'\,|2\rangle = 6\,|2\rangle \qquad H'\,|3\rangle = 12\,|3\rangle$$

would be represented by the matrix

$$\begin{pmatrix} 2 & 0 & 0 \\ 0 & 6 & 0 \\ 0 & 0 & 12 \end{pmatrix}$$

or by the sum over projection operators

$$H' = 2\,|1\rangle\langle 1| + 6\,|2\rangle\langle 2| + 12\,|3\rangle\langle 3|$$

PROBLEMS

2-1 Using bra-ket notation, construct the following projection operators:

(a) The operator that projects three-dimensional space onto a plane that intersects the xz-plane at an angle of 45° with respect to the x-axis, and that intersects the yz-plane along the line $z = 0$;

(b) The operator that projects three-dimensional space onto the line $y = 0$, $x = 3z$.

2-2 The "expectation value" of an operator H' in the state $|v\rangle$ may be defined as $\langle v|\,H'\,|v\rangle$. For a Hermitian H', use a basis where H' is diagonal to calculate $\langle v|\,H'\,|v\rangle$. (Your answer should be in terms of the diagonal elements of the H' matrix, and the components of $|v\rangle$ in this basis.) Interpret your answer to show why it is the "expected value."

2-2 GREEN'S FUNCTIONS AND THE INHOMOGENEOUS EQUATION

The abstract notation discussed in Section 2-1 is particularly helpful in formulating some methods that are used in the theory of linear operators. In this section and in Section 2-3 we will discuss two of these methods, stressing applications to matrix operators. Applications to differential operators are even more important; they will appear scattered through subsequent chapters.

Consider a linear operator H with eigenvectors $|n\rangle$ such that

$$H\,|n\rangle = \lambda_n\,|n\rangle$$

Suppose we are given the inhomogeneous equation

$$H\,|v\rangle - \lambda\,|v\rangle = |w\rangle \tag{2-14}$$

where $|v\rangle$ is a vector to be determined, $|w\rangle$ is a known vector, and λ is a constant not equal to λ_n for any n. Write $|v\rangle = \sum_n v_n\,|n\rangle$.

Then $H\,|v\rangle = \sum_n v_n\lambda_n\,|n\rangle$ and equation (2-14) becomes

$$\sum_n v_n(\lambda_n - \lambda)\,|n\rangle = |w\rangle \tag{2-15}$$

Take the scalar product of both sides with $|m\rangle$.

$$v_m(\lambda_m - \lambda) = \langle m \mid w\rangle \tag{2-16}$$

$$|v\rangle = \sum_n \frac{|n\rangle\langle n \mid w\rangle}{\lambda_n - \lambda} = \left\{\sum_n \frac{|n\rangle\langle n|}{\lambda_n - \lambda}\right\}|w\rangle \tag{2-17}$$

$$|v\rangle = G\,|w\rangle \tag{2-18}$$

The solution $|v\rangle$ can be written as $G\,|w\rangle$, where G is an operator known as the Green's function. Notice that it depends only on the eigenvalues and eigenvectors of the problem at hand. Thus once the eigenvalues and eigenvectors of the operator are known, we can find the answer to equation (2-14) for any $|w\rangle$ without explicitly solving the equation.

Example

$$\begin{pmatrix} 0 & 1 \\ 1 & 0 \end{pmatrix}\begin{pmatrix} v_1 \\ v_2 \end{pmatrix} - \lambda\begin{pmatrix} v_1 \\ v_2 \end{pmatrix} = \begin{pmatrix} 1 \\ 2 \end{pmatrix}$$

We must solve for v_1 and v_2. The eigenvectors and eigenvalues of

$$\begin{pmatrix} 0 & 1 \\ 1 & 0 \end{pmatrix}$$

are

$$\lambda_1 = 1, |1\rangle = 2^{-1/2}\begin{pmatrix} 1 \\ 1 \end{pmatrix}; \qquad \lambda_2 = -1, |2\rangle = 2^{-1/2}\begin{pmatrix} 1 \\ -1 \end{pmatrix}$$

Hence the Green's function is

$$\sum_n \frac{|n\rangle\langle n|}{\lambda_n - \lambda} = \left(\frac{1}{1-\lambda^2}\right)\begin{pmatrix} \lambda & 1 \\ 1 & \lambda \end{pmatrix}$$

and the solution is

$$\begin{pmatrix} v_1 \\ v_2 \end{pmatrix} = \left(\frac{1}{1-\lambda^2}\right)\begin{pmatrix} \lambda + 2 \\ 1 + 2\lambda \end{pmatrix}$$

This can be checked by direct substitution.

Many systems of differential equations can be reduced to algebraic form by an adroit substitution or by transform techniques; the resulting algebra problem can then be solved by methods discussed in this chapter. Fourier and Laplace transform methods, discussed in Section 9-1, provide perhaps the most general framework for reducing differential equations in the time domain to algebraic form. However, often the problem is simple enough that intuition will suffice.

Example

Consider the simple electrical circuit displayed in Figure 2-1. We can write the equations describing potential drops around the loops as

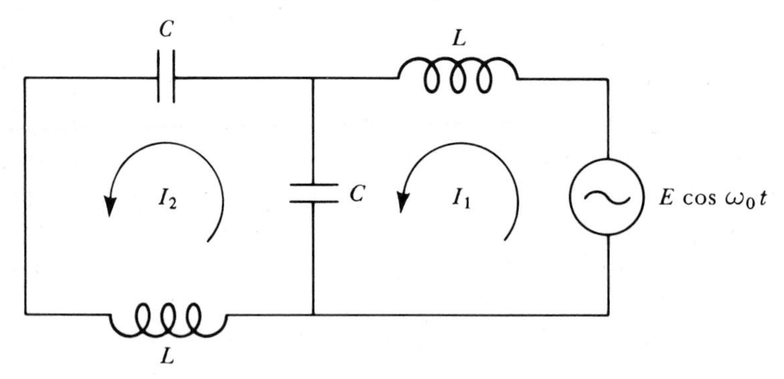

FIGURE 2-1 A simple circuit

$$E\cos\omega_0 t = L\frac{dI_1}{dt} + \frac{1}{C}\left\{\int_0^t [I_1(t') - I_2(t')]\,dt' + Q_1(0)\right\}$$

$$0 = \frac{1}{C}\left\{\int_0^t (I_2(t') - I_1(t')\, dt' + Q_1(0)\right\} + L\frac{dI_2}{dt}$$

$$+ \frac{1}{C}\left\{\int_0^t I_2(t')\, dt' + Q_2(0)\right\}$$

In order to put this into the form of a differential equation, we differentiate both sides:

$$-E\omega_0 \sin \omega_0 t = L\frac{d^2I_1}{dt^2} + \frac{1}{C}(I_1 - I_2)$$

$$0 = \frac{1}{C}(I_2 - I_1) + L\frac{d^2I_2}{dt^2} + \frac{1}{C}I_2$$

We recall that the general solution of an inhomogeneous linear equation is the sum of a particular solution of the inhomogeneous equation with the general solution of the associated homogeneous equation. Hence our currents I_i will take the form $I_i = I_i^P + I_i^G$, where the I_i^P are *any* particular solution of

$$-E\omega_0 \sin \omega_0 t = L\frac{d^2I_1^P}{dt^2} + \frac{1}{C}(I^P - I_2^P)$$

$$0 = \left(\frac{2}{C}I_2^P - \frac{1}{C}I_1^P\right) + L\frac{d^2I_2^P}{dt^2}$$

and the I_i^G are the general solution of

$$0 = L\frac{d^2I_1^G}{dt^2} + \frac{1}{C}(I_1^G - I_2^G)$$

$$0 = \left(\frac{2}{C}I_2^G - \frac{1}{C}I_1^G\right) + L\frac{d^2I_2^G}{dt^2}$$

Any free parameters in the I_i^G are chosen to satisfy the boundary conditions.

We expect that the I_i^P will have oscillatory time behavior with frequency ω_0. In fact, a study of the equations shows that we can have a solution of the form $I_i^P(t) = I_i^P \sin \omega_0 t$. Upon inserting this form, we can remove the $\sin \omega_0 t$ factor, and obtain the algebraic problem

$$\begin{pmatrix} \frac{1}{C} & -\frac{1}{C} \\ -\frac{1}{C} & \frac{2}{C} \end{pmatrix}\begin{pmatrix} I_1^P \\ I_2^P \end{pmatrix} - \omega_0^2 L \begin{pmatrix} I_1^P \\ I_2^P \end{pmatrix} = \begin{pmatrix} -E\omega_0 \\ 0 \end{pmatrix}$$

This has exactly the form displayed in equation (2-14). To solve this equation, we can either invert the matrix

$$\begin{pmatrix} \frac{1}{C} - \omega_0^2 L & -\frac{1}{C} \\ -\frac{1}{C} & \frac{2}{C} - \omega_0^2 L \end{pmatrix}$$

by standard matrix inversion techniques, or we can use the Green's function method.

Let us use the Green's function technique. To apply this, we must examine the solutions of the eigenvector equations

$$\begin{pmatrix} \frac{1}{C} & -\frac{1}{C} \\ -\frac{1}{C} & \frac{2}{C} \end{pmatrix} \begin{pmatrix} I_1 \\ I_2 \end{pmatrix} = \omega^2 L \begin{pmatrix} I_1 \\ I_2 \end{pmatrix}$$

Notice that these are exactly the equations we get for I_i^G, provided we assume the exponential time dependence $e^{-i\omega t}$. Solving for the allowed frequencies, we find that the eigenvectors and eigenvalues are

$$\omega^2 L = \frac{1}{C}\left(\frac{3}{2} + \frac{\sqrt{5}}{2}\right); \begin{pmatrix} I_1^G \\ I_2^G \end{pmatrix} = \begin{pmatrix} 2 \\ -1-\sqrt{5} \end{pmatrix} \frac{1}{\sqrt{10+2\sqrt{5}}}$$

$$\omega^2 L = \frac{1}{C}\left(\frac{3}{2} - \frac{\sqrt{5}}{2}\right); \begin{pmatrix} I_1^G \\ I_2^G \end{pmatrix} = \begin{pmatrix} 2 \\ -1+\sqrt{5} \end{pmatrix} \frac{1}{\sqrt{10-2\sqrt{5}}}$$

The Green's function operator then takes the form

$$G = \frac{|1\rangle\langle 1|}{\lambda_1 - \lambda} + \frac{|2\rangle\langle 2|}{\lambda_2 - \lambda}$$

$$= \left(\frac{1}{10+2\sqrt{5}}\right) \frac{1}{\left[\frac{1}{C}\left(\frac{3}{2}+\frac{\sqrt{5}}{2}\right) - L\omega_0^2\right]} \begin{pmatrix} 4 & -2-2\sqrt{5} \\ -2-2\sqrt{5} & 6+2\sqrt{5} \end{pmatrix}$$

$$+ \left(\frac{1}{10-2\sqrt{5}}\right) \frac{1}{\left[\frac{1}{C}\left(\frac{3}{2}-\frac{\sqrt{5}}{2}\right) - L\omega_0^2\right]} \begin{pmatrix} 4 & -2+2\sqrt{5} \\ -2+2\sqrt{5} & 6-2\sqrt{5} \end{pmatrix}$$

We can use this operator to find the particular solution

$$\begin{pmatrix} I_1^P \\ I_2^P \end{pmatrix} = G \begin{pmatrix} -E\omega_0 \\ 0 \end{pmatrix} = \frac{-E\omega_0}{L^2C\left[\omega_0^4 - \frac{3\omega_0^2}{LC} + \frac{1}{L^2C^2}\right]} \begin{pmatrix} 2 - LC\omega_0^2 \\ 1 \end{pmatrix}$$

$$\begin{pmatrix} I_1(t) \\ I_2(t) \end{pmatrix} = \frac{-E\omega_0 \sin \omega_0 t}{L^2C\left[\omega_0^4 - \frac{3\omega_0^2}{LC} + \frac{1}{L^2C^2}\right]} \begin{pmatrix} 2 - LC\omega_0^2 \\ 1 \end{pmatrix}$$

$$+\frac{1}{\sqrt{10+2\sqrt{5}}}\begin{pmatrix} 2 \\ -1-\sqrt{5} \end{pmatrix}\left[A \sin \sqrt{\frac{1}{LC}\left(\frac{3}{2}+\frac{\sqrt{5}}{2}\right)}\,t\right.$$

$$\left.+ B \cos \sqrt{\frac{1}{LC}\left(\frac{3}{2}+\frac{\sqrt{5}}{2}\right)}\,t\right]$$

$$+\frac{1}{\sqrt{10-2\sqrt{5}}}\begin{pmatrix} 2 \\ -1+\sqrt{5} \end{pmatrix}\left[C \sin \sqrt{\frac{1}{LC}\left(\frac{3}{2}-\frac{\sqrt{5}}{2}\right)}\,t\right.$$

$$\left.+ D \cos \sqrt{\frac{1}{LC}\left(\frac{3}{2}-\frac{\sqrt{5}}{2}\right)}\,t\right]$$

The constants A, B, C, and D are then chosen to satisfy initial conditions.

PROBLEMS

2-3 Consider a system of charged beads connected by springs, and sliding on a rod:

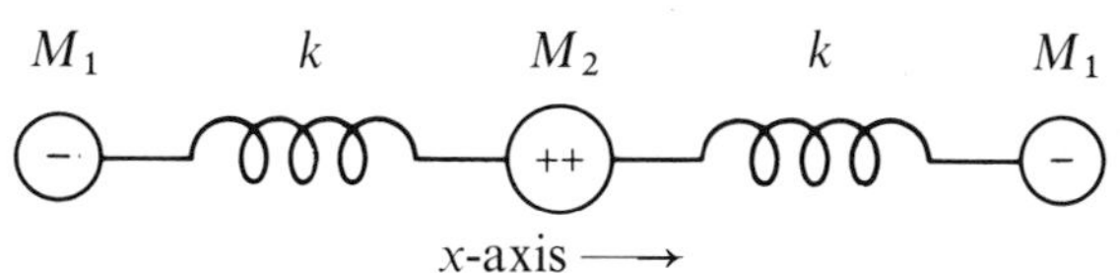

Assume that they are in a uniform electric field $\vec{E} = \hat{e}_x E \cos \omega_0 t$.

(a) How far apart should the masses be before we can neglect the Coulomb force in comparison with the force due to the imposed field?

(b) Neglect the Coulomb force and friction of the beads on the rod. Write the equations of motion for the masses.

(c) Using intuition about the time dependence of the solutions for the inhomogeneous equations found in (b), and their corresponding homogeneous equations, reduce these differential equations to algebraic equations.

Find the solutions of the homogeneous equations. Use these to write the Green's function operator for the inhomogeneous equation. Use this Green's function to find a particular solution of the inhomogeneous equation.

Give your time-dependent overall solutions, expressing all coefficients in terms of the initial positions and velocities of the masses.

2-3 PERTURBATION THEORY

Quite frequently in physics one begins with a simple situation, which is then disturbed slightly. The problem can generally be cast into a form such that the "simple" state is an eigenstate of a known operator, and the "disturbed" state is an eigenstate of a slightly different operator. The new eigenstates can be described in terms of the old eigenstates; it is possible to get a good approximation to the new state because it is only slightly different from the original one.

In this section we discuss several methods for obtaining an expansion of the new eigenstates in terms of the old ones. Again, we phrase our discussion in bra-ket notation although the only problems given here are in terms of matrix operators and column vector eigenfunctions. Application of these techniques to function spaces will be discussed in the problem section of Chapter 12.

BASIC METHOD

Suppose we are to solve the eigenvalue equation

$$H\,|\psi\rangle = \lambda\,|\psi\rangle$$

where $H = H_0 + H'$, and the eigenvectors of H_0 are already known:

$$H_0\,|n^0\rangle = \lambda_n^0\,|n^0\rangle \tag{2-19}$$

The problem is to find eigenvectors $|n\rangle$ and eigenvalues λ_n such that

$$H\,|n\rangle = \lambda_n\,|n\rangle \tag{2-20}$$

If H' is a small perturbation on H_0, we would expect the eigenvectors and eigenvalues of H to be nearly equal to those for H_0. Thus if the new eigenstates are expanded in terms of the old ones

$$|n\rangle = \sum_{m^0} a_{m^0}^n\,|m^0\rangle \tag{2-21}$$

the series should converge rather rapidly. (We expect $a_{n_0}^n$ to be nearly unity and the other coefficients to be small.)

Using this expansion, we find

$$H\,|n\rangle = \lambda_n\,|n\rangle = \lambda_n \sum_{m^0} a_{m^0}^n\,|m^0\rangle \tag{2-22}$$

$$\begin{aligned} H\,|n\rangle = (H_0 + H')\,|n\rangle &= H_0 \sum_{m^0} a_{m^0}^n\,|m^0\rangle + H' \sum_{m^0} a_{m^0}^n\,|m^0\rangle \\ &= \sum_{m^0} a_{m^0}^n \lambda_m^0\,|m^0\rangle + H' \sum_{m^0} a_{m^0}^n\,|m^0\rangle \end{aligned} \tag{2-23}$$

By equating the last expressions in (2-22) and (2-23) and taking the scalar product of the resulting equation with $|m^0\rangle$, we obtain

$$\begin{aligned} \lambda_n a_{m^0}^n &= a_{m^0}^n \lambda_m^0 + \sum_{p^0} a_{p_0}^n \langle m^0|\,H'\,|p^0\rangle \\ &= a_{m^0}^n \lambda_m^0 + a_{m^0}^n \langle m^0|\,H'\,|m^0\rangle + \sum_{p^0 \neq m^0} a_{p^0}^n \langle m^0 \mid H'\,|p^0\rangle \end{aligned} \tag{2-24}$$

Let $|n^0\rangle$ be the eigenvector of H_0 that is the limit of $|n\rangle$ as the perturbation is turned off. Then $a^n_{n^0}$ is large (close to 1) and $a^n_{p^0}$ is expected to be small for $p^0 \neq n^0$. For $m^0 = n^0$ in equation (2-24), one can therefore divide through by $a^n_{n^0}$, yielding

$$\lambda_n = \lambda_n^0 + \langle n^0 | H' | n^0 \rangle + \frac{1}{a^n_{n^0}} \sum_{p^0 \neq n^0} a^n_{p^0} \langle n^0 | H' | p^0 \rangle \tag{2-25}$$

Because

$$\frac{a^n_{p_0}}{a^n_{n^0}} \ll 1 \qquad p_0 \neq n^0 \tag{2-26}$$

the first-order correction to the eigenvalue (that is, the one of first order in H') is $\langle n^0 | H' | n^0 \rangle$ and

$$\lambda_n \approx \lambda_n^0 + \langle n^0 | H' | n^0 \rangle \tag{2-27}$$

In order to compute higher order corrections, one must know $a^n_{p_0}$ and $a^n_{n^0}$. In particular, to obtain a second-order approximation to the eigenvalue, one must have a first-order approximation to $a^n_{p_0}$, etc. The coefficients are found by considering

$$\langle m^0 | H - H^0 | n \rangle = \langle m^0 | H' | n \rangle \tag{2-28}$$

Using equations (2-19) and (2-20), this is just

$$\langle m^0 | \lambda_n - \lambda_m^0 | n \rangle = \langle m^0 | H' | n \rangle \tag{2-29}$$

so

$$a^n_{m^0} = \langle m^0 | n \rangle = \frac{\langle m^0 | H' | n \rangle}{\lambda_n - \lambda_m^0} \tag{2-30}$$

If $m^0 \neq n^0$ and $\lambda_n^0 \neq \lambda_m^0$, then $\lambda_n - \lambda_m^0$ will probably not be a small number. The first-order approximation to $a^n_{m_0}$ is then found by setting

$$\lambda_n \approx \lambda_n^0, \qquad |n\rangle \approx |n^0\rangle$$

in equation (2-30). This leads to

$$a^n_{m^0} \approx \frac{\langle m^0 | H' | n^0 \rangle}{\lambda_n^0 - \lambda_m^0} \qquad m^0 \neq n^0 \tag{2-31}$$

There is no physically interesting first-order correction to $a^n_{n^0}$. This can be seen by noticing that the new eigenstates must be normalized to 1, just as the old ones were. Hence

$$\langle n | n \rangle = 1 = |a^n_{n^0}|^2 + \sum_{m^0 \neq n^0} |a^n_{m^0}|^2 \tag{2-32}$$

The only first-order change in $a^n_{n^0}$ which would satisfy this equation (to first order) would be a pure imaginary term $i\delta$. This can be removed by redefining

$a^n_{n^0} \rightarrow e^{-i\delta} a^n_{n^0}$. As this only corresponds to a change of phase for $|n\rangle$, it is not crucial to the arguments developed here.

This gives enough information to compute λ_n to second order. In practice we would then shuttle back and forth, using the second-order estimate for λ_n to get a second-order estimate of $a^n_{p_0}$, then using this to get a third-order estimate of λ_n, and so on.

Example

Find the eigenvalues and eigenvectors of

$$H = \begin{pmatrix} -.03 & 5.01 \\ 5.01 & -.05 \end{pmatrix}$$

This problem is, of course, so simple that we can solve it exactly. We do this in order to have something to compare with the perturbation theory result. Using a pocket calculator, and rounding, we find

$$\lambda = -5.05001, \qquad \begin{pmatrix} .70640 \\ -.70781 \end{pmatrix}$$

$$\lambda = 4.97001, \qquad \begin{pmatrix} .70781 \\ .70640 \end{pmatrix}$$

We now check this against various approximations in the perturbation calculation.

Split the Hamiltonian as $H = H_0 + H'$,

$$H_0 = \begin{pmatrix} 0 & 5 \\ 5 & 0 \end{pmatrix} \qquad H_1 = \begin{pmatrix} -.03 & .01 \\ .01 & -.05 \end{pmatrix}$$

The eigenvectors and eigenvalues of H_0 are

$$\lambda_1^0 = 5 \qquad |1^0\rangle = \frac{1}{\sqrt{2}} \begin{pmatrix} 1 \\ 1 \end{pmatrix}$$

$$\lambda_2^0 = -5 \qquad |2^0\rangle = \frac{1}{\sqrt{2}} \begin{pmatrix} 1 \\ -1 \end{pmatrix}$$

The first-order corrections to the eigenvalues then are

$$\lambda_1^1 = \lambda_1^0 + \langle 1^0 | H' | 1^0 \rangle = 5 - .03 = 4.97$$

$$\lambda_2^1 = \lambda_2^0 + \langle 2^0 | H' | 2^0 \rangle = -5 - .05 = -5.05$$

and the first-order corrections to the eigenvectors are

$$a^1_{2^0} \approx \frac{\langle 2^0 | H' | 1^0 \rangle}{\lambda_1^0 - \lambda_2^0} = .001$$

$$|1'\rangle = \frac{1}{\sqrt{2}}\begin{pmatrix}1\\1\end{pmatrix} + \frac{10^{-3}}{\sqrt{2}}\begin{pmatrix}1\\-1\end{pmatrix} = \begin{pmatrix}.70781\\.70640\end{pmatrix}$$

$$a^2_{1^0} \approx \frac{\langle 1^0|\, H'\, |2^0\rangle}{\lambda_2^0 - \lambda_1^0} = -.001$$

$$|2'\rangle = \frac{1}{\sqrt{2}}\begin{pmatrix}1\\-1\end{pmatrix} - \frac{10^{-3}}{\sqrt{2}}\begin{pmatrix}1\\1\end{pmatrix} = \begin{pmatrix}.70640\\-.70781\end{pmatrix}$$

so we see that in this case the first-order corrections come very close to the exact answers.

What about the second-order corrections?

$$\lambda_n^2 = \lambda_n^0 + \langle n^0|\, H'\, |n^0\rangle + \sum_{p^0 \neq n^0} a^n_{p^0} \langle n^0|\, H'\, |p^0\rangle$$

so

$$\lambda_1^2 = 4.97 + a^1_{2^0} \langle 1^0|\, H'\, |2^0\rangle = 4.97001$$

$$\lambda_2^2 = -5.05 + a^2_{1^0} \langle 2^0|\, H'\, |1^0\rangle = -5.05001$$

The corrections to the eigenvectors are now calculated by using the first-order corrections:

$$a^n_{m^0} = \frac{\langle m^0|\, H'\, |n\rangle}{\lambda_n - \lambda_m^0} \approx \frac{\langle m^0|\, H'\, |n'\rangle}{\lambda_n^1 - \lambda_m^0} \quad (m^0 \neq n^0)$$

$$a^n_{n^0} = \sqrt{1 - \sum_{(m^0 \neq n^0)} |a^n_{m^0}|^2}$$

This gives (to the same accuracy as above)

$$|1^2\rangle = \frac{1.0}{\sqrt{2}}\begin{pmatrix}1\\1\end{pmatrix} + \frac{.000998}{\sqrt{2}}\begin{pmatrix}1\\-1\end{pmatrix} = \begin{pmatrix}.70781\\.70640\end{pmatrix}$$

$$|2^2\rangle = \frac{-.000998}{\sqrt{2}}\begin{pmatrix}1\\1\end{pmatrix} + \frac{1.0}{\sqrt{2}}\begin{pmatrix}1\\-1\end{pmatrix} = \begin{pmatrix}.70640\\-.70781\end{pmatrix}$$

We see that the iteration procedure converges rapidly.

A MORE ELEGANT FORMALISM

It is possible to cast the equations into a different form which is, for example, easier to program for computer applications. This rearranged series allows iteration to be performed always on the same equation, and eliminates shuttling back and forth between equations.

Begin with equations (2-28), (2-29), and (2-30):

$$\langle m^0|\, H - H_0\, |n\rangle = \langle m^0|\, \lambda_n - \lambda_m^0\, |n\rangle = \langle m^0|\, H'\, |n\rangle$$

and

$$\langle m^0 \mid n \rangle = \frac{\langle m^0 | H' | n \rangle}{\lambda_n - \lambda_m^0}$$

For the case $m^0 = n^0$, this becomes

$$\langle n^0 \mid n \rangle (\lambda_n - \lambda_n^0) = \langle n^0 | H' | n^0 \rangle \langle n^0 \mid n \rangle + \sum_{p^0 \neq n^0} \frac{\langle n^0 | H' | p^0 \rangle \langle p^0 | H' | n \rangle}{\lambda_n - \lambda_p^0} \tag{2-33}$$

if we make use of

$$|n\rangle = |n^0\rangle \langle n^0 \mid n \rangle + \sum_{m^0 \neq n^0} |m^0\rangle \langle m^0 \mid n \rangle$$

$$= |n^0\rangle \langle n^0 \mid n \rangle + \sum_{m^0 \neq n^0} |m^0\rangle \frac{\langle m^0 | H' | n \rangle}{\lambda_n - \lambda_m^0} \tag{2-34}$$

Repeated use of (2-34) in (2-33) yields

$$\langle n^0 \mid n \rangle (\lambda_n - \lambda_n^0) = \langle n^0 | H' | n^0 \rangle \langle n^0 \mid n \rangle + \sum_{p^0 \neq n^0} \frac{\langle n^0 | H' | p^0 \rangle \langle p^0 | H' | n^0 \rangle \langle n^0 \mid n \rangle}{\lambda_n - \lambda_p^0} + \cdots \tag{2-35}$$

Division by $\langle n^0 \mid n \rangle$ leads to

$$\lambda_n = \lambda_n^0 + \langle n^0 | H' | n^0 \rangle + \sum_{p^0 \neq n^0} \frac{\langle n^0 | H' | p^0 \rangle \langle p^0 | H' | n^0 \rangle}{\lambda_n - \lambda_p^0}$$

$$+ \sum_{m^0 \neq n^0} \sum_{p^0 \neq n^0} \frac{\langle n^0 | H' | p^0 \rangle \langle p^0 | H' | m^0 \rangle \langle m^0 | H' | n^0 \rangle}{(\lambda_n - \lambda_p^0)(\lambda_n - \lambda_m^0)} + \cdots \tag{2-36}$$

Everything in equation (2-36) is known except λ_n. The equation must still be solved by iteration, but only the one equation is necessary.

Example

Find the second-order corrections to the eigenvalues of

$$H_0 = \begin{pmatrix} 0 & 5 \\ 5 & 0 \end{pmatrix}$$

due to the perturbation

$$H' = \begin{pmatrix} -.03 & .01 \\ .01 & -.05 \end{pmatrix}$$

Our series (equation 2-36) shows that the second-order corrections can be computed using

$$\lambda_n^2 = \lambda_n^0 + \langle n^0| H' |n^0\rangle + \sum_{p^0 \neq n^0} \frac{\langle n^0| H' |p^0\rangle \langle p^0| H' |n^0\rangle}{\lambda_n^0 - \lambda_p^0}$$

Hence (using $|1^0\rangle$ and $|2^0\rangle$ as given in the previous example),

$$\lambda_1^2 = 5 - .03 + \frac{10^{-2}(10^{-2})}{10} = 4.97001$$

$$\lambda_2^2 = -5 - .05 + \frac{10^{-2}(10^{-2})}{-10} = -5.05001$$

as we found before.

THE CASE OF DEGENERATE EIGENVALUES

If some of the initial eigenvalues (that is, eigenvalues of H_0) are degenerate, the above perturbation expansion is not very useful. This is because some of the values $\lambda_n - \lambda_m^0$ that appear in the denominators of coefficients $\langle m^0 | n\rangle$ will be small (in fact some of them may be zero!). This destroys the utility of equation (2-36) as an expansion with successively smaller terms.

In order that the coefficients $\langle m^0 | n\rangle$ $(m_0 \neq n_0)$ be small, we must pick the eigenstates of H_0 in such a way that they "grow" into the eigenstates of H as the perturbation is turned on. This is automatically true for the eigenstates of non-degenerate eigenvalues (because then $\lambda_n - \lambda_m^0$ is always nonzero for $m^0 \neq n^0$). The only difficulties arise in subspaces where H_0 is degenerate. In this case, however, any vector in the subspace is an eigenvector of H_0 with the same eigenvalue, λ_m^0. We can therefore choose as initial eigenstates those particular ones which will grow into eigenfunctions of H. We shall see that this can be achieved automatically to first order by picking vectors that are eigenfunctions of H' within the degenerate subspace.

In this new basis (which we denote the $|n'\rangle$ basis),

$$\langle m'| H_0 |n'\rangle = \delta_{n'm'}\lambda_{n'}^0$$

$$\langle m'| H' |n'\rangle = \delta_{n'm'}h'_{n'} \quad \text{if } n' \text{ and } m' \text{ are both in the degenerate subspace of } H_0 \tag{2-37}$$

$$\langle m'| H' |n'\rangle = h'_{n'm'} \qquad \text{otherwise}$$

Now both H_0 and H' are diagonal within the degenerate subspace, whereas only H_0 is diagonal outside the subspace. To approximate more closely the situation we had in the nondegenerate case, we separate the diagonal parts of the operator from the nondiagonal ones. Define a new division of the operator $H = H_1 + H_2$, where

$$\langle m'| H_1 |n'\rangle = \delta_{n'm'}\lambda'_{n'}$$

$$\lambda'_{n'} = \begin{cases} \lambda^0_{n'} + h'_{n'} & \text{if } n' \text{ is in the degenerate subspace} \\ \lambda^0_{n'} & \text{otherwise} \end{cases} \tag{2-38}$$

$$\langle m' | \, H_2 \, | n' \rangle = \begin{cases} 0 & \text{if both } n' \text{ and } m' \text{ are in the degenerate subspace} \\ h'_{n'm'} & \text{otherwise} \end{cases} \tag{2-39}$$

Then the states $|n'\rangle$ are eigenfunctions of H_1 with eigenvalues λ'_n and

$$\langle m' \mid n \rangle = \frac{\langle m' | \, H_2 \, | n \rangle}{\lambda_n - \lambda'_{m'}} \tag{2-40}$$

If H_1 is nondegenerate, H_2 can be used as a perturbation in the same way as discussed earlier in the section (that is, equations 2-25, 2-30, and 2-35 apply). Hence the $|n'\rangle$ basis, composed by choosing eigenvectors of H_0 that diagonalize H' within the degenerate subspace, is the set of vectors from which the true eigenvectors grow as $H_2(H')$ is turned on.

If H_1 is still partially degenerate, equation (2-40) can be used to compute coefficients $\langle m' \mid n \rangle$ by the approximation

$$\langle m' \mid n \rangle \approx \frac{\langle m' | \, H_2 \, | n' \rangle}{\lambda'_n - \lambda'_m}$$

only if m' and n' are not both in the remaining degenerate subspace. The other coefficients do not change to first order. (Why?) Higher order calculations with degenerate operators are more complicated and we will not consider them here.

Example

Calculate the first-order change in eigenvectors and eigenvalues of

$$H_0 = \begin{pmatrix} 5 & 0 & 0 \\ 0 & 0 & 5 \\ 0 & 5 & 0 \end{pmatrix}$$

due to the perturbation

$$H' = \begin{pmatrix} 0 & .02 & 0 \\ .02 & -.03 & .01 \\ 0 & .01 & -.05 \end{pmatrix}$$

The unperturbed eigenvectors and eigenvalues are

$$\lambda^0_1 = 5, \, |1^0\rangle = \begin{pmatrix} 1 \\ 0 \\ 0 \end{pmatrix}; \qquad \lambda^0_2 = 5, \, |2^0\rangle = \frac{1}{\sqrt{2}} \begin{pmatrix} 0 \\ 1 \\ 1 \end{pmatrix};$$

$$\lambda^0_3 = -5, \, |3^0\rangle = \frac{1}{\sqrt{2}} \begin{pmatrix} 0 \\ 1 \\ -1 \end{pmatrix}$$

In this basis the operator matrices become

$$H_0 \leftrightarrow \begin{pmatrix} 5 & 0 & 0 \\ 0 & 5 & 0 \\ 0 & 0 & -5 \end{pmatrix} \qquad H' \leftrightarrow \begin{pmatrix} 0 & \frac{.02}{\sqrt{2}} & \frac{.02}{\sqrt{2}} \\ \frac{.02}{\sqrt{2}} & -.03 & .01 \\ \frac{.02}{\sqrt{2}} & .01 & -.05 \end{pmatrix}$$

Next we diagonalize that part of H' which involves only the $|1^0\rangle$, $|2^0\rangle$ subspace. This submatrix,

$$\begin{pmatrix} 0 & \frac{.02}{\sqrt{2}} \\ \frac{.02}{\sqrt{2}} & -.03 \end{pmatrix}$$

has as its eigenvectors and eigenvalues

$$\lambda = .56155 \times 10^{-2}, \begin{pmatrix} .92941 \\ .36905 \end{pmatrix}; \qquad \lambda = -3.56155 \times 10^{-2}, \begin{pmatrix} .36905 \\ -.92941 \end{pmatrix}$$

If we transform to the basis

$$|1'\rangle = .92941\,|1^0\rangle + .36905\,|2^0\rangle$$

$$|2'\rangle = .36905\,|1^0\rangle - .92941\,|2^0\rangle$$

$$|3'\rangle = |3^0\rangle$$

we find

$$H_0 \leftrightarrow \begin{pmatrix} 5 & 0 & 0 \\ 0 & 5 & 0 \\ 0 & 0 & -5 \end{pmatrix}$$

$$H' \leftrightarrow \begin{pmatrix} .56155 \times 10^{-2} & 0 & 1.68343 \times 10^{-2} \\ 0 & -3.56155 \times 10^{-2} & -.40749 \times 10^{-2} \\ 1.68343 \times 10^{-2} & -.40749 \times 10^{-2} & -.05 \end{pmatrix}$$

This allows us to divide $H = H_0 + H'$ into $H = H_1 + H_2$ with matrices in the $|n'\rangle$ basis

$$H_1 \leftrightarrow \begin{pmatrix} 5.00562 & 0 & 0 \\ 0 & 4.96438 & 0 \\ 0 & 0 & -5 \end{pmatrix}$$

$$H_2 \leftrightarrow \begin{pmatrix} 0 & 0 & 1.68343 \times 10^{-2} \\ 0 & 0 & -.40749 \times 10^{-2} \\ 1.68343 \times 10^{-2} & -.40749 \times 10^{-2} & -.05 \end{pmatrix}$$

H_1 is not degenerate, and we can treat H_2 as a perturbation by the method discussed earlier. The first-order eigenvalues are then

$$\lambda_1^{(1)} = 5.00562 \qquad \lambda_2^{(1)} = 4.96438 \qquad \lambda_3^{(1)} = -5.05$$

and the first-order corrections to the eigenvectors are

$$\langle 1' | 2 \rangle \approx \frac{\langle 1' | H_2 | 2' \rangle}{\lambda_2' - \lambda_1'} = 0$$

$$\langle 1' | 3 \rangle \approx \frac{\langle 1' | H_2 | 3' \rangle}{\lambda_3^0 - \lambda_1^1} = -.16825 \times 10^{-2}$$

$$\langle 2' | 1 \rangle \approx 0$$

$$\langle 2' | 3 \rangle \approx .40895 \times 10^{-3}$$

$$\langle 3' | 1 \rangle \approx .16825 \times 10^{-2}$$

$$\langle 3' | 2 \rangle \approx -.40895 \times 10^{-3}$$

We therefore have as our first-order eigenvectors

$$|1\rangle = |1'\rangle + 1.6825 \times 10^{-3} |3'\rangle$$

$$= .92941 |1^0\rangle + .36905 |2^0\rangle + 1.6825 \times 10^{-3} |3^0\rangle$$

$$= (\text{in the } \textit{original} \text{ basis}) \begin{pmatrix} .92941 \\ .26215 \\ .25977 \end{pmatrix}$$

$$|2\rangle = \begin{pmatrix} .36905 \\ -.65748 \\ -.65690 \end{pmatrix} \qquad |3\rangle = \begin{pmatrix} -.00141 \\ .70640 \\ -.70781 \end{pmatrix}$$

It is straightforward to check that $|1\rangle$, $|2\rangle$, and $|3\rangle$ are normalized to 1 to the accuracy given, that they are orthogonal, and that they are indeed eigenvectors of $H = H_0 + H'$ with the eigenvalues given.

The numbers given above were obtained with a pocket calculator. Will more elaborate computing apparatus be necessary to handle the second-order corrections?

PROBLEMS

2-4 For $H = H_0 + H_1$ with

$$H_0 = \begin{pmatrix} E_1^0 & 0 & 0 \\ 0 & E_2^0 & 0 \\ 0 & 0 & E_3^0 \end{pmatrix}$$

and

$$H_1 = \begin{pmatrix} 0 & a & b \\ a^* & c & d \\ b^* & d^* & e \end{pmatrix}$$

find the eigenvalues of H to second order in H_1 and the eigenvectors to first order in H_1.

2-5 Set $E_1^0 = E_2^0$ in the above problem and find the eigenvalues and eigenvectors to the orders specified

2-6 For the operator

$$H = \begin{pmatrix} E & 0 & 0 & 0 \\ 0 & E+2 & \sqrt{2}A & 0 \\ 0 & \sqrt{2}A & E+2 & \sqrt{2}A \\ 0 & 0 & \sqrt{2}A & E+2 \end{pmatrix}$$

consider A to be a small perturbation and find the eigenvectors and eigenvalues to first order in A.

SUMMARY

The bra-ket notation simplifies discussion of vector spaces and allows us to neglect the matrix representation if we desire. This will be particularly helpful later on, when we discuss vector spaces with an infinite number of dimensions. Elements of the original vector space are written as kets $|v\rangle$, elements of the adjoint space are written as bras $\langle v|$, and operators H may be written in the form

$$H = \sum_{n,m} |n\rangle \langle n| H |m\rangle \langle m|$$

where the kets $|n\rangle$ constitute a basis for the vector space, and the bras $\langle m|$ constitute a basis for the adjoint space.

For a given operator H with eigenvectors $|n\rangle$ such that $H|n\rangle = \lambda_n |n\rangle$, the inhomogeneous equation $H|v\rangle - \lambda |v\rangle = |\omega\rangle$ ($\lambda \neq \lambda_n$ for all n) has as its solution

$$|v\rangle = G|\omega\rangle$$

where

$$G = \sum_n \frac{|n\rangle \langle n|}{\lambda_n - \lambda}$$

is an operator known as the Green's function. Thus knowledge of the eigenvectors and eigenvalues of an operator is sufficient to determine solutions to all linear equations involving the operator.

The eigenvector description of an operator also plays a useful role when we consider small changes, or perturbations, in the operator. These should be reflected in small changes in the eigenvalues and eigenvectors; therefore we expand these quantities in series form, with successive terms of the series containing higher orders of the perturbation.

If none of the eigenvalues of the original operator is degenerate, then the expansions may be carried out in a formal manner and we find (for a perturbation H')

$$\lambda_n = \lambda_n^0 + \langle n^0| H' |n^0\rangle + \frac{1}{a_{n^0}^n} \sum_{p^0 \neq n^0} a_{p^0}^n \langle n^0| H' |p^0\rangle$$

where

$$a_{m^0}^n \equiv \langle m^0 \,|\, n\rangle = \frac{\langle m^0| H' |n\rangle}{\lambda_n - \lambda_m^0}$$

These equations must be solved by iteration. The first-order solution is

$$\lambda_n \approx \lambda_n^0 + \langle n^0| H' |n^0\rangle$$

$$a_{n^0}^n \approx 1$$

$$a_{p^0}^n \approx \frac{\langle p^0| H' |n^0\rangle}{\lambda_n^0 - \lambda_p^0} \qquad n^0 \neq p^0$$

The equation for the eigenvalue can be cast into a different form that may be easier to iterate under some circumstances

$$\lambda_n = \lambda_n^0 + \langle n^0| H' |n^0\rangle + \sum_{p^0 \neq n^0} \frac{\langle n^0| H' |p^0\rangle \langle p^0| H' |n^0\rangle}{\lambda_n - \lambda_p^0} + \cdots$$

If two or more of the eigenvalues of the original matrix H_0 are identical (degenerate), then one must choose as basis states $|n'\rangle$ within the degenerate subspace those eigenfunctions of H_0 that are also eigenfunctions of H' (restricted to the degenerate subspace). Using this basis, the first-order corrections are

$$\lambda_n \approx \lambda_n^0 + \langle n'| H' |n'\rangle$$

$a_{m'}^n \approx \delta_{n'm'}$ if n' and m' are both in the degenerate subspace

$a_{p'}^n \approx \dfrac{\langle p'| H' |n^0\rangle}{\lambda_n^0 - \lambda_p'} =$ (to first order) $\dfrac{\langle p'| H' |n^0\rangle}{\lambda_n^0 - \lambda_p^0}$ if p' is in the degenerate subspace and n is not.

$a_{p^0}^n \approx \dfrac{\langle p^0| H' |n^0\rangle}{\lambda_n^0 - \lambda_p^0}$ if neither n^0 nor p^0 is in the degenerate subspace.

CHAPTER 3 Ordinary Differential Equations

For many years mathematical physics courses concentrated almost entirely on the solution of differential equations. Today a number of other methods have become useful, but application of most of them requires a command of differential equations. Hence study of the subject is still vital. This chapter may be divided roughly into two topics: (a) general methods for solving ordinary differential equations, and (b) the properties of solutions of certain physically important equations. In the next chapter we take up a third topic, the application of vector space ideas to these important solutions. All three are essential for an understanding of the "orthogonal function" expansions commonly used in solving physics problems.

Sections 3-1 and 3-2 review the theory of ordinary differential equations, with emphasis on linear equations. Methods for obtaining a solution in closed form, as well as the technique of power series expansions, are illustrated. Special attention is given to the Legendre and Bessel equations.

Additional useful properties of Legendre and Bessel functions are derived in Sections 3-3 and 3-4. Many of these derivations are similar for the two kinds of functions; hence we might suspect that the Legendre and Bessel equations are two examples of some general class of equations, and that we would learn more by studying the entire class. In fact, both equations belong to *two* classes: the class of hypergeometric equations, and the class of Sturm-Liouville equations. Properties of hypergeometric equations are discussed in Section 3-5; Sturm-Liouville equations and their solutions are taken up in the following chapter.

3-1 SOLUTION IN CLOSED FORM

The *order* and *degree* of a differential equation refer to the derivative of highest order after the equation has been rationalized. Thus, the equation

$$\frac{d^3y}{dx^3} + x\sqrt{\frac{dy}{dx}} + x^2y = 0$$

is of third order and second degree, since when rationalized it contains the term $(d^3y/dx^3)^2$.

We first recall some methods that apply particularly to first-order equations. If the equation can be written in the form

$$A(x)\,dx + B(y)\,dy = 0 \tag{3-1}$$

we say the equation is *separable*; the solution is found immediately by integrating.

Example

$$\frac{dy}{dx} + \left[\frac{1-y^2}{1-x^2}\right]^{1/2} = 0$$

$$\frac{dy}{(1-y^2)^{1/2}} + \frac{dx}{(1-x^2)^{1/2}} = 0$$

$$\sin^{-1} y + \sin^{-1} x = C$$

or, taking the sine of both sides,

$$x(1-y^2)^{1/2} + y(1-x^2)^{1/2} = \sin C = C'$$

More generally, it may be possible to integrate immediately an equation of the form

$$A(x, y)\,dx + B(x, y)\,dy = 0 \tag{3-2}$$

If the left side of (3-2) is the differential du of some function $u(x, y)$, we can integrate and obtain the solution

$$u(x, y) = C$$

Such an equation is said to be *exact*. A necessary and sufficient condition that equation (3-2) be exact is

$$\frac{\partial A}{\partial y} = \frac{\partial^2 u}{\partial y\,\partial x} = \frac{\partial^2 u}{\partial x\,\partial y} = \frac{\partial B}{\partial x} \tag{3-3}$$

Example

$$(x + y)\,dx + x\,dy = 0 \tag{3-4}$$

$$A = x + y \qquad B = x$$

$$\frac{\partial A}{\partial y} = \frac{\partial B}{\partial x} = 1$$

Thus (3-4) may be written in the form $du = 0$ for $u = xy + \frac{1}{2}x^2$, and our solution is

$$xy + \tfrac{1}{2}x^2 = C$$

INTEGRATING FACTORS

Sometimes we can find a function $\lambda(x, y)$, such that

$$\lambda(A\,dx + B\,dy)$$

is an exact differential, although $A\,dx + B\,dy$ may not have been. We call such a function λ an *integrating factor*. One can show that such factors always exist (for a first-order equation), but there is no simple way of finding them.

Consider the general linear first-order equation

$$\frac{dy}{dx} + f(x)y = g(x) \tag{3-5}$$

(An equation is *linear* if it contains y and its derivatives to only the first power.) Let us try to find an integrating factor $\lambda(x)$ such that the differential

$$\lambda(x)\,[dy + f(x)y\,dx] = \lambda(x)g(x)\,dx$$

is exact. The right side is immediately integrable, and our criterion (3-3) that the left side be exact is

$$\frac{d\lambda(x)}{dx} = \lambda(x)f(x)$$

This equation is separable, and its solution

$$\lambda(x) = \exp\left[\int^x f(x')\,dx'\right] \tag{3-6}$$

is the integrating factor we were looking for.

Example

$$xy' + (1 + x)y = e^x$$

$$y' + \left(\frac{1 + x}{x}\right)y = \frac{e^x}{x}$$

The integrating factor is $\exp\left\{\int^x [(1 + x')/x']\,dx'\right\} = xe^x$

$$xe^x\left[y' + \left(\frac{1 + x}{x}\right)y\right] = e^{2x}$$

Now our equation is exact; integrating both sides gives

$$xe^x y = \int^x e^{2x'}\,dx' = \frac{1}{2}e^{2x} + C$$

$$y = \frac{e^x}{2x} + \frac{C}{x}e^{-x}$$

CHANGES OF VARIABLE

One can often simplify a differential equation by making a judicious change of variable. For example, the equation

$$y' = f(ax + by + c)$$

becomes separable if one introduces the new dependent variable $v = ax + by + c$. As another example, the so-called *Bernoulli equation*

$$y' + f(x)y = g(x)y^n \tag{3-7}$$

becomes linear if one sets $v = y^{1-n}$. (This substitution becomes "obvious" if the equation is first divided by y^n.)

A function $f(x, y, \ldots)$ in any number of variables is said to be *homogeneous* of degree r in these variables if

$$f(ax, ay, \ldots) = a^r f(x, y, \ldots)$$

A first-order differential equation

$$A(x, y)\,dx + B(x, y)\,dy = 0 \tag{3-8}$$

is said to be homogeneous if A and B are homogeneous functions of the same degree. The substitution $y = vx$ makes the homogeneous equation (3-8) separable, for it becomes

$$A(x, vx)\,dx + B(x, vx)\left[x\,dv + v\,dx\right] = 0$$

$$\left[A(x, vx) + vB(x, vx)\right]dx + xB(x, vx)\,dv = 0$$

Because A and B are homogeneous,

$$A(x, vx) = x^r A(1, v) \qquad B(x, vx) = x^r B(1, v)$$

and (3-8) reduces to

$$\left[A(1, v) + vB(1, v)\right]dx + xB(1, v)\,dv = 0$$

which is obviously separable.

Note that this approach is related to dimensional arguments familiar from physics. A homogeneous function is simply a dimensionally consistent function, if $x, y, \ldots$ are all assigned the same dimension (for example, length). The variable $v = y/x$ is then a "dimensionless" variable.

This suggests a generalization of the idea of homogeneity. Suppose that the equation $A\,dx + B\,dy = 0$ is dimensionally consistent when the dimensionality of y is some power m of the dimensionality of x. That is, suppose

$$\begin{aligned} A(ax, a^m y) &= a^r A(x, y) \\ B(ax, a^m y) &= a^{r-m+1} B(x, y) \end{aligned} \tag{3-9}$$

Such equations are said to be *isobaric*. The substitution $y = vx^m$ reduces the equation to a separable one.

Example

$$xy^2(3y\,dx + x\,dy) - (2y\,dx - x\,dy) = 0 \tag{3-10}$$

Let us test to see if this is isobaric. Give x a "weight" 1 and y a weight m. The first term has weight $3m + 2$, and the second has weight $m + 1$. Therefore, the equation is isobaric with weight $m = -\frac{1}{2}$.

This suggests introducing the "dimensionless" variable $v = yx^{1/2}$. To avoid fractional powers, we instead let

$$v = y^2 x \qquad x = \frac{v}{y^2} \qquad dx = \frac{dv}{y^2} - \frac{2v\,dy}{y^3}$$

Equation (3-10) then reduces to

$$(3v - 2)y\,dv + 5v(1 - v)\,dy = 0$$

which is separable.

An equation of the form

$$(ax + by + c)\,dx + (ex + fy + g)\,dy = 0 \tag{3-11}$$

where $a, \ldots, g$ are constants, may be made homogeneous by a substitution

$$x = X + \alpha \qquad y = Y + \beta$$

where α and β are suitably chosen constants. (This holds provided $af \neq be$; if $af = be$, equation (3-11) is even more trivial.)

CLAIRAUT EQUATION

An equation of the form

$$y - xy' = f(y') \tag{3-12}$$

is known as a *Clairaut equation*. To solve it, differentiate both sides with respect to x. The result is

$$y''[f'(y') + x] = 0$$

We thus have two possibilities. If we set $y'' = 0$, $y = ax + b$, and substitution back into the original equation (3-12) gives $b = f(a)$. Thus $y = ax + f(a)$ is the general solution. However, we also have the possibility $f'(y') + x = 0$. Eliminating y' between this equation and the original differential equation (3-12), we obtain a solution with no arbitrary constants. Such a solution is known as a *singular solution.*

Example

$$y = xy' + (y')^2 \tag{3-13}$$

This is a Clairaut equation with general solution

$$y = cx + c^2$$

However, we must also consider the possibility $2y' + x = 0$, which gives

$$x^2 + 4y = 0$$

This singular solution is an *envelope* of the family of curves given by the general solution, as shown in Figure 3-1. The dotted parabola is the

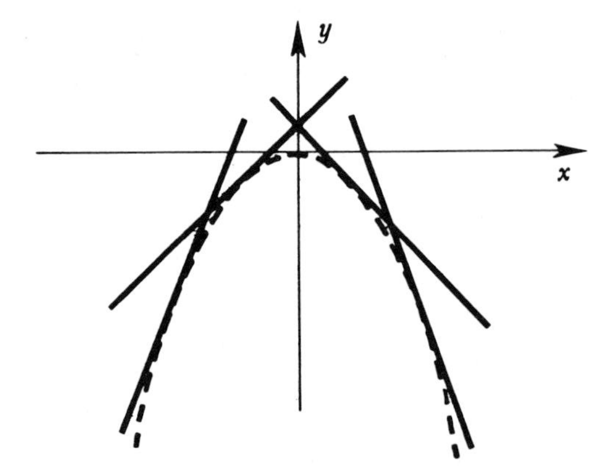

FIGURE 3-1 Solutions of the differential equation (3-13) and their envelope

singular solution, and the straight lines tangent to the parabola are the general solution.

There are various other types of singular solutions, but we shall not discuss them here.

LINEAR EQUATIONS WITH CONSTANT COEFFICIENTS

Next we review some methods that are useful for higher-order differential equations. In contrast to the first-order equations discussed above, where all solutions contained one arbitrary constant, the solution of an nth-order linear equation will in general contain n arbitrary constants. To understand the significance of these constants, consider the linear equation

$$a_n(x)y^{(n)} + a_{n-1}(x)y^{(n-1)} + \cdots + a_1(x)y' + a_0(x)y = f(x)$$

If $f(x) = 0$, the equation is said to be *homogeneous*; otherwise it is *inhomogeneous*. It can be shown that the nth order homogeneous linear equation has n linearly independent solutions. These n functions span the space of solutions for the equation and any solution of the homogeneous equation can be written as a linear combination of them. Thus the n arbitrary constants may be thought of as coordinates in an n-dimensional vector space, which has as basis vectors the n linearly independent solutions. These n constants assume particular values if the solution is required to satisfy certain boundary conditions. The general solution of an inhomogeneous equation is the sum of the general solution of the corresponding homogeneous equation (the so-called *complementary function*) and any solution of the inhomogeneous equation (the so-called *particular integral*).

One important type of higher order equation is the linear equation with constant coefficients

$$a_n y^{(n)} + a_{n-1} y^{(n-1)} + \cdots + a_1 y' + a_0 y = f(x) \qquad \text{(3-14)}$$

A solution of the homogeneous equation (3-14) with $f(x) = 0$ is e^{mx}. Substitution into the homogeneous equation gives

$$a_n m^n + a_{n-1} m^{n-1} + \cdots + a_0 = 0$$

If the n roots of this equation are $m_1, m_2, \ldots, m_n$, the complementary function is

$$c_1 e^{m_1 x} + \cdots + c_n e^{m_n x} \qquad (c_i \text{ are arbitrary constants})$$

Suppose two roots are the same, $m_1 = m_2$. Then we have only $n - 1$ solutions, and we need another. Imagine a limiting procedure in which m_2 approaches m_1. Then

$$\frac{e^{m_2 x} - e^{m_1 x}}{m_2 - m_1}$$

is a solution, and, as m_2 approaches m_1, this solution becomes

$$\left.\frac{d}{dm} e^{mx}\right|_{m=m_1} = x e^{m_1 x}$$

This is our additional solution. If three roots are equal, $m_1 = m_2 = m_3$, then the three solutions are

$$e^{m_1 x}, \qquad x e^{m_1 x}, \qquad x^2 e^{m_1 x}$$

and so on. Arguments involving similar limiting procedures are frequently useful; see, for example, the discussion in Section 3-2 about finding a second solution to Bessel's equation.

A particular integral is generally harder to find. If $f(x)$ has only a finite number of linearly independent derivatives, then the *method of undetermined coefficients* is quite straightforward. This method is commonly used when $f(x)$ is a linear combination of terms of the form x^n, e^{ax}, $\sin kx$, $\cos kx$, or, more generally,

$$x^n e^{mx} \cos \alpha x \qquad x^n e^{mx} \sin \alpha x$$

Take for $y(x)$ a linear combination of $f(x)$ and its independent derivatives and determine the coefficients by requiring that $y(x)$ obey the differential equation.

Example

$$y'' + 3y' + 2y = e^x \tag{3-15}$$

To find the complementary function, substitute $y = e^{mx}$:

$$m^2 + 3m + 2 = 0$$

$$m = -1, -2$$

$$y = c_1 e^{-x} + c_2 e^{-2x}$$

To find a particular integral, try $y = Ae^x$. Substitution into the differential equation (3-15) gives

$$6A = 1 \qquad A = \tfrac{1}{6}$$

Thus, the general solution is

$$y = \tfrac{1}{6}e^x + c_1 e^{-x} + c_2 e^{-2x}$$

If $f(x)$, or a term in $f(x)$, is also part of the complementary function, the particular integral may contain this term and its derivatives multiplied by some power of x. To see how this works, solve the above example (3-15) with the right-hand side e^x replaced by e^{-x}.

There are several formal devices for obtaining particular integrals. If D means d/dx, then we can write our equation (3-14) as

$$(D - m_1)(D - m_2)\dots(D - m_n)y = f(x) \tag{3-16}$$

A formal solution of (3-16) is

$$y = \frac{f(x)}{(D - m_1)\dots(D - m_n)}$$

or, expanding by partial fraction techniques,

$$y = A_1 \frac{f(x)}{D - m_1} + \cdots + A_n \frac{f(x)}{D - m_n}$$

What does $f(x)/(D - m)$ mean? It is the solution of $(D - m)y = f(x)$, which is a first-order linear equation whose solution is trivial. This method reduces the solution of an nth-order differential equation to the solving of n first-order equations.

Alternatively, we can just peel off the factors in (3-16) one at a time. That is,

$$(D - m_2)(D - m_3)\dots(D - m_n)y = \frac{f(x)}{D - m_1}$$

We evaluate the right side, divide by $D - m_2$, evaluate again, and so on.

Example

The equation

$$\frac{d^2y}{dx^2} - 3\frac{dy}{dx} + 2y = x$$

can be factored as $(D - 2)(D - 1)y = x$, hence

$$y = \frac{x}{(D - 2)(D - 1)} = \frac{x}{D - 2} - \frac{x}{D - 1}$$

The equation $(D - 2)g_1(x) = x$ has the solution $g_1(x) = -x/2 - \frac{1}{4}$. Likewise the equation $(D - 1)g_2(x) = x$ has the solution $g_2(x) = -x - 1$. Thus a particular solution is

$$y = g_1 - g_2 = \frac{x}{2} + \frac{3}{4}$$

VARIATION OF PARAMETERS

Next, we consider the very important method known as *variation of parameters* for obtaining a particular integral. This method applies equally well to linear equations with nonconstant coefficients. Before giving a general discussion of the method and applying it to an example, we shall digress briefly on the subject of *osculating parameters.*

Suppose we are given two linearly independent functions $y_1(x)$ and $y_2(x)$. By means of these we can define the two-parameter family of functions

$$c_1y_1(x) + c_2y_2(x) \tag{3-17}$$

Now consider some arbitrary function $y(x)$. Can we represent it by an appropriate choice of c_1 and c_2 in (3-17)? Clearly, the answer in general is no. Let us try the more modest approach of *approximating* $y(x)$ in the neighborhood of some fixed point $x = x_0$ by a curve of the family (3-17). Since there are two parameters at our disposal, a natural choice is to fit the *value* $y(x_0)$ and *slope* $y'(x_0)$ exactly. That is, c_1 and c_2 are determined from the two simultaneous equations

$$\begin{aligned} y(x_0) &= c_1y_1(x_0) + c_2y_2(x_0) \\ y'(x_0) &= c_1y_1'(x_0) + c_2y_2'(x_0) \end{aligned} \tag{3-18}$$

The c_1 and c_2 obtained in this way vary from point to point (that is, as x_0 varies) along the curve $y(x)$. They are called *osculating parameters* (from the Latin, *osculari*, to kiss) because the curve they determine fits the curve $y(x)$ as closely as possible at the point in question.

One can, of course, generalize to an arbitrary number N of functions y_i and parameters c_i. One chooses the c_i to reproduce the function $y(x)$ and its first $N - 1$ derivatives at the point x_0.

We now return to the problem of solving linear differential equations. For simplicity, we shall restrict ourselves to second-order equations. Consider the inhomogeneous equation

$$p(x)y'' + q(x)y' + r(x)y = s(x) \tag{3-19}$$

and suppose we know the complementary function to be $c_1y_1(x) + c_2y_2(x)$. Let us seek a solution of (3-19) of the form

$$y = u_1(x)y_1(x) + u_2(x)y_2(x) \tag{3-20}$$

where the $u_i(x)$ are functions to be determined. In order to substitute (3-20) into (3-19), we must evaluate y' and y''. From (3-20),

$$y' = u_1y_1' + u_2y_2' + u_1'y_1 + u_2'y_2 \tag{3-21}$$

Before going on to calculate y'', we observe that it would be convenient to impose the condition that the sum of the last two terms in (3-21) vanish, that is,

$$u_1'y_1 + u_2'y_2 = 0 \tag{3-22}$$

This prevents second derivatives of the u_i from appearing, for now

$$y' = u_1y_1' + u_2y_2' \tag{3-23}$$

and differentiating gives

$$y'' = u_1y_1'' + u_2y_2'' + u_1'y_1' + u_2'y_2' \tag{3-24}$$

Note that the condition (3-22) not only simplifies the subsequent algebra but also ensures that u_1 and u_2 are exactly the osculating parameters discussed above; compare (3-20) and (3-23) with (3-18). [Note also that since we have replaced the single dependent variable, $g(x)$, by two, $u_1(x)$ and $u_2(x)$, in (3-20), we should expect to apply some auxiliary condition such as (3-22).]

The rest of the procedure is straightforward. If (3-20), (3-23), and (3-24) are substituted into the original differential equation (3-19) and we use the fact that y_1 and y_2 are solutions of the homogeneous equation, we obtain

$$p(x)(u_1'y_1' + u_2'y_2') = s(x) \tag{3-25}$$

Hence (3-22) and (3-25) are simultaneous linear equations for the u_i', and can be solved in a straightforward manner. The student is urged to complete the solution.

Example

Consider the differential equation

$$x^2y'' - 2y = x \tag{3-26}$$

The complementary function, that is, the general solution of

$$x^2y'' - 2y = 0 \tag{3-27}$$

is most easily found by noting that $y = x^m$ is a natural trial solution. (In fact, $y = x^m$ is an obvious trial solution for any linear differential equation of the form $c_n x^n y^{(n)} + c_{n-1}x^{n-1}y^{n-1} + \cdots + c_1 xy' + c_0 y = 0$.) Substitution into (3-27) shows that $m = 2$ or -1, so that the complementary function is

$$c_1 x^2 + \frac{c_2}{x}$$

Therefore, we are led to try to find a solution of (3-26) of the form

$$y = u_1 x^2 + \frac{u_2}{x} \tag{3-28}$$

We differentiate, obtaining

$$y' = 2xu_1 - \frac{1}{x^2}u_2 + x^2 u_1' + \frac{1}{x}u_2'$$

and impose the condition

$$x^2 u_1' + \frac{u_2'}{x} = 0 \tag{3-29}$$

Then

$$y' = 2xu_1 - \frac{1}{x^2}u_2$$

and

$$y'' = 2u_1 + \frac{2}{x^3}u_2 + 2xu_1' - \frac{1}{x^2}u_2' \tag{3-30}$$

Substituting (3-28) and (3-30) back into the differential equation (3-26), we obtain

$$2x^3 u_1' - u_2' = x$$

Solving this together with (3-29) gives

$$u_1' = \frac{1}{3x^2} \qquad u_2' = -\frac{x}{3}$$

$$u_1 = -\frac{1}{3x} + c_1 \qquad u_2 = -\frac{x^2}{6} + c_2$$

Then the general solution of (3-26) is

$$y = -\frac{x}{2} + c_1x^2 + \frac{c_2}{x}$$

OTHER DEVICES

Changes of variable sometimes help. We shall discuss one general transformation that is particularly useful. Consider the second-order linear equation

$$y'' + f(x)y' + g(x)y = 0 \tag{3-31}$$

The substitution

$$y = v(x)p(x) \tag{3-32}$$

will give another linear equation in $v(x)$ that may be easier to solve for certain choices of $p(x)$. The resulting equation is

$$v'' + \left(\frac{2p'}{p} + f\right)v' + \left(\frac{p'' + fp' + gp}{p}\right)v = 0 \tag{3-33}$$

If we know one solution of the original equation (3-31) we can take p to be that solution and thereby eliminate the term in v from (3-33). This is very useful because we can then find the general solution by two straightforward integrations.

Alternatively, we may choose

$$p = \exp\left[-\tfrac{1}{2}\int f(x)\,dx\right] \tag{3-34}$$

and eliminate the first-derivative term of (3-33). This procedure is helpful as an aid to recognizing equations, and it is especially useful in connection with approximate methods of solution.

In conclusion, we shall briefly mention some other devices for solving differential equations. For details, refer to a text on differential equations such as Burkill (5), Ince (19), or Forsyth (14).

If the dependent variable y is absent, let $y' = p$ be the new dependent variable. This lowers the order of the equation by one.

If the independent variable x is absent, let y be the new independent variable and $y' = p$ be the new dependent variable. This also lowers the order of the equation by one.

If the equation is homogeneous in y, let $v = \log y$ be a new dependent variable. The resulting equation will not contain v, and the substitution $v' = p$ will then reduce the order of the equation by one.

If the equation is isobaric when x is given the weight 1 and y the weight m, the change in dependent variable $y = vx^m$, followed by the change in independent variable $u = \log x$, gives an equation in which the new independent variable u is absent.

Always watch for the possibility that an equation is exact. Also consider the possibility of finding an integrating factor. For example, the commonly occurring equation $y'' = f(y)$ can be integrated immediately if both sides are multiplied by y'.

Laplace and Fourier transform techniques are discussed in Section 9-1.

PROBLEMS

Find the general solutions of Problems 3-1 to 3-20:

3-1 $x^2y' + y^2 = xyy'$

3-2 $y' = \dfrac{x\sqrt{1 + y^2}}{y\sqrt{1 + x^2}}$

3-3 $y' = \dfrac{a^2}{(x + y)^2}$

3-4 $y' + y \cos x = \dfrac{1}{2} \sin 2x$

3-5 $(1 - x^2)y' - xy = xy^2$

3-6 $2x^3y' = 1 + \sqrt{1 + 4x^2y}$

3-7 $y'' + y'^2 + 1 = 0$

3-8 $y'' = e^y$

3-9 $x(1 - x)y'' + 4y' + 2y = 0$

3-10 $(1 - x)y^2\, dx - x^3\, dy = 0$

3-11 $xy' + y + x^4y^4e^x = 0$

3-12 $(1 + x^2)y' + y = \tan^{-1} x$

3-13 $x^2y'^2 - 2(xy - 4)y' + y^2 = 0$ (general solution *and* singular solution)

3-14 $yy'' - y'^2 - 6xy^2 = 0$

3-15 $x^4yy'' + x^4y'^2 + 3x^3yy' - 1 = 0$

3-16 $x^2y'' - 2y = x$

3-17 $y''' - 2y'' - y' + 2y = \sin x$

3-18 $y^{iv} + 2y'' + y = \cos x$

3-19 $y'' + 3y' + 26 = \exp[e^x]$

3-20 $a^2 y''^2 = (1 + y'^2)^3$

3-21 The differential equation obeyed by the charge q on a capacitor C connected in series with a resistance R to a voltage

$$V = V_0 \left(\frac{t}{\tau}\right)^2 e^{-t/\tau}$$

is

$$R\frac{dq}{dt} + \frac{1}{C}q = V_0 \left(\frac{t}{\tau}\right)^2 e^{-t/\tau}$$

Find $q(t)$ if $q(0) = 0$.

3-22 In the activation of an indium foil by a constant slow neutron flux, the number N of radioactive atoms obeys the equation

$$\frac{dN}{dt} = \lambda N_S - \lambda n$$

where N_S is the constant number after "saturation." Find $N(t)$ if $N(0) = 0$.

3-23 Find the general solution of

$$A(x)y''(x) + A'(x)y'(x) + \frac{y(x)}{A(x)} = 0$$

where $A(x)$ is a known function and $y(x)$ is unknown.

3-24 Find the general solution of

$$xy'' + 2y' + n^2 xy = \sin \omega x$$

Hint: Eliminate the first derivative term.

3-25 Note that $y = x$ would be a solution of

$$(1 - x)y'' + xy' - y = (1 - x)^2$$

if the right side were zero. Use this fact to obtain the general solution of the equation as given.

3-26 Consider the differential equation

$$y'' + p(x)y' + q(x)y = 0$$

on the interval $a \le x \le b$. Suppose we know two solutions, $y_1(x)$ and $y_2(x)$, such that

$$y_1(a) = 0 \qquad y_2(a) \neq 0$$

$$y_1(b) \neq 0 \qquad y_2(b) = 0$$

Give the solution of the equation

$$y'' + p(x)y' + q(x)y = f(x)$$

which obeys the conditions $y(a) = y(b) = 0$, in the form

$$y(x) = \int_a^b G(x, x')f(x')\,dx'$$

where $G(x, x')$, the Green's function, involves only the solutions y_1 and y_2 and assumes different functional forms for $x' < x$ and $x' > x$.

Illustrate by solving

$$y'' + k^2 y = f(x)$$

$$y(a) = y(b) = 0$$

3-27 Find the general solution of the differential equation

$$xy'' + \frac{3}{x}y = 1 + x^3$$

in real form (no i's in answer).

3-2 POWER SERIES SOLUTIONS

Before discussing series solutions in general, consider a simple (although nonlinear) example:

$$y'' = x - y^2 \tag{3-35}$$

Try $y = c_0 + c_1 x + c_2 x^2 + \cdots$. Then (3-35) becomes

$$2c_2 + 6c_3 x + 12c_4 x^2 + \cdots = x - c_0^2 - 2c_0 c_1 x - (c_1^2 + 2c_0 c_2)x^2 - \cdots$$

Equating coefficients of equal powers of x gives

$$c_2 = -\tfrac{1}{2}c_0^2$$

$$c_3 = \tfrac{1}{6} - \tfrac{1}{3}c_0 c_1$$

$$c_4 = -\tfrac{1}{12}c_1^2 + \tfrac{1}{12}c_0^3 \qquad \text{etc.}$$

Suppose, for example, we want the solution with $y = 0$, $y' = 1$ at $x = 0$. Then

$$c_0 = 0, \qquad c_1 = 1, \qquad c_2 = 0, \qquad c_3 = \tfrac{1}{6}, \qquad c_4 = -\tfrac{1}{12}, \ldots$$

This method of solution is a very useful one, but we have been careless about justifying the method, establishing convergence of the series, and so on. We now briefly outline the general theory of series solutions of *linear* differential equations.

Consider the equation

$$\frac{d^n y}{dx^n} + f_{n-1}(x)\frac{d^{n-1}y}{dx^{n-1}} + \cdots + f_0(x)y = 0 \tag{3-36}$$

If $f_0(x), \ldots, f_{n-1}(x)$ are regular (in the sense of complex variable theory; see Chapter 7) at a point $x = x_0$, x_0 is said to be an *ordinary point* of the differential equation. In practice, this means that the $f_i(x)$ behave near $x = x_0$ like $(x - x_0)^{m_i}$, where m_i is a positive integer or zero.

Near an ordinary point, the general solution of a differential equation can be written as a Taylor series; the Taylor series is an ordinary power series

$$y = \sum_{m=0}^{\infty} c_m(x - x_0)^m \tag{3-37}$$

whose coefficients c_m may be conveniently found by substitution in the differential equation (3-36) as in the example above. This series will converge (see Chapter 5) for all x such that $|x - x_0| < |x_s - x_0|$ where x_s is the nearest singularity; a *singularity* is any point which is not an ordinary point.

If x_0 is not an ordinary point, but $(x - x_0)f_{n-1}(x)$, $(x - x_0)^2 f_{n-2}(x), \ldots,$ $(x - x_0)^n f_0(x)$ are regular at x_0, x_0 is said to be a *regular singular point* of the differential equation. It can be shown that near a regular singular point, we can always find at least one solution of the form

$$y = (x - x_0)^s \sum_{m=0}^{\infty} c_m(x - x_0)^m \qquad \text{with } c_0 \neq 0 \tag{3-38}$$

where the exponent s of the leading term will not necessarily be an integer. The series again converges in any circle that includes no singularities except for x_0.

The algebraic work involved in substituting the series (3-37) or (3-38) in a differential equation is simplified if $x_0 = 0$. Thus it is usually convenient to first make a translation to x_0 as origin, that is, to rewrite the equation in terms of a new independent variable, $z = x - x_0$.

If a point is neither an ordinary point nor a regular singularity, it is an *irregular singular point*.

Example: Solutions of Legendre's equation

We first consider an example of expansion about an ordinary point: *Legendre's differential equation* is

$$(1 - x^2)y'' - 2xy' + n(n + 1)y = 0 \tag{3-39}$$

The points $x = \pm 1$ are regular singular points. We shall expand about the ordinary point $x = 0$. Try

$$y = \sum_{m=0}^{\infty} c_m x^m = c_0 + c_1 x + c_2 x^2 + \cdots$$

Substitution into the differential equation (3-39) gives

$$c_2 = -\frac{n(n+1)}{2}c_0$$

$$c_3 = \frac{2-n(n+1)}{6}c_1$$

$$c_4 = -\frac{n(n+1)[6-n(n+1)]}{24}c_0$$

$$c_5 = \frac{[2-n(n+1)][12-n(n+1)]}{120}c_1$$

$$\vdots$$

The general recursion relation is

$$\frac{c_{i+2}}{c_i} = \frac{i(i+1)-n(n+1)}{(i+1)(i+2)} = \frac{(i+n+1)(i-n)}{(i+1)(i+2)} \tag{3-40}$$

Note that this is a two-term recursion relation; that is, it relates two coefficients only. Their indices differ by two. These facts could have been noticed at the beginning by inspection of the differential equation.

The general solution of the differential equation (3-39) is therefore

$$y = c_0\left[1 - n(n+1)\frac{x^2}{2!} + n(n+1)(n-2)(n+3)\frac{x^4}{4!} - + \cdots\right]$$
$$+ c_1\left[x - (n-1)(n+2)\frac{x^3}{3!} + (n-1)(n+2)(n-3)(n+4)\frac{x^5}{5!} - + \cdots\right] \tag{3-41}$$

Legendre's equation is frequently encountered in boundary value problems for quantum wave functions or electromagnetic fields. In these situations, the variable x stands for $\cos\theta$, with θ the angle from the z-axis in spherical polar coordinates. Hence, for physical applications, it is necessary to consider the behavior of the infinite series in (3-41) near the singular points $x = \pm 1$. If we let $i \to \infty$ in the recursion relation (3-40), we see that

$$\frac{c_{i+2}}{c_i} \to 1$$

Thus, if we go further and further out in either series, we find it more and more resembling a geometric series with x^2 being the ratio of successive terms. The sum of such a series approaches infinity as x^2 approaches 1. [This discussion is incomplete. Infinite series are discussed in more detail in Chapter 5, and the series of (3-41) is studied as a special case in Section 5-1. Their sums go to infinity logarithmically ($y \sim \ln(1-x^2)$).]

In cases of physical interest, however, we often require solutions $y(x)$ that are finite for $-1 \leq x \leq +1$. There are two ways we can arrange this:

1. Let $c_1 = 0$, and choose n to be one of the integers $\ldots, -5, -3, -1, 0, 2, 4, \ldots$. Then the first series in (3-41) terminates and the second is absent.

2. Let $c_0 = 0$, and choose n to be one of the integers $\ldots, -6, -4, -2, 1, 3, 5, \ldots$. Then the second series in (3-41) terminates, and the first is absent.

We see that if we wish a solution of the Legendre differential equation (3-39) that is finite on the interval $-1 \leq x \leq 1$, n must be an integer. The resulting solution $y(x)$ is a polynomial which, when normalized by the condition $y(1) = 1$, is known as a Legendre polynomial $P_n(x)$. [Note that $P_n(x) = P_{-n-1}(x)$.]

Suppose we had started by expanding about the regular singular point $x = 1$. The change of variables $t = x - 1$ in equation (3-39) leads to

$$t(t + 2)y'' + 2(t + 1)y' - n(n + 1)y = 0 \tag{3-42}$$

Since $t = 0$ is a regular singular point, substitution of the power series

$$y = t^s \sum_{m=0}^{\infty} c_m t^m$$

into (3-42) will give us a solution. The coefficient of t^{s-1} in the resulting equation is

$$2c_0 s^2 = 0$$

or

$$s^2 = 0$$

This is called the *indicial equation*. Its only root is $s = 0$. Setting $s = 0$ in the recursion relations for the coefficients leads to

$$c_{m+1} = \frac{-c_m[m(m + 1) - n(n + 1)]}{2(m + 1)^2} \tag{3-43}$$

In order that we have a finite polynomial solution, $c_{m+1} = 0$ for some m; hence, n must be a positive integer. Calculation of the solutions for $n = 0, 1, 2, \ldots$ etc. shows that this expansion gives the same polynomials as our previous expansion about $x = 0$.

There is, however, one difficulty about (3-43). For each n, we are forced to a single solution. Yet we know that a second-order equation should have two independent solutions. Clearly our method is adequate for finding only one of these; some refinement of the technique is necessary to find the other linearly independent solution.

FINDING A SECOND SOLUTION; APPLICATION TO BESSEL'S EQUATION

To illustrate the refined technique, consider *Bessel's equation* (another equation which commonly appears in electromagnetic boundary value problems)

$$x^2y'' + xy' + (x^2 - m^2)y = 0 \tag{3-44}$$

This has a regular singular point at $x = 0$, and thus the existence of a solution of the form

$$y = x^s \sum_{n=0}^{\infty} c_n x^n \qquad (c_0 \neq 0)$$

is guaranteed. Again we can see from the differential equation that a two-term recursion relation will be obtained. If we substitute into the differential equation, the coefficient of x^s is $c_0(s^2 - m^2) = 0$, which gives

$$s^2 - m^2 = 0 \tag{3-45}$$

as the indicial equation. Its roots are $s = \pm m$. Next consider the coefficient of x^{s+1}. It is $c_1[(s+1)^2 - m^2] = 0$. Thus $c_1 = 0$ except in the single case $m = \frac{1}{2}$, $s = -m = -\frac{1}{2}$; and in that case we can set $c_1 = 0$ since the terms thereby omitted are equivalent to those that make up the other solution, with $s = +m = +\frac{1}{2}$.

We therefore confine ourselves to even values of n in the sum, writing

$$y = x^{\pm m}(c_0 + c_2x^2 + c_4x^4 + \cdots)$$

The recursion relation is easily found to be

$$\frac{c_{n+2}}{c_n} = \frac{-1}{(s+n+2)^2 - m^2} = \frac{-1}{(n+2)(2s+n+2)} \tag{3-46}$$

Thus our solution is

$$y = c_0 x^s \left[1 - \frac{x^2}{4(s+1)} + \frac{x^4}{4 \cdot 8(s+1)(s+2)} - + \cdots\right] \tag{3-47}$$

Suitably normalized, this series is called a Bessel function; we shall discuss Bessel functions in more detail in Section 3-4.

If m is not an integer, we have two independent solutions to our equation, namely (3-47) with $s = \pm m$. If m is an integer (which we may assume positive or zero), we can only choose $s = +m$; for $s = -m$, all denominators in (3-47) beyond a certain term vanish. If we multiply through by $(s + m)$ before setting $s = -m$, to cancel the offending factors, we get a multiple of the solution with $s = +m$, as may be readily verified.

This is a situation reminiscent of our difficulty in Section 3-1 in obtaining independent solutions of a differential equation with constant coefficients. The resolution of the difficulty is quite analogous. We begin by letting $y(x, s)$ denote the series

$$y(x, s) = x^s \sum_{n=0}^{\infty} c_n x^n$$

with $c_0 = 1, c_1 = 0$, and all other coefficients given by (3-46), but without making use of the indicial equation (3-45). That is, we set

$$y(x, s) = x^s \left[1 - \frac{x^2}{(s+2+m)(s+2-m)} + \frac{x^4}{(s+2+m)(s+2-m)(s+4+m)(s+4-m)} - + \cdots \right] \tag{3-48}$$

If we abbreviate the Bessel differential operator as L,

$$L = x^2 \frac{d^2}{dx^2} + x \frac{d}{dx} + (x^2 - m^2)$$

then the series $Ly(x, s)$ will contain only a term in x^s, because the recursion relation (3-46) makes the coefficients of all higher powers vanish. We obtain [compare (3-45)]

$$Ly(x, s) = (s - m)(s + m)x^s$$

Again we see that s must equal $\pm m$ in order that $y(x, s)$ be a solution of Bessel's equation $Ly = 0$. However, if m is a positive integer, we have remarked that

$$[(s + m)y(x, s)]_{s=-m}$$

is a constant multiple of $y(x, m)$, and we must find a second solution. To do so, consider the result obtained by substituting $(s + m)y(x, s)$ into Bessel's equation. We get

$$L[(s + m)y(x, s)] = (s + m)^2(s - m)x^s$$

The derivative of the right side with respect to s vanishes at $s = -m$. Therefore,

$$\left\{ \frac{\partial}{\partial s} [(s + m)y(x, s)] \right\}_{s=-m}$$

is a solution of Bessel's equation, and is in fact the second solution we were looking for.

Example

Let us find a second solution of Bessel's equation for $m = 2$.

$$y(x, s) = x^s \left[1 - \frac{x^2}{s(s+4)} + \frac{x^4}{s(s+2)(s+4)(s+6)} - + \cdots \right]$$

$$(s+2)y(x,s) = x^2\left[(s+2) - \frac{(s+2)}{s(s+4)}x^2 + \frac{x^4}{s(s+4)(s+6)} - + \cdots\right] \tag{3-49}$$

Remembering that $d/ds\, x^s = x^s \ln x$ and

$$\frac{d}{dx}\left(\frac{uv\ldots}{w\ldots}\right) = \frac{uv\ldots}{w\ldots}\left(\frac{u'}{u} + \frac{v'}{v} + \cdots - \frac{w'}{w} - \cdots\right)$$

we obtain

$$\begin{aligned}\frac{\partial}{\partial s}[(s+2)y(x,s)] &= (s+2)y(x,s)\ln x \\ &\quad + x^s\left\{1 - \frac{(s+2)}{s(s+4)}\left[\frac{1}{(s+2)} - \frac{1}{s} - \frac{1}{s+4}\right]x^2\right. \\ &\quad \left. + \frac{1}{s(s+4)(s+6)}\left(-\frac{1}{s} - \frac{1}{s+4} - \frac{1}{s+6}\right)x^4 - + \ldots\right\}\end{aligned}$$

Setting $s = -2$ and observing that $[(s+2)y(x,s)]_{s=-2} = -\frac{1}{16}y(x,2)$, we obtain

$$\left\{\frac{\partial}{\partial s}[(s+2)y(x,s)]\right\}_{s=-2} = -\frac{1}{16}y(x,2)\ln x + x^{-2}\left(1 + \frac{x^2}{4} + \frac{x^4}{64} + \cdots\right) \tag{3-50}$$

This solution and $y(x, 2)$ are two independent solutions of Bessel's equation with $m = 2$.

Similar methods can be used to find the second independent solution to (3-42).

FACTORING OUT BEHAVIOR NEAR A SINGULARITY

Let us now consider the differential equation

$$\frac{d^2\psi}{dx^2} + (E - x^2)\psi = 0 \tag{3-51}$$

This is the Schrödinger equation for a one-dimensional quantum-mechanical harmonic oscillator, in appropriate units. If one tries a direct power series solution of (3-51) about $x = 0$, one obtains a three-term recursion relation. These are a little inconvenient, so that it is advantageous to look for a transformation of variables that will lead to a simpler equation.

A trick that often helps in such a situation is to "factor out" the behavior near some singularity or singularities. Where are the singularities of this equation? There are none in the finite z-plane, but we must digress for a moment on *singularities at infinity*.

Consider the differential equation

$$y'' + P(x)y' + Q(x)y = 0 \tag{3-52}$$

Recall that $x = 0$ is an ordinary point if P and Q are regular there, whereas $x = 0$ is a regular singular point if xP and x^2Q are regular there. Let $z = 1/x$. The equation becomes

$$\frac{d^2y}{dz^2} + \left[\frac{2}{z} - \frac{1}{z^2}\,p(z)\right]\frac{dy}{dz} + \frac{1}{z^4}\,q(z)y = 0 \tag{3-53}$$

where

$$p(z) = P(x) \qquad \text{and} \qquad q(z) = Q(x)$$

Then $x = \infty$ is an ordinary point or a singularity of equation (3-52) depending on whether $z = 0$ is an ordinary point or a singularity of equation (3-53). That is, $x = \infty$ is an ordinary point if $2x - x^2P(x)$ and $x^4Q(x)$ are regular there, and $x = \infty$ is a regular singular point if $xP(x)$ and $x^2Q(x)$ are regular there. This concludes our digression.

Using these criteria, we see that our differential equation (3-51) is highly singular at $x = \infty$. For large x, the equation is approximately

$$\frac{d^2\psi}{dx^2} - x^2\psi = 0$$

and the solutions are roughly

$$\psi \sim e^{\pm x^2/2} \tag{3-54}$$

That is, if we substitute either function (3-54) into the differential equation (3-51), the terms that are dominant at infinity cancel.

In quantum mechanics, physically acceptable solutions must not become infinite at $x \to \infty$; therefore, try

$$\psi = ye^{-x^2/2} \tag{3-55}$$

This change of variable does not ensure the desired behavior at infinity, of course, and we shall have to select solutions $y(x)$ that give this behavior. In fact, we may expect in general $y(x) \to e^{x^2}$ so as to give the divergent behavior $\psi \to e^{+x^2/2}$

The differential equation (3-51) becomes

$$y'' - 2xy' + (E - 1)y = 0 \tag{3-56}$$

(If we write $E - 1 = 2n$, (3-56) is known as the *Hermite differential equation.*) We can obtain the general solution of this equation in the form of a power series that will converge everywhere, and the recursion relation for the coefficients will contain only two terms. The recursion relation is

$$\frac{c_{m+2}}{c_m} = \frac{2m + 1 - E}{(m + 1)(m + 2)}$$

and the solution is

$$y = c_0\left[1 + (1 - E)\frac{x^2}{2!} + (1 - E)(5 - E)\frac{x^4}{4!} + \cdots\right]$$

$$+ c_1\left[x + (3 - E)\frac{x^3}{3!} + (3 - E)(7 - E)\frac{x^5}{5!} + \cdots\right] \tag{3-57}$$

If $E = 2n + 1$ where n is an integer, one series terminates after the term in x^n (the even or odd series, depending on n). The resulting polynomial, suitably normalized, is called a *Hermite polynomial* of order n, $H_n(x)$. The other series can be eliminated by setting its coefficient, c_0 or c_1, equal to zero, and the resulting solution (3-55) will go to zero at infinity.

If either series of (3-57) does not terminate, its behavior at large x is determined by the terms far out in the series where the recursion relation is approximately

$$\frac{c_{m+2}}{c_m} \approx \frac{2}{m}$$

Thus either series behaves for large x like e^{x^2}, and $\psi \to e^{x^2/2}$ as expected. A solution ψ, which remains bounded as $x \to \pm\infty$, is therefore only possible if $E = 2n + 1$ with integral n; that is

$$\psi = \psi_n(x) = H_n(x)e^{-x^2/2}$$

for

$$E = E_n = 2n + 1 \tag{3-58}$$

This is another example of how boundary conditions may impose restrictions on the acceptable values for a constant appearing in a differential equation. The acceptable solutions ψ_n are called *eigenfunctions* of the differential operator $-d^2/dx^2 + x^2$ belonging to the *eigenvalues* E_n.

As another example of series solutions, we consider briefly the *associated Legendre equation*

$$(1 - x^2)y'' - 2xy' + \left[n(n + 1) - \frac{m^2}{1 - x^2}\right]y = 0 \tag{3-59}$$

Note that when $m = 0$, this reduces to Legendre's equation. The origin is an ordinary point, and $x = \pm 1$ are regular singularities.

A straightforward attempt at a power series solution about $x = 0$ again yields a three-term recursion relation. Let us try to factor out the behavior at $x = \pm 1$. Let $x = \pm 1 + z$ and write the approximate form of the differential equation valid for $|z| \ll 1$. Near either singularity it is

$$zy'' + y' - \frac{m^2}{4z}y = 0$$

This has solutions $y = z^{+m/2}$ and $z^{-m/2}$, the former being well behaved if $m \geq 0$, as we may assume without loss of generality.

We are therefore led to make the change of variable $y = v(1 - x^2)^{m/2}$ in (3-59), thus factoring out the behavior at both singularities simultaneously. The equation becomes

$$(1 - x^2)v'' - 2(m + 1)xv' + [n(n + 1) - m(m + 1)]v = 0 \tag{3-60}$$

with a straightforward series solution about $x = 0$. The recursion relation for the coefficients is

$$\frac{c_{r+2}}{c_r} = \frac{(r + m)(r + m + 1) - n(n + 1)}{(r + 1)(r + 2)} = \frac{(r + m - n)(r + m + n + 1)}{(r + 1)(r + 2)} \tag{3-61}$$

Again we obtain even and odd series solutions for $v(x)$, both of which behave like $(1 - x^2)^{-m}$ near $x = \pm 1$ if they do not terminate. A bounded solution exists only if n and m are such that one series terminates after some term x^r. From (3-61) the condition is

$$(n - m) = r = \text{an integer} \geq 0$$

Usually in physical applications n and m are both integers. It may be verified that y is then simply a constant times

$$(1 - x^2)^{m/2} \left(\frac{d}{dx}\right)^m P_n(x) \tag{3-62}$$

which is called an *associated Legendre function*. These functions will be discussed in more detail in Section 4-3.

PROBLEMS

3-28 Consider the equation

$$\frac{d^2y}{dx^2} + \frac{2}{x}\frac{dy}{dx} + \left\{K + \frac{2}{x} - \frac{l(l + 1)}{x^2}\right\}y = 0$$

$$l = \text{nonnegative integer } 0 \leq x \leq \infty$$

Find all values of the constant K that can give a solution that is finite on the entire range of x (including ∞). An equation like this arises in solving the Schrödinger equation for the hydrogen atom.
Hint: Let $y = v/x$, then "factor out" the behavior at infinity.

3-29 For what values of the constant K does the differential equation

$$y'' - \left(\frac{1}{4} + \frac{K}{x}\right)y = 0 \qquad (0 < x < \infty)$$

have a nontrivial solution vanishing at $x = 0$ and $x = \infty$?

3-30 Find the values of the constant k for which the equation

$$xy'' - 2xy' + (k - 3x)y = 0$$

has a solution bounded on the range $0 \leq x < \infty$.

3-31 A solution of the differential equation

$$xy'' + 2y' + (E - x)y = 0$$

is desired such that $y(0) = 1$, $y(\infty) = 0$. For what values of E is this possible?

3-32 By considering equation (3-61) for $r \to \infty$, verify that $v(x)$ indeed behaves like $(1 - x^2)^{-m}$ as $x \to \pm 1$.

3-33 Bessel's equation for $m = 0$ is

$$x^2 y'' + xy' + x^2 y = 0$$

We have found one solution

$$J_0(x) = 1 - \frac{x^2}{4} + \frac{x^4}{64} - + \cdots$$

Show that a second solution exists of the form

$$J_0(x) \ln x + Ax^2 + Bx^4 + Cx^6 + \cdots$$

and find the first three coefficients A, B, C.

3-34 Consider the differential equation

$$xy'' + (2 - x)y' - 2y = 0$$

Give two solutions, one regular at the origin and having the value 1 there, the other of the form

$$\frac{1}{x} + A(x) \ln x + B(x)$$

where $A(x)$ and $B(x)$ are regular at the origin. Three terms of each series will suffice.

3-35 Our study of particular examples has demonstrated (for those cases) that the behavior of the "second" solution near a regular singular point is at worst like $(x - x_0)^{-m}$ or $(x - x_0)^{-m} \ln (x - x_0)$. Consider the equation

$$x^2 y' + y = 0$$

which has $x = 0$ as an irregular singular point. Find the solution and discuss its behavior near $x = 0$.

3-3 LEGENDRE FUNCTIONS

In our study of the Legendre differential equation (3-39)

$$(1 - x^2)\frac{d^2y}{dx^2} - 2x\frac{dy}{dx} + n(n+1)y = 0$$

we found two independent series solutions [equation (3-41)], such that if n is an integer, one or the other solution is simply a polynomial in x. Let us study these polynomials in somewhat greater detail.

A straightforward rewriting by means of factorials enables us to express the solution (3-41) in the form

$$y = c_0 \sum_k (-1)^k \frac{\left(\frac{n}{2}\right)!}{\left(\frac{n}{2}+k\right)!} \frac{\left(\frac{n}{2}\right)!}{\left(\frac{n}{2}-k\right)!} \frac{(n+2k)!}{n!} \frac{x^{2k}}{(2k)!}$$

$$+ c_1 \sum_k (-1)^k \frac{\left(\frac{n-1}{2}\right)!}{\left(\frac{n-1}{2}+k\right)!} \frac{\left(\frac{n-1}{2}\right)!}{\left(\frac{n-1}{2}-k\right)!} \frac{(n+2k)!}{n!} \frac{x^{2k+1}}{(2k+1)!}$$

If $n = 0, 2, 4, \ldots$, the first series becomes a polynomial. We shall introduce the new summation index $r = n/2 - k$, and write this solution as

$$c_0 \frac{\left[\left(\frac{n}{2}\right)!\right]^2}{n!}(-1)^{n/2} \sum_r (-1)^r \frac{(2n-2r)!}{(n-r)!\,r!} \frac{x^{n-2r}}{(n-2r)!}$$

If, on the other hand, $n = 1, 3, 5, \ldots$, it is the second series that concerns us. In this series we define $r = (n-1)/2 - k$, and obtain the solution

$$\frac{c_1}{2} \frac{\left[\left(\frac{n-1}{2}\right)!\right]^2}{n!}(-1)^{(n-1)/2} \sum_r (-1)^r \frac{(2n-2r)!}{(n-r)!\,r!} \frac{x^{n-2r}}{(n-2r)!}$$

It is now clear that for any non-negative integer n, even or odd, the polynomial

$$P_n(x) = K_n \sum_r (-1)^r \frac{(2n-2r)!}{(n-r)!\,r!} \frac{x^{n-2r}}{(n-2r)!} \qquad (K_n \text{ arbitrary})$$

is a solution of Legendre's equation.

Before fixing the normalization constant K_n, we shall engage in some more factorial manipulation, and write

$$P_n(x) = K_n \sum_r \frac{(-1)^r}{(n-r)!\,r!} \left(\frac{d}{dx}\right)^n x^{2n-2r}$$

$$= \frac{K_n}{n!} \left(\frac{d}{dx}\right)^n \sum_r (-1)^r \frac{n!}{r!(n-r)!} x^{2n-2r}$$

$$= \frac{K_n}{n!} \left(\frac{d}{dx}\right)^n (x^2-1)^n$$

As mentioned earlier when we discussed the solution of Legendre's equation, it is convenient to impose the normalization $P_n(1) = 1$. The resulting value of K_n may be found by inspection, since $P_n(1)$ equals $K_n/n!$ times the nth derivative of the product

$$\overbrace{(x-1)(x-1)\ldots(x-1)}^{n \text{ factors}} \quad \overbrace{(x+1)(x+1)\ldots(x+1)}^{n \text{ factors}}$$

evaluated at $x = 1$. If a single factor $(x - 1)$ survives after the n differentiations, setting $x = 1$ will cause that term to vanish. Thus the only nonvanishing contributions to $P_n(1)$ occur when each of the n differentiations turns one factor $(x - 1)$ into 1. There are $n!$ ways this can happen, and setting $x = 1$ gives $(1 + 1)^n = 2^n$ each time; the result is

$$P_n(1) = \frac{K_n}{n!} n!\, 2^n = 2^n K_n$$

and we choose, therefore,

$$K_n = \frac{1}{2^n}$$

The resulting expression

$$P_n(x) = \frac{1}{2^n\, n!} \left(\frac{d}{dx}\right)^n (x^2-1)^n \tag{3-63}$$

is called *Rodrigues' formula* for the *Legendre polynomials*. The first few are

$$\begin{aligned} P_0(x) &= 1 & P_3(x) &= \tfrac{1}{2}(5x^3 - 3x) \\ P_1(x) &= x & P_4(x) &= \tfrac{1}{8}(35x^4 - 30x^2 + 3) \\ P_2(x) &= \tfrac{1}{2}(3x^2 - 1). & P_5(x) &= \tfrac{1}{8}(63x^5 - 70x^3 + 15x) \end{aligned} \tag{3-64}$$

Notice that the P_k for even k are even functions of x, the P_k for odd k are odd functions of x, and $P_k(1) = 1$ for all k, as arranged.

These polynomials have important orthonormality properties. To derive them we use Rodrigues' expression (3-63). Consider the integral

$$I_{mn} = \int_{-1}^{1} P_m(x)P_n(x)\,dx \qquad (m < n)$$

$$= \frac{1}{2^{m+n}}\frac{1}{m!\,n!}\int_{-1}^{1}\left[\left(\frac{d}{dx}\right)^m (x^2-1)^m\right]\left[\left(\frac{d}{dx}\right)^n (x^2-1)^n\right]dx$$

We integrate by parts n times:

$$I_{mn} = -\frac{1}{2^{m+n}}\frac{1}{m!\,n!}\int_{-1}^{1}\left[\left(\frac{d}{dx}\right)^{m+1} (x^2-1)^m\right]\left[\left(\frac{d}{dx}\right)^{n-1} (x^2-1)^n\right]dx$$

$$= \frac{(-1)^n}{2^{m+n}m!\,n!}\int_{-1}^{1}\left[\left(\frac{d}{dx}\right)^{m+n} (x^2-1)^m\right](x^2-1)^n\,dx$$

Therefore,

$$I_{mn} = 0 \quad \text{since} \quad \left(\frac{d}{dx}\right)^{m+n}(x^2-1)^m = 0 \qquad \text{if } m < n$$

Suppose $m = n$. Then

$$I_{nn} = \int_{-1}^{1} [P_n(x)]^2\,dx = \frac{1}{2^{2n}(n!)^2}\int_{-1}^{1}\left[\left(\frac{d}{dx}\right)^n (x^2-1)^n\right]^2 dx$$

and n integrations by parts now give

$$I_{nn} = \frac{(-1)^n}{2^{2n}(n!)^2}\int_{-1}^{1}(x^2-1)^n\left(\frac{d}{dx}\right)^{2n}(x^2-1)^n\,\mathrm{d}x$$

$$= \frac{(2n)!}{2^{2n}(n!)^2}\int_{-1}^{1}(1-x^2)^n\,dx$$

Let $x = 2u - 1$.

$$I_{nn} = \frac{2(2n)!}{(n!)^2}\int_0^1 du(1-u)^n u^n$$

We will show in Chapter 10 that

$$\int_0^1 du(1-u)^n u^n = B(n+1, n+1) = \frac{n!\,n!}{(2n+1)!}$$

Therefore

$$I_{nn} = \frac{2}{2n+1}$$

and

$$\int_{-1}^{1} P_m(x)P_n(x)\,dx = \frac{2}{2n+1}\,\delta_{mn} \tag{3-65}$$

What is the second linearly independent solution of Legendre's equation, besides $P_n(x)$? We noted in the discussion following (3-42) that it cannot be expanded in a power series about $x = 1$. Also, our derivation of the second solution for Bessel's equation [equation (3-50)] would lead us to guess that the corresponding solution of Legendre's equation has a logarithmic singularity at $x = 1$. Let us check this by deriving it explicitly.

If we make the change of variable $y = vP_n(x)$ in Legendre's equation, we obtain the equation

$$(1 - x^2)v'' P_n(x) + v'[2(1 - x^2)P_n'(x) - 2xP_n(x)] = 0$$

The solution of this equation is straightforward and gives

$$v = c\int^x \frac{dx'}{(1 - x'^2)[P_n(x')]^2} + c'$$

Thus we have found a second solution; the conventional definition is

$$Q_n(z) = -P_n(z)\int_{+\infty}^{z} \frac{dz'}{(z'^2 - 1)[P_n(z')]^2} \tag{3-66}$$

Notice that $Q_n(z)$ approaches zero like $z^{-(n+1)}$ as $z \to \infty$. It can be shown [see Copson (8), p. 287, for example], that

$$Q_n(z) = \frac{1}{2}P_n(z)\ln\left(\frac{z+1}{z-1}\right) + f_{n-1}(z) \tag{3-67}$$

where $f_{n-1}(z)$ is a polynomial of degree $n - 1$ in z.

The polynomial $f_{n-1}(z)$ is determined by the condition that $Q_n(z) \to 0$ like $z^{-(n+1)}$ as $z \to \infty$. For example,

$$Q_0(z) = \frac{1}{2}\ln\left(\frac{z+1}{z-1}\right) = \frac{1}{z} + \frac{1}{3z^3} + \frac{1}{5z^5} + \cdots$$

$$\begin{aligned} Q_1(z) &= \frac{1}{2}z\ln\left(\frac{z+1}{z-1}\right) + f_0 \\ &= z\left(\frac{1}{z} + \frac{1}{3z^3} + \cdots\right) + f_0 \end{aligned}$$

Clearly, $f_0 = -1$ and

$$Q_1(z) = \frac{1}{2}z\ln\left(\frac{z+1}{z-1}\right) - 1, \text{ etc.}$$

PROBLEMS

3-36 Using the techniques illustrated in Section 3-2 by Bessel's equation, find the second solution for Legendre's equation by expansion in powers about $x = 1$ and compare with the Q_n's.

3-37 The *Wronskian* of two functions $f_1(x)$ and $f_2(x)$ is defined as the determinant

$$W(x) = \begin{vmatrix} f_1(x) & f_2(x) \\ f_1'(x) & f_2'(x) \end{vmatrix}$$

Find the Wronskian of $P_K(x)$ and $Q_K(x)$ (K an integer).

3-4 BESSEL FUNCTIONS

We have already encountered Bessel's equation (3-44)

$$x^2y'' + xy' + (x^2 - m^2)y = 0 \tag{3-68}$$

and the series solution

$$y(x) = x^m\left[1 - \frac{1}{m+1}\left(\frac{x}{2}\right)^2 + \frac{1}{(m+1)(m+2)}\frac{1}{2!}\left(\frac{x}{2}\right)^4 - + \cdots\right]$$

We define the Bessel function $J_m(x)$ as

$$\begin{aligned} J_m(x) &= \frac{1}{m!}\left(\frac{x}{2}\right)^m\left[1 - \frac{1}{m+1}\left(\frac{x}{2}\right)^2 + \frac{1}{(m+1)(m+2)}\frac{1}{2!}\left(\frac{x}{2}\right)^4 - + \cdots\right] \\ &= \sum_{r=0}^{\infty}\frac{(-1)^r}{r!\,\Gamma(m+r+1)}\left(\frac{x}{2}\right)^{m+2r} \end{aligned} \tag{3-69}$$

The *gamma function* $\Gamma(m+1) = m!$ may be used to define the factorial symbol for non-integer m (see Section 10-2).

Useful connections between adjacent Bessel functions and their derivatives, known as recursion relations, can be deduced immediately from the power series (3-69);

$$\begin{aligned} &J_{m-1}(x) + J_{m+1}(x) \\ &= \sum_{r=0}^{\infty}\left[\frac{(-1)^r}{r!\,\Gamma(m+r)}\left(\frac{x}{2}\right)^{m+2r-1} + \frac{(-1)^r}{r!\,\Gamma(m+r+2)}\left(\frac{x}{2}\right)^{m+2r+1}\right] \\ &= \left(\frac{x}{2}\right)^{m-1}\left\{\frac{1}{\Gamma(m)} + \sum_{r=1}^{\infty}\left[\frac{(-1)^r}{r!\,\Gamma(m+r)}\left(\frac{x}{2}\right)^{2r} - \frac{(-1)^r}{(r-1)!\,\Gamma(m+r+1)}\left(\frac{x}{2}\right)^{2r}\right]\right. \\ &= \left(\frac{x}{2}\right)^{m-1}\left[\frac{1}{\Gamma(m)} + \sum_{r=1}^{\infty}(-1)^r\left(\frac{x}{2}\right)^{2r}\frac{m}{r!\,\Gamma(m+r+1)}\right] \\ &= m\sum_{r=0}^{\infty}\frac{(-1)^r}{r!\,\Gamma(m+r+1)}\left(\frac{x}{2}\right)^{m+2r-1} \end{aligned}$$

Therefore,

$$J_{m-1}(x) + J_{m+1}(x) = \frac{2m}{x}J_m(x) \tag{3-70}$$

Similarly,

$$J_{m-1}(x) - J_{m+1}(x) = 2J'_m(x) \tag{3-71}$$

Adding and subtracting (3-70) and (3-71) gives

$$J_{m-1}(x) = \frac{m}{x} J_m(x) + J'_m(x) \tag{3-72}$$

$$J_{m+1}(x) = \frac{m}{x} J_m(x) - J'_m(x) \tag{3-73}$$

The first three Bessel functions of integral order are shown in Figure 3-2. Notice that J_0 is the only one that is nonzero at $x = 0$.

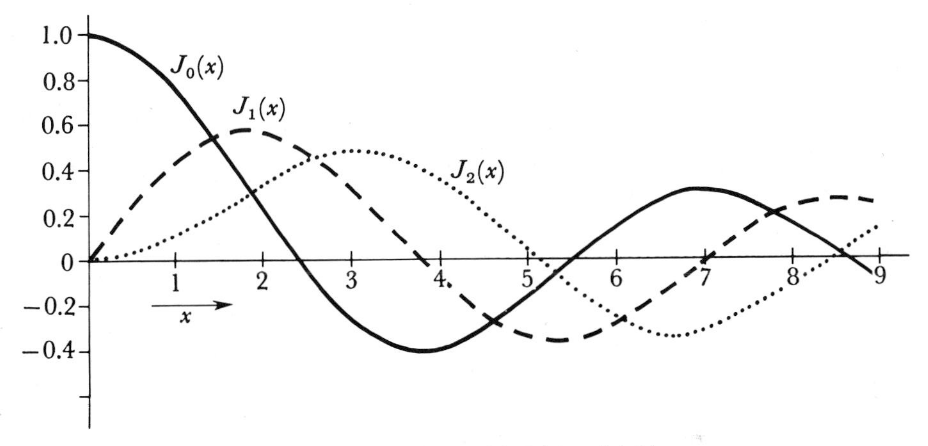

FIGURE 3-2 The Bessel functions $J_0(x)$, $J_1(x)$, and $J_2(x)$

Bessel functions of half-integral order may be simply expressed in terms of trigonometric functions. From the series solutions, one finds

$$J_{1/2}(x) = \left(\frac{2}{\pi x}\right)^{1/2} \sin x \qquad J_{-1/2}(x) = \left(\frac{2}{\pi x}\right)^{1/2} \cos x$$

Then, from the recursion relations (3-73) and (3-72),

$$J_{3/2}(x) = \sqrt{\frac{2}{\pi x}} \left(\frac{1}{x} \sin x - \cos x\right)$$

$$J_{-3/2}(x) = \sqrt{\frac{2}{\pi x}} \left(-\frac{1}{x} \cos x - \sin x\right)$$

$J_{\pm 5/2}$, $J_{\pm 7/2}$, etc, can be obtained similarly.

When m is not an integer, two independent solutions of Bessel's equation are given by $J_m(x)$ and $J_{-m}(x)$. These are not independent, however, when m is an integer, as discussed in Section 3-2. Therefore, one defines

$$Y_m(x) = \frac{\cos(m\pi)J_m(x) - J_{-m}(x)}{\sin m\pi} \tag{3-74}$$

Notice that if m is an integer, this is defined by the limiting process used in Sections 3-1 and 3-2. $J_m(x)$ and $Y_m(x)$ are always an independent pair of solutions. To show this, one may calculate the Wronskian of J_m and Y_m and verify that it is nonzero for all values of m.

Some useful integral relations are found as follows. From

$$J_0'(x) = -J_1(x)$$

we obtain

$$\int^x J_1(x')\,dx' = -J_0(x)$$

From (3-72),

$$[x^n J_n(x)]' = x^n\left[J_n'(x) + \frac{n}{x}J_n(x)\right]$$

$$= x^n J_{n-1}(x)$$

Therefore,

$$\int x^n J_{n-1}(x)\,dx = x^n J_n(x) \tag{3-75}$$

Also,

$$[x^{-n}J_n(x)]' = x^{-n}\left[J_n'(x) - \frac{n}{x}J_n(x)\right]$$

$$= -x^{-n}J_{n+1}(x)$$

Therefore

$$\int x^{-n}J_{n+1}(x)\,dx = -x^{-n}J_n(x) \tag{3-76}$$

Bessel functions possess an orthogonality property analogous to that of the Legendre polynomials. Let

$$J_m(kx) = f(x) \qquad J_m(lx) = g(x)$$

Then

$$f'' + \frac{1}{x}f' + \left(k^2 - \frac{m^2}{x^2}\right)f = 0$$

$$g'' + \frac{1}{x}g' + \left(l^2 - \frac{m^2}{x^2}\right)g = 0$$

Multiplying the second by xf, and subtracting xg times the first gives

$$[x(fg' - gf')]' = (k^2 - l^2)xfg$$

$$\int xf(x)g(x)\,dx = \frac{x}{k^2 - l^2}\,[f(x)g'(x) - g(x)f'(x)]$$

$$\int_a^b J_m(kx)\,J_m(lx)x\,dx = \frac{1}{(k^2 - 1^2)}\,[lxJ_m(kx)J_m'(lx) - kxJ_m(lx)J_m'(kx)]_a^b$$

If $J_m(kx)$ and $J_m(lx)$ vanish at a and b, or if $J_m'(kx)$ and $J_m'(lx)$ vanish at a and b, or under more general conditions (for example, the two functions may vanish at a and the two derivatives at b, we obtain

$$\int_a^b J_m(kx)J_m(lx)x\,dx = 0 \qquad \text{provided } k \neq 1 \tag{3-77}$$

What if $k = l$? We must evaluate

$$\int J_m^2(kx)x\,dx = \frac{1}{k^2}\int J_m^2(y)y\,dy \qquad \text{where } y = kx$$

Integrate once by parts

$$I = \int J_m^2(y)y\,dy = \frac{1}{2}y^2J_m^2(y) - \int J_m(y)J_m'(y)y^2\,dy$$

But from Bessel's equation,

$$y^2J_m(y) = m^2J_m(y) - yJ_m'(y) - y^2J_m''(y)$$

Therefore,

$$I = \frac{1}{2}y^2J_m^2(y) - \int J_m'(y)\,[m^2J_m(y) - yJ_m'(y) - y^2J_m''(y)]\,dy$$

$$= \frac{1}{2}y^2J_m^2(y) - \frac{1}{2}m^2J_m^2(y) + \frac{1}{2}y^2\,[J_m'(y)]^2$$

Thus we have derived the "normalization integral,"

$$\int J_m^2(kx)x\,dx = \frac{1}{2}\left(x^2 - \frac{m^2}{k^2}\right)J_m^2(kx) + \frac{1}{2}\,x^2[J_m'(kx)]^2$$

If, for example, $J_m(kx)$ vanishes at $x = a$ and $x = b$,

$$\int_a^b J_m^2(kx)x\,dx = \frac{x^2}{2}\,[J_m'(kx)]^2\Big|_a^b$$

$$= \frac{x^2}{2}\,[J_{m+1}(kx)]^2\Big|_a^b \tag{3-78}$$

PROBLEMS

3-38 Let $f(x)$ and $g(x)$ be two solutions of Bessel's equation (with the same m). Show that their Wronskian is of the form

$$fg' - gf' = \frac{\text{constant}}{x}$$

3-39 Find the Wronskian of $J_m(x)$ and $J_{-m}(x)$ (m arbitrary).

3-40 Find the Wronskian of $J_m(x)$ and $Y_m(x)$ (m arbitrary).

3-41 Demonstrate that the definition $J_{-m}(x) = (-1)^m J_m(x)$ for integral m is consistent with the results one would obtain from the recursion relations (3-70) and (3-71). When we study the properties of the gamma function in Section 10-2, we will find that (3-69) also gives this relationship.

3-42 Verify directly that

$$J_{1/2}(x) = \left(\frac{2}{\pi x}\right)^{1/2} \sin x$$

and

$$J_{-1/2}(x) = \left(\frac{2}{\pi x}\right)^{1/2} \cos x$$

3-43 A function $f(x)$ on the interval $0 < x < a$ can be expanded in a series

$$f(x) = \sum_n c_n J_m(k_n x)$$

where the k_n are chosen so that $J_m(k_n a) = 0$, and m is arbitrary. By considering the limit $a \to \infty$, derive the formulas for the *Hankel transform*

$$f(x) = \int_0^\infty g(y) J_m(xy) y\, dy$$

$$g(y) = \int_0^\infty f(x) J_m(xy) x\, dx$$

3-44 Show that

$$Y_n(x) = \frac{1}{\pi}\left[\frac{\partial J_n(x)}{\partial n} - (-)^n \frac{\partial J_{-n}(x)}{\partial n}\right]$$

if n is an integer.

3-45 What linear homogeneous second-order differential equation has

$$x^\alpha J_{\pm m}(\beta x^\gamma)$$

as solutions? Give the general solution of

$$y'' + x^2 y = 0$$

3-5 HYPERGEOMETRIC FUNCTIONS

Consider the general second-order linear differential equation with three regular singularities. Let the three singularities be at ξ, η, ρ, and the differential equation be

$$y'' + P(z)y' + Q(z)y = 0 \tag{3-79}$$

We know from Section 3-2 that P and Q have the form

$$P = \frac{\text{(polynomial)}}{(z-\xi)(z-\eta)(z-\rho)} \qquad Q = \frac{\text{(polynomial)}}{(z-\xi)^2(z-\eta)^2(z-\rho)^2} \tag{3-80}$$

(We are using the fact that a function that has no singularities in the finite complex plane, and does not have an essential singularity at infinity, must be a polynomial.) Imposing the condition that the point at infinity is an ordinary point gives

$$P = \frac{A}{z-\xi} + \frac{B}{z-\eta} + \frac{C}{z-\rho} \qquad A + B + C = 2$$

$$Q = \frac{1}{(z-\xi)(z-\eta)(z-\rho)}\left(\frac{D}{z-\xi} + \frac{E}{z-\eta} + \frac{F}{z-\rho}\right)$$

Let us now look for a solution $y \approx (z-\xi)^\alpha$ near $z = \xi$ by the method of Section 3-2. The indicial equation is

$$\alpha^2 + (A-1)\alpha + \frac{D}{(\xi-\eta)(\xi-\rho)} = 0$$

Let α_1 and α_2 denote the two roots. Then

$$\alpha_1 + \alpha_2 = 1 - A$$

$$\alpha_1\alpha_2 = \frac{D}{(\xi-\eta)(\xi-\rho)}$$

If these two equations are solved for A and D in terms of α_1 and α_2, and a similar investigation is made at the other two singularities, the result is that our differential equation (3-79) can be written

$$y'' + \left(\frac{1-\alpha_1-\alpha_2}{z-\xi} + \frac{1-\beta_1-\beta_2}{z-\eta} + \frac{1-\gamma_1-\gamma_2}{z-\rho}\right)y' - \frac{(\xi-\eta)(\eta-\rho)(\rho-\xi)}{(z-\xi)(z-\eta)(z-\rho)}$$
$$\times\left[\frac{\alpha_1\alpha_2}{(z-\xi)(\eta-\rho)} + \frac{\beta_1\beta_2}{(z-\eta)(\rho-\xi)} + \frac{\gamma_1\gamma_2}{(z-\rho)(\xi-\eta)}\right]y = 0 \tag{3-81}$$

We have used β and γ to denote the exponents at η and ρ, respectively.

The fact that y is a solution of this equation (3-81) is often written

$$y = P\begin{Bmatrix} \xi & \eta & \rho \\ \alpha_1 & \beta_1 & \gamma_1 & z \\ \alpha_2 & \beta_2 & \gamma_2 \end{Bmatrix} \quad (\textit{Riemann P symbol}) \tag{3-82}$$

Note that we must have $\alpha_1 + \alpha_2 + \beta_1 + \beta_2 + \gamma_1 + \gamma_2 = 1$ in order that $z = \infty$ be an ordinary point.

The Riemann P symbol represents the *general* solution of equation (3-81). From it we can read out six different *special* solutions that behave near the singularities in the manner indicated by the exponents. For example, one of these six is the solution that behaves for z near η like $(z-\eta)^{\beta_1}$. These six special solutions are not independent, of course.

This solution (3-82) can be related fairly easily to functions with different behavior near the singular points. For example,

$$\frac{(z-\xi)^\lambda}{(z-\eta)^\lambda} P\begin{Bmatrix} \xi & \eta & \rho \\ \alpha_1 & \beta_1 & \gamma_1 & z \\ \alpha_2 & \beta_2 & \gamma_2 \end{Bmatrix} = P\begin{Bmatrix} \xi & \eta & \rho \\ \alpha_1 + \lambda & \beta_1 - \lambda & \gamma_1 & z \\ \alpha_2 + \lambda & \beta_2 - \lambda & \lambda_2 \end{Bmatrix} \tag{3-83}$$

[This can be checked by substitution into (3-81)]. Hence

$$\frac{(z-\rho)^{\alpha_1+\beta_1}}{(z-\xi)^{\alpha_1}(z-\eta)^{\beta_1}} P\begin{Bmatrix} \xi & \eta & \rho \\ \alpha_1 & \beta_1 & \gamma_1 & z \\ \alpha_2 & \beta_2 & \gamma_2 \end{Bmatrix}$$

$$= P\begin{Bmatrix} \xi & \eta & \rho \\ 0 & 0 & \gamma_1 + \alpha_1 + \beta_1 & z \\ \alpha_2 - \alpha_1 & \beta_2 - \beta_1 & \gamma_2 + \alpha_1 + \beta_1 \end{Bmatrix} \tag{3-84}$$

It is often convenient to put our singularities at 0, 1, ∞. This is accomplished by a *homographic transformation*

$$w = \frac{(z-\xi)(\eta-\rho)}{(z-\rho)(\eta-\xi)} \tag{3-85}$$

which takes the points $z = \xi, \eta, \rho$ into the points $w = 0, 1, \infty$. It then turns out (reasonably enough!) that

$$y = P\begin{Bmatrix} \xi & \eta & \rho \\ \alpha_1 & \beta_1 & \gamma_1 & z \\ \alpha_2 & \beta_2 & \gamma_2 \end{Bmatrix} = P\begin{Bmatrix} 0 & 1 & \infty \\ \alpha_1 & \beta_1 & \gamma_1 & w \\ \alpha_2 & \beta_2 & \gamma_2 \end{Bmatrix} \tag{3-86}$$

In other words, if the independent variable in the differential equation (3-81) is transformed according to (3-85), the resulting differential equation is of the same form, with the replacements $z \to w$, $\xi \to 0$, $\eta \to 1$, $\rho \to \infty$.

By combining (3-84) with (3-86), we find that

$$P\begin{Bmatrix} \xi & \eta & \rho & \\ \alpha_1 & \beta_1 & \gamma_1 & z \\ \alpha_2 & \beta_2 & \gamma_2 & \end{Bmatrix}$$

$$= \frac{(z-\xi)^{\alpha_1}(z-\eta)^{\beta_1}}{(z-\rho)^{\alpha_1+\beta_1}} P\begin{Bmatrix} \xi & \eta & \rho & \\ 0 & 0 & \gamma_1+\alpha_1+\beta_1 & z \\ \alpha_2-\alpha_1 & \beta_2-\beta_1 & \gamma_2+\alpha_1+\beta_1 & \end{Bmatrix} \tag{3-87}$$

$$= \frac{(z-\xi)^{\alpha_1}(z-\eta)^{\beta_1}}{(z-\rho)^{\alpha_1+\beta_1}} P\begin{Bmatrix} 0 & 1 & \infty & \\ 0 & 0 & \gamma_1+\alpha_1+\beta_1 & \dfrac{(z-\xi)(\eta-\rho)}{(z-\rho)(\eta-\xi)} \\ \alpha_2-\alpha_1 & \beta_2-\beta_1 & \gamma_2+\alpha_1+\beta_1 & \end{Bmatrix}$$

Hence every solution (3-82) to a second-order linear differential equation with three regular singular points can be related to a function of the form

$$f(x) = P\begin{Bmatrix} 0 & 1 & \infty & \\ 0 & 0 & a & x \\ 1-c & c-a-b & b & \end{Bmatrix} \tag{3-88}$$

which is a solution of the equation

$$f''(x) + \left[\frac{c}{x} + \frac{1-c+a+b}{x-1}\right]f'(x) + \frac{ab}{x(x-1)}f(x) = 0 \tag{3-89}$$

that is,

$$x(1-x)f'' + [c-(a+b+1)x]f' - abf = 0 \tag{3-90}$$

Equation (3-90) is called the *hypergeometric equation.*

Thus the complete theory of second-order linear equations with three regular singular points may be obtained by study of hypergeometric functions. We will not make a study here, but instead will limit ourselves to a survey of some physical applications.

One solution of (3-90) (the one that behaves like a constant near $x = 0$) is the *hypergeometric series*

$$\begin{aligned} {}_2F_1(a, b; c; x) &= 1 + \frac{ab}{c}\frac{x}{1!} + \frac{a(a+1)b(b+1)}{c(c+1)}\frac{x^2}{2!} + \cdots \\ &= \frac{\Gamma(c)}{\Gamma(a)\Gamma(b)} \sum_{n=0} \frac{\Gamma(a+n)\Gamma(b+n)}{\Gamma(c+n)}\frac{x^n}{n!} \end{aligned} \tag{3-91}$$

The notation ${}_2F_1$ signifies that there are two "numerator parameters" a and b, and one "denominator parameter" c. Many of the functions encountered in mathematical physics are special cases of ${}_2F_1(a, b; c; x)$.

For example, Legendre's differential equation (3-39) has regular singular points at $x = \pm 1$, $x = \infty$. It can be shown by simple change of variables that

$$P_n(x) = (\text{constant})\ {}_2F_1\left(1 + n, -n; 1; \frac{x+1}{2}\right) \tag{3-92}$$

Substitution into (3-91) provides us with a convenient expansion for $P_n(x)$ about $x = -1$, which complements the expansions about $x = 0$ and $x = 1$ found in Section 3-2. Furthermore, because a, b, and c may assume any values in (3-90), equation (3-92) gives a convenient definition for $P_n(x)$ if n takes on non-integer values (a situation in which the Rodrigues formula is useless).

More generally, if a or b is a negative integer, the hypergeometric series becomes a polynomial, known as a *Jacobi polynomial.* Jacobi polynomials occur in the study of the transformation properties of spherical harmonics under coordinate rotations.

CONFLUENT HYPERGEOMETRIC FUNCTIONS

There is a close relative of the hypergeometric function, called the *confluent hypergeometric function,* which includes as special cases many more functions of physical interest. To obtain these functions, we begin with the differential equation for ordinary hypergeometric functions

$$x(1-x)y'' + [c - (a+b+1)x]y' - aby = 0$$

with regular singular points $x = 0, 1, \infty$. If we set $x = z/b$, the singularities move to $z = 0, b, \infty$. As $b \to \infty$, the equation becomes

$$zy'' + (c - z)y' - ay = 0 \tag{3-93}$$

This confluent hypergeometric equation [equation (3-93)] has a regular singular point at $z = 0$ and an essential singularity at $z = \infty$, which arose from the confluence $b \to \infty$.

By taking the same limit ($b \to \infty$) in the solutions of (3-90), we can find solutions of (3-93). These solutions now have one "numerator parameter" and one "denominator parameter," so we may write, for example,

$$\lim_{b\to\infty} {}_2F_1\left(a, b; c; \frac{z}{b}\right) \equiv {}_1F_1(a; c; z) \tag{3-94}$$

The function ${}_1F_1(a; c; z)$ is called a *confluent hypergeometric function* or *Kummer function*; its power series expansion takes the form

$${}_1F_1(a; c; z) = \frac{\Gamma(c)}{\Gamma(a)} \sum_{n=0}^{\infty} \frac{\Gamma(a+n)}{\Gamma(c+n)} \frac{z^n}{n!} = 1 + \frac{a}{c}\frac{z}{1!} + \frac{a(a+1)}{c(c+1)}\frac{z^2}{2!} + \cdots$$

Many frequently occurring functions are special cases of confluent hypergeometric functions. Some examples are

$$e^z = {}_1F_1(a; a; z) \qquad (a \text{ arbitrary})$$

$$J_n(z) = \frac{1}{\Gamma(n+1)}\left(\frac{z}{2}\right)^n e^{iz}\ {}_1F_1\left(n + \frac{1}{2}; 2n+1; -2iz\right)$$

$$\operatorname{erf} z = \frac{2z}{\sqrt{\pi}}\,{}_1F_1\left(\frac{1}{2};\frac{3}{2};-z^2\right)$$

$H_n(z) = n$th Hermite polynomial [compare (3-57)]

$$= 2^n\left[\frac{\Gamma\left(-\frac{1}{2}\right)}{\Gamma\left(-\frac{n}{2}\right)}\,z\,{}_1F_1\left(\frac{1-n}{2};\frac{3}{2};z^2\right) + \frac{\Gamma\left(\frac{1}{2}\right)}{\Gamma\left(\frac{1-n}{2}\right)}\,{}_1F_1\left(\frac{-n}{2};\frac{1}{2};z^2\right)\right]$$

If a is a negative integer, ${}_1F_1(a; c; z)$ becomes a polynomial which, when suitably normalized, is a Laguerre polynomial. Specifically, it is usually written (for $n = 0, 1, 2, \ldots$)

$$\Gamma_n^{\alpha}(z) = \frac{\Gamma(\alpha + n + 1)}{n!\,\Gamma(\alpha + 1)}\,{}_1F_1(-n; \alpha + 1; z)$$

PROBLEMS

3-46 Consider the differential equation

$$z^2(z^2 - 1)y'' + z(2z^2 - 1)y' + \tfrac{1}{16}y = 0$$

(a) Give the general solution, involving hypergeometric functions with argument

$$\frac{2z}{z+1}$$

(b) Give the general solution, involving hypergeometric functions with argument

$$\frac{z-1}{2z}$$

3-47 Express in terms of elementary functions:

(a) ${}_2F_1(1, a; 2; z)$
(b) ${}_2F_1(1, 1; 2; z)$

3-48 Give two solutions of the ordinary hypergeometric equation that are useful near $z = 1$.

3-49 Write the most general linear homogeneous second-order differential equation with just two regular singularities, at ξ and η, in terms of ξ, η, and the exponents at ξ and η. What conditions are there on the exponents? Consider the relation

$$\left(\frac{z-\xi}{z-\eta}\right)^{\lambda} P\begin{Bmatrix} \xi & \eta & \\ \alpha_1 & \beta_1 & z \\ \alpha_2 & \beta_2 & \end{Bmatrix} = P\begin{Bmatrix} \xi & \eta & \\ \alpha_1+\lambda & \beta_1-\lambda & z \\ \alpha_2+\lambda & \beta_2-\lambda & \end{Bmatrix}$$

If the exponents on the left obey the conditions referred to above, show that the new exponents on the right also obey these conditions.

3-50 What is the general solution of the differential equation

$$y'' + \frac{2z-1}{z(z-1)}y' - \frac{2}{9}\frac{1}{z^2(z-1)(z-2)}y = 0?$$

3-51 Express the general solution of

$$z^2(z^2-1)^2y'' + z(z^2-1)(2z^2-1)y' - [(3\alpha^2 - \tfrac{1}{4})z^2 + \alpha^2]y = 0$$

near $z = 1$ in terms of hypergeometric functions ${}_2F_1$.

3-52 Find the Wronskian of ${}_1F_1(a; c; z)$ and $z^{1-c}{}_1F_1(1+a-c; 2-c; z)$.

SUMMARY

In solving first-order differential equations, one should look to see whether the variables can be separated in the form

$$A(x)\,dx + B(y)\,dy = 0$$

If so, the equation can be immediately integrated. Another common device is to search for an integrating factor $\lambda(x, y)$ such that the equation becomes immediately integrable after both sides have been multiplied by λ.

The general solution of an nth-order linear differential equation is the sum of the general solution of the corresponding *homogeneous* equation (the complementary function) with a particular solution of the *inhomogeneous* equation. The nth-order homogeneous equation has n independent solutions; hence the complementary function contains n arbitrary constants. These may be fixed by boundary conditions of the problem. Of the many methods for finding particular solutions, the method of variation of parameters is one of the most generally successful. In this method one takes as a trial function the sum of solutions of the homogeneous equation, multiplied by arbitrary functions instead of arbitrary constants. The equation is then solved for these new functions.

Often solutions can be found in the form of power series expansions about some point. For a linear homogeneous equation of the form

$$\frac{d^ny}{dx^n} + f_{n-1}(x)\frac{d^{n-1}y}{dx^{n-1}} + \cdots + f_0(x)y = 0$$

the possibility of a power series expansion of y about the point x_0 depends on the behavior of the functions $f_i(x)$ at this point. If none of them is singular there, then y may be expanded in a series about the point. Frequently, however, the f_i do have singularities at points of physical interest. In this case it is useful to know that if $(x - x_0)f_{n-1}(x)$, $(x - x_0)^2 f_{n-2}(x), \ldots, (x - x_0)^n f_0(x)$ are regular at x_0, then there is always at least one solution of the form

$$y = (x - x_0)^s \sum_{m=0}^{\infty} c_m (x - x_0)^m$$

If an attempt to make a power series expansion yields a three-term recursion relation for the coefficients, factoring out the desired behavior near some singularity may simplify the expansion of the remainder.

Solutions of Legendre's equation are normally classified into two kinds: the Legendre polynomials $P_n(x)$, which are finite at $x = \pm 1$ and orthogonal on the interval $(-1, +1)$ with weight function 1; and the Legendre functions of the second kind, $Q_n(x)$, which have logarithmic singularities at $x = \pm 1$. The $Q_n(x)$ behave like x^{-n-1} as x approaches infinity.

Solutions of Bessel's equation are likewise classified according to their behavior at $x = 0$: the ordinary Bessel functions $J_n(x)$ are finite at this point, whereas the Bessel functions of the second kind have a logarithmic singularity here. The $J_n(k_i x)$ for a given n but different k_i are orthogonal on the interval $(0, a)$ with weight function x, provided the numbers k_i are chosen such that $J_n(k_i a) = 0$.

The complete theory of second-order linear equations with three regular singular points may be obtained by study of hypergeometric functions. Likewise solutions of many second-order equations with one regular singular point and one essential singularity can be related to confluent hypergeometric functions. Practically all of the functions commonly used in physics are special cases of hypergeometric functions.

CHAPTER 4 Function Spaces and Expansions in Orthogonal Sets

Ordinary differential equations such as those discussed in Chapter 3 often describe the spatial variation of some physical system. These differential operators in space frequently play a role analogous to the matrix operators of Chapter 1, expressing the same information for the continuous system that matrices give for finite discrete systems. It is not surprising, therefore, to find that certain differential operators are Hermitian, with real eigenvalues, and that their eigenfunctions can be used as basis functions for expansions. This chapter is concerned with various aspects of this analogy.

In Section 4-1 we display the transition from a finite system to a continuous system, showing how differential operators in the continuous case play the same role as matrix operators in the finite case. Section 4-2 is devoted to properties of Sturm-Liouville operators, showing that they act in the same way as Hermitian matrices. Expansions in terms of eigenfunctions of Sturm-Liouville operators are displayed. In Section 4-3 we extend these ideas to the case of continuous eigenvalues, with emphasis on Fourier transforms.

With this background we can appreciate some more general ideas about function spaces, given in Section 4-4, and specific tricks such as the ladder operator method discussed in Section 4-5.

4-1 TRANSITION FROM THE DISCRETE CASE

Consider a system of three beads connected by springs and sliding on a rod (Figure 4-1). The equations of motion for this system are easily found to be

$$\begin{aligned} m\ddot{\bar{x}}_1 &= k(\bar{x}_2 - \bar{x}_1) \\ m\ddot{\bar{x}}_2 &= -k(\bar{x}_2 - \bar{x}_1) + k(\bar{x}_3 - \bar{x}_2) \qquad (4\text{-}1) \\ m\ddot{\bar{x}}_3 &= -k(\bar{x}_3 - \bar{x}_2) \end{aligned}$$

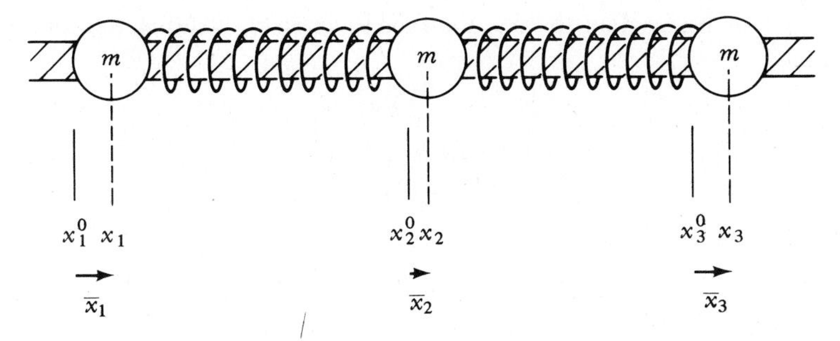

FIGURE 4-1 Beads connected by springs

where $\bar{x}_i$ is the deviation of the ith bead from its equilibrium position. Similarly for a system of n beads, we get

$$\begin{aligned} m\ddot{\bar{x}}_1 &= k(\bar{x}_2 - \bar{x}_1) \\ m\ddot{\bar{x}}_n &= -k(\bar{x}_n - \bar{x}_{n-1}) \end{aligned} \tag{4-2}$$

and for $i \neq n$,

$$m\ddot{\bar{x}}_i = -k(\bar{x}_i - \bar{x}_{i-1}) + k(\bar{x}_{i+1} - \bar{x}_i)$$

As n gets very large, we can neglect the end beads. An infinite array of beads may then be described by the set of equations

$$m\ddot{\bar{x}}_i = -k(\bar{x}_i - \bar{x}_{i-1}) + k(\bar{x}_{i+1} - \bar{x}_i) \tag{4-3}$$

It is convenient to use a different notation to discuss the transition to the continuous case: Let $\bar{x}_i = U(i)$ be the displacement of the ith bead from its equilibrium position. Then we have

$$m \frac{d^2}{dt^2} U(i) = -k[U(i) - U(i-1)] + k[U(i+1) - U(i)]$$

This is a difference equation. To go over to a differential equation, we now must make a transition from the discrete parameter i to a continuous parameter x. Instead of parametrizing the masses by their number i in the array, let us use their equilibrium position $x = ia$. The difference $U(i) - U(i-1)$ is the change in U per unit change in i, and if we use x as a parameter we can write this as

$$U(ia) - U[(i-1)a] = \frac{\Delta U(ia)}{\Delta i} = \frac{\Delta U(x)}{\Delta x} \cdot a \tag{4-4}$$

We therefore can rewrite our equation as

$$\begin{aligned} m \frac{d^2 U(ia)}{dt^2} &= ak \left[\frac{\Delta U[(i+1)a]}{\Delta(ia)} - \frac{\Delta U(ia)}{\Delta(ia)} \right] = \frac{ak\,\Delta\left(\dfrac{\Delta U(ia)}{\Delta ia}\right)}{\Delta i} = a^2 k \frac{\Delta^2 U(ia)}{\Delta(ia)^2} \\ &= a^2 k \frac{\Delta^2 U(x)}{(\Delta x)^2} \end{aligned} \tag{4-5}$$

This may now be applied to very small changes in x, yielding the differential equation

$$\frac{d^2U}{dx^2} - \frac{m}{a^2k}\frac{d^2U}{dt^2} = 0 \tag{4-6}$$

Notice that a^2k/m has the dimensions of (velocity)^2, so we can replace this quantity by the new parameter c^2. We have thus derived the equation for longitudinal oscillations of a rod tapped at one end.

In this example the equations of motion, which looked like

$$m\frac{d^2}{dt^2}\begin{pmatrix} \bar{x}_1 \\ \bar{x}_2 \\ \bar{x}_3 \end{pmatrix} = \begin{pmatrix} -k & k & 0 \\ k & -2k & k \\ 0 & k & -k \end{pmatrix}\begin{pmatrix} \bar{x}_1 \\ \bar{x}_2 \\ \bar{x}_3 \end{pmatrix}$$

in the discrete case, have gone over to

$$m\frac{d^2}{dt^2}U(x) = a^2k\frac{d^2}{dx^2}U(x)$$

in the continuous case. That is, the matrix operator was replaced by a differential operator, the index i which labeled the elements of the column vector was replaced by the running index x, and the vector itself was replaced by a density $U(x)$. Although this was a particularly simple case, we will have similar equations whenever we have to describe the behavior at all points in space.

Since the equations go over so simply from the matrix difference form to the differential form, we should expect that differential operators would have many of the properties of matrices. In particular, we might hope for differential operators that would act like Hermitian matrices; the eigenvectors of these operators could then be used for doing expansions, and Green's functions could be used to advantage in solving the inhomogeneous equations.

4-2 THE STURM-LIOUVILLE EQUATION

Consider the differential equation

$$p(x)\frac{d^2\psi_\lambda}{dx^2} + \frac{dp(x)}{dx}\frac{d\psi_\lambda}{dx} + q(x)\psi_\lambda = -\lambda r(x)\psi_\lambda \tag{4-7}$$

or

$$\frac{d}{dx}\left[p(x)\frac{d\psi_\lambda}{dx}\right] + q(x)\psi_\lambda = -\lambda r(x)\psi_\lambda$$

Here λ is a parameter that identifies the solution. For this type of equation, the Sturm-Liouville equation, we can prove a theorem:

Provided certain types of boundary conditions (to be specified below) *are satisfied on the interval* (a, b), *the integral*

$$(\lambda^* - \lambda')\int_a^b r(x)\psi_\lambda^*(x)\psi_{\lambda'}(x)\,dx$$

is zero.

Proof: We begin by writing

$$\frac{d}{dx}\left[p(x)\frac{d\psi_\lambda}{dx}\right] + q(x)\psi_\lambda = -\lambda r(x)\psi_\lambda \tag{4-8}$$

$$\frac{d}{dx}\left[p(x)\frac{d\psi_{\lambda'}}{dx}\right] + q(x)\psi_{\lambda'} = -\lambda' r(x)\psi_{\lambda'} \tag{4-9}$$

Multiply (4-8) by $\psi_{\lambda'}^*$ and (4-9) by ψ_λ^* and integrate both sides of each equation over the interval (a, b),

$$\int_a^b \psi_{\lambda'}^* \frac{d}{dx}\left[p(x)\frac{d\psi_\lambda}{dx}\right]dx + \int_a^b q(x)\psi_{\lambda'}^*\psi_\lambda\,dx = -\lambda\int_a^b r(x)\psi_{\lambda'}^*\psi_\lambda\,dx \tag{4-10}$$

$$\int_a^b \psi_\lambda^* \frac{d}{dx}\left[p(x)\frac{d\psi_{\lambda'}}{dx}\right]dx + \int_a^b q(x)\psi_\lambda^*\psi_{\lambda'}\,dx = -\lambda'\int_a^b r(x)\psi_\lambda^*\psi_{\lambda'}\,dx \tag{4-11}$$

Using integration by parts, we can rewrite these as

$$\begin{aligned} -\int_a^b \frac{d\psi_{\lambda'}^*}{dx}p(x)\frac{d\psi_\lambda}{dx}dx &+ \int_a^b q(x)\psi_{\lambda'}^*\psi_\lambda\,dx \\ &= -\left[\psi_{\lambda'}^* p(x)\frac{d\psi_\lambda}{dx}\right]_a^b - \lambda\int_a^b r(x)\psi_{\lambda'}^*\psi_\lambda\,dx \end{aligned} \tag{4-12}$$

and

$$\begin{aligned} -\int_a^b \frac{d\psi_\lambda^*}{dx}p(x)\frac{d\psi_{\lambda'}}{dx}dx &+ \int_a^b q(x)\psi_\lambda^*\psi_{\lambda'}\,dx \\ &= -\left[\psi_\lambda^* p(x)\frac{d\psi_{\lambda'}}{dx}\right]_a^b - \lambda'\int_a^b r(x)\psi_\lambda^*\psi_{\lambda'}\,dx \end{aligned} \tag{4-13}$$

For real $p(x)$ and $q(x)$, the left-hand side of (4-12) is just the complex conjugate of the left-hand side of (4-13). If $r(x)$ is also real, we arrive at

$$(\lambda^* - \lambda')\int_a^b r(x)\psi_\lambda^*\psi_{\lambda'}\,dx = \left[p(x)\left(\psi_\lambda^*\frac{d\psi_{\lambda'}}{dx} - \psi_{\lambda'}\frac{d\psi_\lambda^*}{dx}\right)\right]_a^b \tag{4-14}$$

Therefore

$$(\lambda^* - \lambda')\int_a^b r(x)\psi_\lambda^*\psi_{\lambda'}\,dx = 0 \tag{4-15}$$

provided ψ_λ and $\psi_{\lambda'}$ satisfy boundary conditions such that

$$\left[p(x)\left(\psi_\lambda^* \frac{d\psi_{\lambda'}}{dx} - \psi_{\lambda'} \frac{d\psi_\lambda^*}{dx}\right)\right]_a^b = 0$$

Two major conclusions may be drawn from (4-15):

1. If $r(x)$ does not change sign on the interval (a, b), then

$$\int_a^b r(x)\,|\psi_\lambda|^2\,dx \neq 0 \qquad \text{and hence } \lambda^* = \lambda$$

2. For $\lambda' \neq \lambda$, $\displaystyle\int_a^b r(x)\psi_\lambda^*\psi_{\lambda'}\,dx = 0$

Many of the functions important in physics satisfy Sturm-Liouville equations. For example Legendre's equation is a Sturm-Liouville equation with $\lambda = n(n+1)$, $p(x) = 1 - x^2$, $q(x) = 0$, and $r(x) = 1$. The associated Legendre equation is of the same form, with $\lambda = n(n+1)$, $p(x) = 1 - x^2$, $q(x) = -m^2/(1 - x^2)$, and $r(x) = 1$. Hence we know without further study that the associated Legendre functions P_n^m and P_k^m with n and k integers must obey

$$\int_{-1}^{1} P_n^m(x)P_k^m(x)\,dx = 0 \qquad k \neq n$$

In fact, we can show by the same means we used to get (3-65)

$$\int_{-1}^{1} P_l^m(x)P_{l'}^m(x)\,dx = \frac{(l+m)!}{(l-m)!}\,\frac{2}{2l+1}\,\delta_{ll'} \tag{4-16}$$

Bessel's equation may be cast into Sturm-Liouville form by dividing by x:

$$xy'' + y' + \left(xk^2 - \frac{m^2}{x}\right)y = 0$$

Here $p(x) = x$, $q(x) = -m^2/x$, $r(x) = x$, and $\lambda = k^2$. Thus it is not surprising that Bessel functions of a given m but different k are orthogonal with weight function x, as was proved in Section 3-4.

A similar analysis can be made for the equation of the quantum harmonic oscillator. Even the equation

$$\frac{d^2\Psi}{dx^2} + \omega^2\Psi = 0$$

for simple harmonic motion is of Sturm-Liouville type. Hence our study of the Sturm-Liouville equation provides a unified point of view from which to approach a large number of problems.

STURM-LIOUVILLE OPERATORS AS HERMITIAN OPERATORS

Conclusion 1 of our theorem tells us that the eigenvalues $-\lambda$ of

$$S = \frac{1}{r(x)}\frac{d}{dx}\left[p(x)\frac{d}{dx}\right] + \frac{q(x)}{r(x)} \tag{4-17}$$

are real; conclusion 2 tells us that the eigenfunctions Ψ_λ and $\Psi_{\lambda'}$ for different eigenvalues are orthogonal with respect to a scalar product

$$\langle \Psi_\lambda | \Psi_{\lambda'} \rangle_S = \int_a^b r(x)\Psi_\lambda^*(x)\Psi_{\lambda'}(x)\,dx \tag{4-18}$$

These are two properties that we proved for Hermitian matrix operators in Section 1-4. We might hope, therefore, that the Sturm-Liouville operator also possesses the property of self-adjointness, which we defined for matrix operators by

$$\langle \alpha | H | \beta \rangle^* = \langle \beta | H | \alpha \rangle$$

The orthogonality relationship tells us the form of the scalar product

$$\langle \Psi | \phi \rangle_S = \int_a^b r(x)\Psi^*(x)\phi(x)\,dx \tag{4-19}$$

Therefore we should use this product in translating the bra-ket notation. This gives

$$\begin{aligned}\langle \Psi | S | \phi \rangle_S &= \int_a^b r(x)\Psi^*(x)\left\{\frac{1}{r(x)}\frac{d}{dx}\left[p(x)\frac{d\phi(x)}{dx}\right] + \frac{q(x)}{r(x)}\phi(x)\right\}dx \\ &= \int_a^b \Psi^*(x)\left\{\frac{d}{dx}\left[p(x)\frac{d\phi(x)}{dx}\right] + q(x)\phi(x)\right\}dx\end{aligned}$$

Likewise, we obtain

$$\begin{aligned}\langle \varphi | S | \psi \rangle_S &= \int_a^b \varphi^*(x)\left\{\frac{d}{dx}\left[p(x)\frac{d\psi(x)}{dx}\right] + q(x)\psi(x)\right\}dx \\ &= \int_a^b \left\{\psi(x)\frac{d}{dx}\left[p(x)\frac{d\varphi^*}{dx}\right] + \psi(x)\varphi^*(x)q(x)\right\}dx \\ &\quad + \left[p(x)\left(\varphi^*(x)\frac{d\psi(x)}{dx} - \psi(x)\frac{d\varphi^*}{dx}(x)\right)\right]\Bigg|_a^b\end{aligned}$$

where we have used integration by parts. If φ and ψ satisfy the same boundary conditions as were imposed on the eigenfunctions to obtain (4-15), then $p(x)$ times the Wronskian of φ and ψ is zero, and

$$\langle \varphi | S | \psi \rangle = \langle \psi | S | \varphi \rangle^* \tag{4-20}$$

for our differential operator as well.

We conclude that Sturm-Liouville differential operators satisfy the same relations as Hermitian matrix operators, provided the bra-ket notation is interpreted properly (that is, that the space of vectors used is spanned by the eigenvectors of S, and a scalar product is defined that makes these eigenvectors mutually orthogonal).

Expansions of functions in the space in terms of eigenfunctions of S are particularly easy if we normalize the basis such that

$$\langle \psi_m \mid \psi_n \rangle = \delta_{mn}$$

Then

$$|f\rangle = \sum_n a_n \, |\psi_n\rangle$$

$$a_n = \langle \psi_n \mid f \rangle_S = \int_a^b \psi_n^*(x) r(x) f(x) \, dx$$

Such expansions often simplify a problem. Hence, it is useful to know the following fact: if an interval (a, b) is "natural" for a particular Sturm-Liouville operator S in the sense that eigenfunctions ψ_λ of S obey the condition

$$\left[p(x) \psi_\lambda(x) \frac{d\psi_\lambda(x)}{dx} \right]_a^b = 0 \tag{4-21}$$

then the eigenfunctions of S are complete on (a, b); that is, every infinitely differentiable function on (a, b) can be expanded in terms of them. [For more details, consult Morse and Feshbach (27), Chapter 6.] In fact, certain classes of functions with discontinuities can also be expanded. Let us list some examples:

Example

Given a function $f(\theta)$ defined on the interval $(0, 2\pi)$ such that $f(0) = f(2\pi)$, what set of functions should we use as an expansion basis? Consider the Sturm-Liouville equation

$$\frac{d^2 y}{d\theta^2} + m^2 y = 0 \tag{4-22}$$

For $m =$ integer, this has eigenfunctions $\sin m\theta$, $\cos m\theta$ that obey $f(0) = f(2\pi)$. Hence they provide an ideal basis for expansion and we may write

$$f(\theta) = \sum_{n=1} a_n \sin n\theta + \sum_{m=0} b_m \cos m\theta \tag{4-23}$$

The eigenfunctions are orthogonal with $r(\theta) = 1$, and may be normalized by using

$$\int_0^{2\pi} \sin^2 n\theta \, d\theta = \int_0^{2\pi} \cos^2 n\theta \, d\theta = \pi \qquad n \neq 0$$

Hence

$$a_n = \frac{1}{\pi}\int_0^{2\pi} f(\theta) \sin n\theta \, d\theta$$

$$b_n = \frac{1}{\pi}\int_0^{2\pi} f(\theta) \cos n\theta \, d\theta \qquad n \neq 0 \tag{4-24}$$

$$b_0 = \frac{1}{2\pi}\int_0^{2\pi} f(\theta) \, d\theta$$

This expansion is usually referred to as a Fourier series.

Example

Any reasonable function $f(x)$ on the interval $-1 \leq x \leq 1$ can be expanded in a series of Legendre polynomials

$$f(x) = \sum_{n=0}^{\infty} c_n P_n(x) \tag{4-25}$$

Here the c_n may be thought of as the components of a vector in a coordinate system with base vectors P_n. Multiplying both sides of (4-25) by $P_m(x)$ and integrating from -1 to $+1$ gives

$$c_m = \frac{(2m+1)}{2}\int_{-1}^{1} P_m(x) f(x) \, dx \tag{4-26}$$

MORE EIGENFUNCTION EXPANSIONS

Associated Legendre functions, with m fixed, are also a complete set of functions, in that an arbitrary reasonable function on the interval $-1 \leq x \leq +1$ may be expanded in a series of the form

$$f(x) = \sum_{n=m}^{\infty} c_n P_n^m(x) \tag{4-27}$$

Combining (4-27) and the idea of a Fourier series already discussed, we see that a function $f(\Omega)$, where Ω is an abbreviation for the polar angles θ, φ, may be expanded in a series

$$f(\Omega) = \sum_{n=0}^{\infty} \sum_{m=-n}^{n} A_{mn} P_n^{|m|}(\cos\theta) e^{im\varphi} \tag{4-28}$$

Although the physics behind the usefulness of (4-28) will not be discussed until Chapter 12, we shall continue briefly on this topic. It is convenient to define spherical harmonics

$$Y_{lm}(\Omega) = \left[\frac{(2l+1)}{4\pi}\frac{(l-|m|)!}{(l+|m|)!}\right]^{1/2} P_l^{|m|}(\cos\theta)e^{im\varphi} \times \begin{cases}(-1)^m & m \geq 0 \\ 1 & m < 0\end{cases} \tag{4-29}$$

We leave it to the student to verify that (4-29) may also be written

$$Y_{lm}(\Omega) = \frac{1}{2^l l!}\left[\frac{(2l+1)}{4\pi}\frac{(l-m)!}{(l+m)!}\right]^{1/2} e^{im\varphi}(-\sin\theta)^m \frac{d^{l+m}}{d(\cos\theta)^{l+m}}(\cos^2\theta - 1)^l \tag{4-30}$$

valid for m positive or negative. It follows from either (4-29) or (4-30) that

$$Y_{l,-m}(\Omega) = (-1)^m Y_{lm}^*(\Omega) \tag{4-31}$$

The normalization constant in (4-29) has been chosen so that

$$\int d\Omega\, Y_{lm}^*(\Omega)\, Y_{l'm'}(\Omega) = \delta_{ll'}\,\delta_{mm'} \tag{4-32}$$

The expansion (4-28) may now be written

$$f(\Omega) = \sum_{l=0}^{\infty}\sum_{m=-l}^{l} B_{lm} Y_{lm}(\Omega) \tag{4-33}$$

where the B_{lm} are easily found from (4-32) to be

$$B_{lm} = \int d\Omega\, Y_{lm}^*(\Omega) f(\Omega) \tag{4-34}$$

We conclude this section by listing a few spherical harmonics.

$$\begin{aligned}
Y_{0,0} &= \sqrt{\frac{1}{4\pi}} & Y_{2,\pm 2} &= \sqrt{\frac{15}{32\pi}}\sin^2\theta\, e^{\pm 2i\varphi} \\
Y_{1,\pm 1} &= \mp\sqrt{\frac{3}{8\pi}}\sin\theta\, e^{\pm i\varphi} & Y_{2,\pm 1} &= \mp\sqrt{\frac{15}{8\pi}}\sin\theta\cos\theta\, e^{\pm i\varphi} \\
Y_{1,0} &= \sqrt{\frac{3}{4\pi}}\cos\theta & Y_{2,0} &= \sqrt{\frac{5}{16\pi}}(3\cos^2\theta - 1)
\end{aligned} \tag{4-35}$$

Example

Consider a function $f(x)$ on the interval $0 < x < a$ where $f(0) = 0 = f(a)$. We can write

$$f(x) = \sum_{n=1}^{\infty} c_n J_m(k_n x) \qquad m \neq 0 \tag{4-36}$$

where the k_n are chosen so that $J_m(k_n a) = 0$. Then, since

$$\int_0^a x J_m(k_n x) J_m(k_p x)\, dx = \delta_{np}\frac{a^2}{2}[J_{m+1}(k_p a)]^2 \tag{4-37}$$

we can easily obtain

$$c_n = \frac{\int_0^a f(x) J_m(k_n x) x \, dx}{\frac{a^2}{2} [J_{m+1}(k_n a)]^2} \tag{4-38}$$

If we wish to represent a function on some more general interval $a < x < b$, we may use, for example, functions of the form

$$J_m(kx) Y_m(ka) - Y_m(kx) J_m(ka) \tag{4-39}$$

with k chosen so that the functions vanish at $x = b$.

Expansion in terms of eigenfunctions is also helpful in finding solutions of the inhomogeneous Sturm-Liouville equation

$$S\psi(x) + \lambda\psi(x) = f(x) \tag{4-40}$$

Following the methods outlined in Section 2-2, we write

$$\psi(x) = \sum_n a_n \psi_n(x)$$

Substitution into (4-40) gives

$$\sum_n a_n \psi_n(x)(\lambda - \lambda_n) = f(x)$$

Using

$$\int_a^b \psi_m^*(x) r(x) \psi_n(x) \, dx = \delta_{nm}$$

we find

$$a_n(\lambda - \lambda_n) = \int_a^b \psi_m^*(x) r(x) f(x) \, dx$$

Thus

$$\begin{aligned} \psi(x) &= \sum_n \frac{\psi_n(x)}{\lambda - \lambda_n} \int_a^b \psi_n^*(x') r(x') f(x') \, dx' \\ &= \int_a^b r(x') G(x, x') f(x') \, dx' \end{aligned} \tag{4-41}$$

where

$$G(x, x') = \sum_n \frac{\psi_n(x) \psi_n^*(x')}{\lambda - \lambda_n} \tag{4-42}$$

is the Green's function for the problem. We might also write (4-41) as

$$|\psi\rangle = G\,|f\rangle$$

where

$$G = \sum_n \frac{|\psi_n\rangle\langle\psi_n|}{\lambda - \lambda_n}$$

and it is understood that all scalar products between functions are defined by (4-19). Relations (4-42) and (4-41) can be used to derive two important properties of these Green's functions:

1. $G(x, x')^* = G(x', x)$, the so-called *reciprocity relation.*
2. $G(x, a)$ is the solution $\psi(x)$ for an inhomogeneous term

$$f(x) = \frac{\delta(x-a)}{r(x)}$$

where $\delta(x - x')$, the delta function, is defined by its properties

$$\delta(x - x') = 0 \qquad x \neq x'$$
$$\int_{-\infty}^{\infty} \delta(x - x')\,dx = 1 \tag{4-43}$$

PROBLEMS

4-1 A function $f(x)$ equals e^{-x} for $0 < x < 1$.
(a) Expand $f(x)$ as a Fourier series of the form $\sum_n B_n \sin n\pi x$.
(b) Expand $f(x)$ as a Fourier series of period 1.

4-2
$$f(x) = \begin{cases} +1 & 0 < x < 1 \\ -1 & -1 < x < 0 \end{cases}$$

Expand $f(x)$ as an infinite series of Legendre polynomials $P_l(x)$.

4-3 Verify the representation (4-30) for the spherical harmonic $Y_{lm}(\Omega)$.

4-4 What is the solution of

$$\frac{d^2y}{dx^2} + \frac{2}{x}\frac{dy}{dx} - \frac{l(l+1)}{x^2}y = \delta(x-a) \qquad (a > 0)$$

on the interval $0 < x < \infty$, subject to the boundary conditions $y(0) = y(\infty) = 0$? l is a positive integer.

4-5 $f_n(x)$ is a polynomial of order $n (n = 0, 1, 2, \ldots)$ and these polynomials are mutually orthogonal on the range 0 to ∞, with weight function e^{-x}; that is,

$$\int_0^{\infty} e^{-x} f_n(x) f_m(x)\,dx = 0 \quad \text{if} \quad m \neq n.$$

Find a differential equation satisfied by $f_n(x)$ of the form

$$x\frac{d^2 f_n}{dx^2} + g(x)\frac{df_n}{dx} + \lambda_n f_n(x) = 0$$

4-6 The solution of

$$y'' + \omega^2 y = g(x) \qquad 0 \le x \le 2\pi$$

subject to the boundary conditions

$$y(0) = y(2\pi) \qquad y'(0) = y'(2\pi)$$

can be written in the form

$$y(x) = \int_0^{2\pi} G(x, x', \omega) g(x')\, dx'$$

Find the Green's function $G(x, x', \omega)$ in closed form. This Green's function is of practical importance in treating the effects of magnet errors on the periodic orbits in a synchrotron.

4-7 We have shown (4-42) that the Green's function for a Hermitian differential operator L satisfies the symmetry relation

$$G(x, x') = [G(x', x)]^*$$

Show this by a more direct method than used in the text, beginning with the differential equations for $G(x, x')$ and $G(x, x'')$; that is,

$$LG(x, x') = \frac{\delta(x - x')}{r(x)}$$

$$LG(x, x'') = \frac{\delta(x - x'')}{r(x)}$$

4-8 Show that

$$\int_a^b f(x)\, \delta[g(x)]\, dx = \frac{f(x_0)}{|g'(x_0)|}$$

provided $g(x) = 0$ has a single root x_0 in the interval $a < x < b$.

4-9 Evaluate

$$\int_0^{\pi} dx \int_1^2 dy\, \delta(\sin x)\, \delta(x^2 - y^2)$$

4-3 USE OF UNCOUNTABLE BASIS SETS—FOURIER TRANSFORMS

We pointed out in the previous section that the equation

$$\frac{d^2\Psi}{dx^2} = -\omega^2\Psi$$

for simple harmonic motion is of Sturm-Liouville type. When the interval (a, b) is finite, say $(-T/2, T/2)$, the orthogonal set of eigenfunctions is $e^{2\pi i n x/T}$, indexed by the real numbers n. Any set that can be put into one-to-one correspondence with the integers is said to be countable. Hence our Fourier series provides an expansion in a countable basis for functions that are periodic with period T. (We must specify that the functions be periodic in order that

$$\left(\Psi_1^* \frac{d\psi_2}{dx} - \Psi_2 \frac{d\psi_1^*}{dx}\right)\Bigg|_{-T/2}^{T/2} = 0$$

so the properties discussed in Section 4-2 hold.)

What happens when the interval (a, b) is extended to include all the real numbers? We can explore this case by studying the behavior of expressions for the Fourier series in the limit as $T \to \infty$. The complex Fourier series

$$\begin{aligned} f(x) &= \sum_{n=-\infty}^{\infty} a_n e^{-2\pi i n x/T} \\ a_n &= \frac{1}{T} \int_{-T/2}^{T/2} f(x) e^{-2\pi i m x/T} \, dx \end{aligned} \tag{4-44}$$

can be converted into an integral by writing

$$\omega = \frac{2\pi m}{T}$$

$$\begin{aligned} \sum_n a_n e^{2\pi i m x/T} &= \int dn \, a_n e^{2\pi i m x/T} = \int_{-\infty}^{\infty} \frac{T}{2\pi} (d\omega) a(\omega) e^{i\omega x} \\ &= \frac{T}{2\pi} \int a(\omega) e^{i\omega x} \, d\omega \end{aligned} \tag{4-45}$$

We give $Ta(\omega)$ the new name $g(\omega)$. Then the formulae (4-44) can be rewritten as

$$f(x) = \frac{1}{2\pi} \int_{-\infty}^{\infty} g(\omega) e^{i\omega x} \, d\omega \tag{4-46}$$

$$g(\omega) = \int_{-\infty}^{\infty} f(x) e^{-i\omega x} \, dx \tag{4-47}$$

$g(\omega)$ is called the Fourier transform of $f(x)$, or vice versa. We see that $g(\omega)$ can be interpreted as the amount of frequency ω in the function $f(x)$ and that all real values of frequency ω are included in the description of $f(x)$. This represents $f(x)$ in terms of the basis functions $e^{i\omega x}$; there are an uncountable number of such functions.

Formula (4-46) can be rewritten in our bra-ket notation. We label the eigenfunctions of d^2/dx^2 by their eigenvalues; thus $e^{i\omega x}$ is the function whose ket is labeled by ω, $e^{i\omega x} \leftrightarrow |\omega\rangle$. Then

$$|f\rangle = \int_{-\infty}^{\infty} d\omega \frac{g(\omega)}{2\pi} |\omega\rangle \tag{4-48}$$

expresses our $f(x)$ in terms of these elementary functions.

Consider the unit operator in this space. Intuitively we try to write it as

$$1 = \int_{-\infty}^{\infty} \frac{|\omega'\rangle \langle\omega'| \, d\omega'}{N(\omega')} \tag{4-49}$$

where $[N(\omega')]^{-1/2} |\omega'\rangle$ is the normalized eigenvector. In order that this object act like the identity, we must have

$$1 \, |\omega\rangle = \int_{-\infty}^{\infty} \frac{|\omega'\rangle \langle\omega' \, | \, \omega\rangle}{N(\omega')} \, d\omega' = |\omega\rangle$$

and hence $\langle\omega' \, | \, \omega\rangle = N(\omega')\delta(\omega' - \omega)$. We see that the normalization of states labeled by continuous eigenvalues is more singular than when the states are labeled by discrete eigenvalues, but that the same general ideas apply. The value of $N(\omega)$ will be deduced below.

Using our rules for Sturm-Liouville operators, the scalar product of two functions will be

$$\mathbf{f}_1 \cdot \mathbf{f}_2 = \int_{-\infty}^{\infty} f_1^*(x) f_2(x) \, dx \tag{4-50}$$

Hence the scalar product between $|\omega\rangle$ and $|\omega'\rangle$ is

$$\langle\omega' \, | \, \omega\rangle = \int_{-\infty}^{\infty} e^{-i\omega' x} e^{i\omega x} \, dx = N(\omega)\delta(\omega - \omega')$$

However, by inserting equation (4-47) into equation (4-46), we obtain

$$\begin{aligned} f(x) &= \frac{1}{2\pi} \int_{-\infty}^{\infty} e^{i\omega' x} \, d\omega' \int_{-\infty}^{\infty} f(x') e^{-i\omega' x'} \, dx' \\ &= \int_{-\infty}^{\infty} f(x') \left[\frac{1}{2\pi} \int_{-\infty}^{\infty} e^{i\omega' x - i\omega' x'} \, d\omega' \right] dx' \end{aligned}$$

Clearly

$$\frac{1}{2\pi} \int_{-\infty}^{\infty} e^{i\omega' x - i\omega' x'} d\omega' = \delta(x - x') \tag{4-51}$$

and $N(\omega) = 2\pi$.

A similar interesting interpretation can be made of equation (4-47).

$$g(\omega) = \int_{-\infty}^{\infty} f(x) e^{-i\omega x}\, dx$$

If we think of $g(\omega)$ and $e^{-i\omega x}$ as functions of ω, then this can be represented by

$$|g\rangle = \int_{-\infty}^{\infty} f(x)\, |x\rangle\, dx \tag{4-52}$$

where $|x\rangle$ is the eigenfunction $e^{-i\omega x}$ of $d^2/d\omega^2$. The vectors $|x\rangle$ are very handy for calculation.

Consider the scalar product

$$\langle f_1 \,|\, f_2 \rangle = \int_{-\infty}^{\infty} f_1^*(x) f_2(x)\, dx$$

By inserting the unit operator

$$1 = \int \frac{|x\rangle\langle x|\, dx}{N(x)}$$

we obtain

$$\langle f_1 \,|\, f_2 \rangle = \int_{-\infty}^{\infty} dx \frac{\langle f_1 \,|\, x\rangle\langle x \,|\, f_2\rangle}{N(x)}$$

and comparison shows us that

$$\begin{aligned} \langle x \,|f_2\rangle &= \sqrt{N(x)} f_2(x) = \sqrt{2\pi} f_2(x) \\ \langle f_1 \,|\, x\rangle &= \sqrt{N(x)} f_1^*(x) = \sqrt{2\pi} f_1^*(x) \end{aligned} \tag{4-53}$$

Any function may therefore be expanded in terms of its values as

$$|f\rangle = \int dx \frac{|x\rangle\langle x \,|\, f\rangle}{2\pi} = \int_{-\infty}^{\infty} dx \frac{f(x)}{\sqrt{2\pi}} |x\rangle \tag{4-54}$$

In a very real sense, the vector $|x\rangle$ picks out the value of the function at point x.

Use of the two alternative representations

$$|f\rangle = \int_{-\infty}^{\infty} dx \frac{f(x)}{\sqrt{2\pi}} |x\rangle$$

and

$$|f\rangle = \int_{-\infty}^{\infty} d\omega \frac{g(\omega)}{2\pi} |\omega\rangle$$

allows us two different methods for writing the scalar product between two functions:

$$\langle f_1 \mid f_2 \rangle = \int_{-\infty}^{\infty} dx' \, \langle x' | \frac{f_1^*(x')}{\sqrt{2\pi}} \int_{-\infty}^{\infty} dx \frac{f_2(x)|x\rangle}{\sqrt{2\pi}}$$
$$= \int_{-\infty}^{\infty} dx \, f_1^*(x) f_2(x) \tag{4-55}$$

and

$$\langle f_1 \mid f_2 \rangle = \int_{-\infty}^{\infty} d\omega' \, \langle \omega' | \frac{g_1^*(\omega')}{2\pi} \int_{-\infty}^{\infty} d\omega \frac{g_2(\omega)|\omega\rangle}{2\pi}$$
$$= \frac{1}{2\pi} \int_{-\infty}^{\infty} d\omega \, g_1^*(\omega) g_2(\omega) \tag{4-56}$$

Equation (4-55) is the standard Sturm-Liouville type scalar product, whereas (4-56) is the usual sum over coefficients of the eigenvectors. The continuous eigenvalues make these two forms look similar for this case. (See Section 9-1 for a more extensive discussion of Fourier transforms.)

4-4 SOME MORE GENERAL IDEAS ABOUT FUNCTION SPACES

In Section 4-2 we defined our scalar product between functions as

$$\varphi \cdot \psi \equiv \langle \varphi \mid \psi \rangle_S = \int_a^b \varphi^*(x) r(x) \psi(x) \, dx \tag{4-57}$$

(for functions φ in the space spanned by eigenfunctions of some Sturm-Liouville operator) because this particular definition of the scalar product made the eigenfunctions of our operator orthogonal. At first glance this looks quite different from the types of scalar products defined in Chapter 1 between vectors in finite dimensional vector spaces. Actually, however, equation (4-57) is a simple translation into the continuous case of the formulae expressed in Chapter 1. It is worthwhile to explore this analogy in detail.

Let us begin with the case $r(x) = 1$. The scalar product integral

$$\int_a^b \varphi^*(x) \psi(x) \, dx$$

is then a summing of values of $\varphi^*(x)\psi(x)$ over the x's in the interval. In the case of a finite vector space, the comparable scalar product

$$\sum_i e_i^* f_i$$

is a summing of values of $e_i^* f_i$ over the available i's. The arguments of Section 4-1, applied to this case, show that the sum goes over into the integral provided we index the values φ by x rather than by i. It is convenient to use *this* scalar product to define the dual space vector $\langle f|$ corresponding to our original vector

$|f\rangle$; we simply have

$$\langle e \mid f\rangle = \sum_n \langle e \mid n\rangle \langle n \mid f\rangle = \sum_n e_n^* f_n \tag{4-58}$$

In the continuous case the corresponding relation is

$$\langle \varphi \mid \psi\rangle = \int_a^b \frac{\langle \varphi \mid x\rangle \langle x \mid \psi\rangle}{2\pi} = \int_a^b \varphi(x)^* \psi(x)\, dx \tag{4-59}$$

However, not all scalar products in the discrete case need take the simple form (4-58). We can satisfy all the requirements of the scalar product with a much more general expression. If M_{ij} is a matrix such that $M_{ij}^* = M_{ji}$, then we can define

$$\langle e \mid f\rangle_M = \sum_{nm} e_n^* M_{nm} f_m \tag{4-60}$$

This is linear in both the spaces, and has the required property that $\langle e \mid f\rangle_M = \langle f \mid e\rangle_M^*$. When M is such that the positivity constraint (1-35) is also satisfied, then $\langle e \mid f\rangle_M = \langle e|\, M\, |f\rangle$ is a scalar product of the form discussed in Section 1-3. If we follow through our discussion of using the scalar product to assign dual space vectors, then this new scalar product makes the assignment $|e\rangle \to \langle e|_M = \langle e|\, M$; whereas $|e\rangle \to \langle e|$ is the assignment made by our simplest scalar product (4-58).

In the continuous case, we have the analogous situation: allow the dual space vector $\langle f|$ to be defined by the simple scalar product (4-59); then a more general scalar product corresponds to the mapping $|f\rangle \to \langle f|\, \varphi$ so that

$$\mathbf{f} \cdot \mathbf{g} = \langle f|\, \theta\, |g\rangle \tag{4-61}$$

The scalar product between two functions may then be expressed in terms of their values by inserting the unit operator into this formula:

$$\begin{aligned} \mathbf{f} \cdot \mathbf{g} &= \langle f| \int_a^b \frac{|x\rangle \langle x|}{2\pi}\, dx\, \theta \int_a^b \frac{|x'\rangle \langle x'|}{2\pi}\, dx'\, |g\rangle \\ &= \int_a^b \int_a^b dx\, dx'\, \frac{\langle f \mid x\rangle}{2\pi} \langle x|\, \theta\, |x'\rangle \frac{\langle x' \mid g\rangle}{2\pi} \\ &= \int_a^b \int_a^b dx\, dx'\, f(x^*)\, \frac{\langle x|\, \theta\, |x'\rangle}{2\pi}\, g(x') \end{aligned} \tag{4-62}$$

Ordinarily the linear operator θ is local; i.e., $\langle x|\, \theta\, |x'\rangle = \theta(x)\delta(x - x')2\pi$. Then the scalar product reduces to

$$\mathbf{f} \cdot \mathbf{g} = \int_a^b dx\, f^*(x)\theta(x)g(x)$$

exactly the form we have been using.

In the case of a finite vector space, of course, if faced with a scalar product like (4-60), we generally would not continue to work in the original basis $|n\rangle$

—this is too clumsy. Instead we would transform to a new basis $|\tilde{n}\rangle$ such that $\langle\tilde{m}\,|\,\tilde{n}\rangle_M = \delta_{\tilde{n}\tilde{m}}$. In this basis, ($1 = \sum_{\tilde{m}} |\tilde{m}\rangle\langle\tilde{m}|$ if all scalar products are $\langle\ |\ \rangle_M$)

$$\langle e\,|\,f\rangle_M = \langle e|\,M\,|f\rangle = \sum_{\tilde{n}\tilde{m}} \langle e\,|\,\tilde{m}\rangle_M \langle\tilde{m}|\,M\,|\tilde{n}\rangle\langle\tilde{n}\,|\,f\rangle_M$$

$$= \sum_{\tilde{n}} \langle e\,|\,\tilde{m}\rangle_M \langle\tilde{m}\,|\,f\rangle_M = \sum_{\tilde{m}} f_{\tilde{m}} e^*_{\tilde{m}}$$

In the continuous case, we may continue to work in the $|x\rangle$ basis because of our familiarity with calculations there. However if the function $\theta(x)$ is chosen so that basis functions $|e_n\rangle$ are orthonormal

$$\mathbf{e}_n \cdot \mathbf{e}_m = \int_a^b e_n^*(x)\theta(x)e_m(x)\,dx = \delta_{nm} \tag{4-63}$$

then the scalar product of two functions $|f\rangle = \sum_n f_n\,|e_n\rangle$ and $|g\rangle = \sum_n g_n\,|e_n\rangle$ becomes

$$\mathbf{f} \cdot \mathbf{g} = \int_a^b f^*(x)\theta(x)g(x)\,dx = \sum_n f_n^* g_n \tag{4-64}$$

just as we had in the finite case.

PROBLEMS

4-10 In a two-dimensional vector space, consider the scalar product $\langle a\,|\,b\rangle_M \equiv \langle a|\,M\,|b\rangle$ defined by the matrix

$$M = \begin{pmatrix} 2 & 0 \\ 0 & 3 \end{pmatrix}$$

Find a basis $|\tilde{n}\rangle$ that is orthonormal with respect to this scalar product, and demonstrate explicitly that for any two vectors in the space

$$\langle e\,|\,f\rangle_M = \sum_{\tilde{m}} f_{\tilde{m}} e^*_{\tilde{m}}$$

4-11 In Section 4-2 we discussed expansion of a function on the interval $(0, a)$ by use of $J_m(k_n x)$, where $J_m(k_n a) = 0$. This is an expansion in terms of a countable basis. Suppose we have the interval $(0, \infty)$ instead. If the scalar product to be used is defined by Bessel's equation, what set of expansion functions is appropriate? Give their orthogonality properties. By analogy with the case of Fourier transforms, the use of this uncountable basis set is called a Fourier-Bessel transform.

4-5 LADDER OPERATORS FOR FUNCTION SPACES

For certain types of Sturm-Liouville equations, there is a particularly elegant method for obtaining information about the eigenvalues and eigenfunctions. This method, the factorization of the differential equation into raising and lowering operators, has formal applications in quantum mechanics. As an introduction to these ideas, we will use them to find recursion relations between adjacent eigenfunctions of some differential operators.

Let us begin with the equation for the nth eigenfunction of a quantum oscillator

$$\frac{d^2\psi_n}{dx^2} + (2n + 1 - x^2)\psi_n = 0 \tag{4-65}$$

This can be rewritten as

$$\left(\frac{d}{dx} - x\right)\left(\frac{d}{dx} + x\right)\psi_n = -2n\psi_n \tag{4-66}$$

If we now define

$$R_n = \left(\frac{d}{dx} - x\right)\psi_n$$

simple substitution can be used to show that

$$\left(\frac{d}{dx} - x\right)\left(\frac{d}{dx} + x\right)R_n = -2(n + 1)R_n$$

But this is simply equation (4-66) with n replaced by $n + 1$. Hence R_n must be proportional to ψ_{n+1}, and we see that $d/dx - x$ is a raising operator that takes us from a solution to the one with next higher n. Likewise, one can show that $d/dx + x$ is a lowering operator that takes ψ_{n+1} into a function proportional to ψ_n.

This means that it is only necessary to solve equation (4-65) once, for $n = 0$. Subsequent eigenfunctions can be obtained (up to a normalization constant) simply by acting on this fundamental one with the raising operator $d/dx - x$.

The oscillator problem is a particularly simple one. For some equations, the raising and lowering operators change with n. Consider for example the Legendre equation

$$(x^2 - 1)\frac{d^2y}{dx^2} + 2x\frac{dy}{dx} - n(n + 1)y = 0 \tag{4-67}$$

After the substitution $x = \cos z$, this becomes

$$\frac{d^2y}{dz^2} + \frac{\cos z}{\sin z}\frac{dy}{dz} + n(n + 1)y = 0$$

which can be further simplified by the change of variables $t = \ln[\tan z/2]$

$$\frac{d^2y}{dt^2} + \frac{n(n + 1)}{\cosh^2 t}y = 0 \tag{4-68}$$

Equation (4-68) can be rewritten as follows:

$$\left[n \tanh t - \frac{d}{dt}\right]\left[n \tanh t + \frac{d}{dt}\right]\phi_n = n^2 \phi_n \tag{4-69}$$

or

$$\left[(n+1) \tanh t + \frac{d}{dt}\right]\left[(n+1) \tanh t - \frac{d}{dt}\right]\phi_n = (n+1)^2 \phi_n \tag{4-70}$$

Suppose we have found a solution ϕ_n to (4-69). Define

$$R_n = \left[(n+1) \tanh t - \frac{d}{dt}\right]\phi_n$$

Then, using (4-70), we find

$$\left[(n+1) \tanh t - \frac{d}{dt}\right]\left[(n+1) \tanh t + \frac{d}{dt}\right]R_n = (n+1)^2 R_n$$

and R_n must be proportional to ϕ_{n+1}. Hence we can again generate the solutions corresponding to higher values of n just from the one for $n = 0$.

Equation (4-68) assumes a particularly simple form for $n = 0$:

$$\frac{d^2\phi_0}{dt^2} = 0$$

with solution

$$\phi_0 = \alpha t + \beta = \alpha \ln\left[\tan \frac{\cos^{-1} x}{2}\right] + \beta$$

$$= \frac{\alpha}{2} \ln\left[\frac{1-x}{1+x}\right] + \beta$$

Clearly, the values $\alpha = 0$, $\beta = 1$ give $\phi_0 = P_0(x)$ whereas $\alpha = -1$, $\beta = 0$ give $\phi_0 = Q_0(x)$. We then get something proportional to ϕ_1 by acting on $\phi_0 = \alpha t + \beta$ with $(n+1) \tanh t - d/dt = \tanh t - d/dt$. This gives $\phi_1 = (\tanh t)(\alpha t + \beta) - \alpha$. Choice of $\alpha = 0$, $\beta = 1$ gives

$$\phi_1 = -x = -P_1(x)$$

whereas choice of $\alpha = +1$, $\beta = 0$ gives $\phi_1 = Q_1(x)$.

If we invert our change of variables in the raising operator, we obtain a recursion relation

$$\phi_{n+1}(x) = (n+1)x\phi_n + (x^2 - 1)\frac{d\phi_n}{dx}$$

A similar recursion relation can be derived from the lowering operator. [Note that the functions derived in this way must still be normalized according to the convention preferred, usually $P_n(1) = 1$.]

PROBLEMS

4-12 Find P_2 and P_3 by the technique outlined in the text.

4-13 Find Q_2 and Q_3 by the ladder operator technique.

4-14 Show that

$$(n+1)\tanh t + d/dt$$

is a "lowering operator" that takes ϕ_{n+1} to ϕ_n in (4-69).

4-15 After separation of angular and radial variables, the quantum mechanical Hamiltonian for the three-dimensional isotropic harmonic oscillator takes the form

$$H_l = \frac{1}{2m}\pi^2 + \frac{l(l+1)}{2mr^2} + \frac{m\omega^2 r^2}{2}$$

in the subspace with angular momentum l. The Hermitian operator π is defined by

$$\pi = \frac{1}{i}\frac{d}{dr} - \frac{i}{r}.$$

By defining operators

$$b_l = \frac{i\pi}{\sqrt{2m}} - \frac{(l+1)}{\sqrt{2m}\,r} + \frac{r\omega\sqrt{m}}{\sqrt{2}}$$

$$b_l^+ = -\frac{i\pi}{\sqrt{2m}} - \frac{(l+1)}{\sqrt{2m}\,r} + \frac{r\omega\sqrt{m}}{\sqrt{2}}$$

we can "factorize" the Hamiltonian as

$$b_l b_l^+ = H_{l+1} - \omega(l + \tfrac{1}{2})$$

$$b_l^+ b_l = H_l - (l + \tfrac{3}{2})\omega$$

Note that $[\pi, r] = -i$ (i.e., $\pi r\psi - r\pi\psi = -i\psi$ for any function ψ)

(a) Prove that if $|\psi_l\rangle$ is an eigenstate of H_l with eigenenergy E then $b_l|\psi_l\rangle$ is an eigenstate of H_{l+1} with eigenenergy $E - \omega$. What are the properties of $b_l^+|\psi_{l+1}\rangle$?

(b) Define $|\chi\rangle = b_l|\psi_l\rangle$. From the fact that $\langle\chi|\chi\rangle \geq 0$, prove that the energies of states $|\psi_l\rangle$ must be $\geq (l + \tfrac{3}{2})\omega$.

(c) For each l, start with the state of lowest energy and "ladder down" in l using b^+ operators to sweep out the allowed spectrum of states.

(d) You may also find it interesting to explore the factorization in terms of another set of operators

$$a_l = \frac{i\pi}{\sqrt{2m}} + \frac{l}{\sqrt{2m}\,r} + \frac{r\omega\sqrt{m}}{\sqrt{2}}$$

$$a_l^+ = -\frac{i\pi}{\sqrt{2m}} + \frac{l}{\sqrt{2m}\,r} + \frac{r\omega\sqrt{m}}{\sqrt{2}}$$

SUMMARY

If a linear differential operator is of Sturm-Liouville type,

$$\frac{1}{r(x)}\frac{d}{dx}\left[p(x)\frac{d\psi}{dx}\right] + \frac{q(x)}{r(x)}\psi = S\psi$$

and if its eigenfunctions ψ_λ satisfy boundary conditions such that

$$\left[p(x)\left(\psi_\lambda^*\frac{d\psi_{\lambda'}}{dx} - \psi_{\lambda'}\frac{d\psi_\lambda^*}{dx}\right)\right]_a^b = 0$$

then the eigenvalues of the operator are real, eigenfunctions belonging to different eigenvalues are orthogonal on the interval (a, b) with a weight function $r(x)$, and every infinitely differentiable function on (a, b) can be expanded in terms of these eigenfunctions. This allows us to think of the set of all infinitely differentiable functions on (a, b) as a vector space with basis vectors given by the eigenvectors of the Sturm-Liouville operator. Since Bessel's equation and Legendre's equation are of Sturm-Liouville type, this gives us another way of understanding expansions in Bessel functions and Legendre polynomials.

In some cases (for example Legendre polynomials), the set of eigenfunctions of the operator is countable; in other cases it is uncountable and we have an expansion similar to a Fourier transform. If the basis vectors are uncountable, the overlap $\langle\omega\,|\,\omega'\rangle$ will be proportional to a delta function $\delta(\omega - \omega')$ rather than the Kronecker delta we have in the countable case. All the basic ideas are the same, however, whether the set used for expansion is countable or uncountable.

Some Sturm-Liouville operators may be "factored" into raising and lowering operators that take one solution of the eigenvalue equation into another. In these cases, one can frequently discover the properties of the spectrum of the operator without explicitly solving the differential equation. The ladder operators may also be used to generate "higher" solutions once some "basic" solution has been found.

CHAPTER 5 Infinite Series

In the preceding chapters we have frequently encountered expansions of functions in series of other functions: power series solutions for differential equations, expansions of functions on an interval in terms of eigenfunctions of some Sturm-Liouville operator, etc. In some cases, we even defined functions (such as J_n) by giving infinite series. Thus far, however, we have paid very little attention to the inverse problem: given a series, can we sum its terms in a unique way to get a well-defined answer? If this is not possible, the series is not of much use; thus it is important to be able to tell whether a series will converge. It is equally important, once a series is recognized as convergent, to be able to sum it into closed form for evaluation (if this is possible).

The first section of this chapter deals with criteria for determining the convergence properties of a given series. We then go on to list a number of common series that the student should be able to recognize, and finally we discuss a number of methods by which unfamiliar series may (sometimes) be related to more familiar ones.

5-1 CONVERGENCE

An infinite series

$$\sum_{n=1}^{\infty} a_n = a_1 + a_2 + a_3 + \cdots$$

is said to *converge* to the sum S provided the sequence of partial sums has the limit S; that is, provided for every real number ε, there exists an N_0 such that

$$\left| S - \sum_{n=1}^{N} a_n \right| < \varepsilon \tag{5-1}$$

for all N greater than N_0. If the series $\sum_{n=1}^{\infty} a_n$ converges to the sum S, we will simply write this as an equality:

$$\sum_{n=1}^{\infty} a_n = S \tag{5-2}$$

The series $\sum_{n=1}^{\infty} a_n$ is said to converge *absolutely* if the related series $\sum_{n=1}^{\infty} |a_n|$ converges. Absolute convergence implies convergence, but not vice versa; for example, the series

$$1 - \tfrac{1}{2} + \tfrac{1}{3} - + \cdots$$

converges (to the sum ln 2), but it does not converge absolutely, because

$$1 + \tfrac{1}{2} + \tfrac{1}{3} + \cdots$$

does not converge (see Problem 5-1).

It should be emphasized that the numbers a_n may be *complex* numbers. We must, of course, give the symbol $|a_n|$ of the preceding paragraph its usual meaning when a_n is complex:

$$|a_n| = \sqrt{(\text{Re } a_n)^2 + (\text{Im } a_n)^2}$$

That is, $|a_n|$ denotes the absolute value (or modulus) of a_n.

Example

For any x and any N, we can always write

$$\frac{1}{1-x} = 1 + x + x^2 + \cdots + x^N + \frac{x^{N+1}}{1-x} \tag{5-3}$$

If $|x| < 1$, then given any ε we can always find an N such that

$$\left| \frac{x^{N+1}}{1-x} \right| < \varepsilon$$

Hence for $|x| < 1$, the sequence $S_n = \sum_1^N x^n$ converges to $1/(1-x)$ as $N \to \infty$. We say then that $\sum_{n=1}^{\infty} x^n$ converges to $1/(1-x)$. If $|x| > 1$, the sequence of partial sums $\sum_1^N x^n$ clearly does not converge.

Suppose that we know that a series $\sum_{n=1}^{\infty} a_n$ converges absolutely; i.e., $\sum_{n=1}^{\infty} |a_n|$ converges. Consider the series $\sum_{n=1}^{\infty} b_n$, where $|b_n| < |a_n|$. It is easy to show that $\sum b_n$ also converges absolutely, as follows: Given any ε, we can find an N_0 such that $\sum_N^{\infty} |a_n| < \varepsilon$ for all N higher than N_0. Since $|b_n| < |a_n|$, we also have $\sum_N^{\infty} |b_n| < \varepsilon$ for all N higher than N_0, and hence $\sum_1^{\infty} b_n$ converges absolutely.

Comparison of a series with the convergent series $\sum_0^{\infty} x^n$ (for $|x| < 1$) allows us to derive the *ratio test*: If the ratio $|a_{n+1}/a_n|$ of the absolute magnitude of successive terms in the infinite series

$$a_0 + a_1 + a_2 + \cdots$$

has as its limit a number less than one as $n \to \infty$, the series converges absolutely. To see why this is true, consider $\sum_N^\infty a_n \leq \sum_N^\infty |a_n|$. This can be rewritten in the form

$$|a_N| \left\{ 1 + \left| \frac{a_{N+1}}{a_N} \right| + \left| \frac{a_{N+2}}{a_{N+1}} \right| \left| \frac{a_{N+1}}{a_N} \right| + \cdots \right\}$$

Because the limit of $|a_{n+1}/a_n|$ is $K < 1$ as $n \to \infty$, given any $\varepsilon < 1 - K$ we can find an N_0 such that $|a_{n+1}/a_n| \leq K' = K + \varepsilon < 1$ for $n > N_0$. Thus for $N > N_0$, $\sum_N^\infty a_n$ converges absolutely by comparison with the convergent series $|a_N| \sum_{n=0}^\infty (K')^n$, and hence $\sum_0^\infty a_n$ converges.

It is easy to see that if the limit as $n \to \infty$ of $|a_{n+1}/a_n|$ is greater than one, $\sum_0^\infty a_n$ cannot converge, because the partial sums $\sum_0^N a_n$ do not converge as a sequence. If $|a_{n+1}/a_n|$ has no limit, or if the limit is one, we must investigate further.

A second criterion is comparison with an infinite integral. The series

$$f(1) + f(2) + f(3) + \cdots$$

converges or diverges with the infinite integral

$$\int^\infty f(x)\, dx$$

provided $f(x)$ is monotonically decreasing. For if f is monotonically decreasing, $f(n+1)$ lies below $f(x)$ in the interval $n < x < n + 1$, and hence

$$\sum_{n=b}^\infty f(n) = f(b)(1) + f(b+1)(1) + \cdots \leq \int_{b-1}^\infty f(x)\, dx$$

Clearly convergence of the integral implies convergence of the series. On the other hand, $f(n)$ lies above $f(x)$ in the interval $n < x < n + 1$, so

$$\sum_{n=b}^\infty f(n) \geq \int_b^\infty f(x)\, dx$$

We then see that divergence of the integral implies divergence of the sum.

For example, consider the series for the *Riemann zeta function*

$$\zeta(s) = 1 + \frac{1}{2^s} + \frac{1}{3^s} + \frac{1}{4^s} + \cdots \tag{5-4}$$

The ratio of successive terms is

$$\frac{a_{n+1}}{a_n} = \left(\frac{n}{n+1}\right)^s = \left(1 + \frac{1}{n}\right)^{-s} \underset{n \to \infty}{\sim} 1 - \frac{s}{n} + \cdots$$

The ratio approaches one as $n \to \infty$. Thus, comparison with the geometric series fails. However

$$\int \frac{dx}{x^s} = -\frac{1}{s-1}\frac{1}{x^{s-1}}$$

so that the criterion for convergence of the zeta function series is $s > 1$. This enables us to sharpen our ratio test; if

$$\left|\frac{a_{n+1}}{a_n}\right| \to 1 - \frac{s}{n} \tag{5-5}$$

with s greater than 1, the series converges absolutely. (If s is complex, the condition is Re $s > 1$.)

Example

Consider the hypergeometric series

$$F(a, b; c; x) = 1 + \frac{ab}{c}\frac{x}{1!} + \frac{a(a+1)b(b+1)}{c(c+1)}\frac{x^2}{2!} + \cdots$$

The ratio of successive terms is

$$\begin{aligned}\frac{a_{n+1}}{a_n} &= \frac{(a+n)(b+n)}{(c+n)(n+1)}x \\ &= \frac{\left(1+\frac{a}{n}\right)\left(1+\frac{b}{n}\right)}{\left(1+\frac{c}{n}\right)\left(1+\frac{1}{n}\right)}x \\ &= \left(1 + \frac{a+b-c-1}{n} + \cdots\right)x\end{aligned}$$

Thus the series converges if $|x| < 1$, or, when $|x| = 1$, if

$$a + b - c < 0 \tag{5-6}$$

[If we allow a, b, c to be complex numbers, condition (5-6) becomes

$$\text{Re}(a + b - c) < 0]$$

We can sharpen our ratio test even more by considering a very slowly converging series such as

$$\sum_{n=2}^{\infty} \frac{1}{n(\ln n)^s} = \frac{1}{2(\ln 2)^s} + \frac{1}{3(\ln 3)^s} + \cdots \tag{5-7}$$

The ratio of successive terms is

$$\begin{aligned}\frac{a_{n+1}}{a_n} &= \left(\frac{n}{n+1}\right)\left[\frac{\ln n}{\ln(n+1)}\right]^s \\ &= \left(1 - \frac{1}{n} + \cdots\right)\left[\frac{\ln n}{\ln n + \ln\left(1 + \frac{1}{n}\right)}\right]^s \\ &= \left(1 - \frac{1}{n} + \cdots\right)\left[1 + \frac{1}{n \ln n} + \cdots\right]^{-s} \\ &= 1 - \frac{1}{n} - \frac{s}{n \ln n} + \cdots\end{aligned} \tag{5-8}$$

and hence the series would not have passed the criterion of equation (5-5). By comparing with the integral

$$\int \frac{dx}{x(\ln x)^s} = -\frac{1}{s-1}\frac{1}{(\ln x)^{s-1}}$$

we see that the series (5-7) converges provided $s > 1$.

The series of course diverges if $s = 0$. This leads us to suspect that if $(a_{n+1}/a_n) \sim 1 - (1/n) + f(n)$, $f(n)$ should fall less rapidly than $1/n^2$ if the series is to converge (that is, some sort of logarithmic dependence should be present). We can in fact prove that if

$$\frac{a_{n+1}}{a_n} \sim 1 - \frac{1}{n} + \frac{K}{n^2}$$

the series diverges. For consider the series

$$\sum_n \frac{1}{n+K-1}$$

This diverges by the integral test. The ratio of its successive terms is

$$\frac{a_{n+1}}{a_n} = \frac{n+K-1}{n+K} = 1 - \frac{1}{n+K} \sim 1 - \frac{1}{n} + \frac{K}{n^2}$$

exactly the behavior under consideration.

Example

Consider the series solution (3-41) of Legendre's equation

$$1 - n(n+1)\frac{x^2}{2!} + n(n+1)(n-2)(n+3)\frac{x^4}{4!} - + \cdots$$

The ratio of successive terms is

$$\frac{a_{i-1}}{a_{i-2}} = -\frac{(n-2i+4)(n+2i-3)}{(2i-3)(2i-2)}x^2$$

For i very large

$$\frac{a_{i-1}}{a_{i-2}} \approx \left[1 - \frac{1}{i} + \mathcal{O}\left(\frac{1}{i^2}\right)\right]x^2$$

Thus, if $x^2 = 1$, the series diverges.

There are, of course, other convergence criteria; for example, if the signs of the a_n alternate and a_n approaches zero monotonically, then the series $\sum_n a_n$ converges (but not necessarily absolutely).

PROBLEMS

5-1 $\sum_{n=1}^{\infty} 1/n$ is the area under the squared curve in Figure 5-1; this is greater

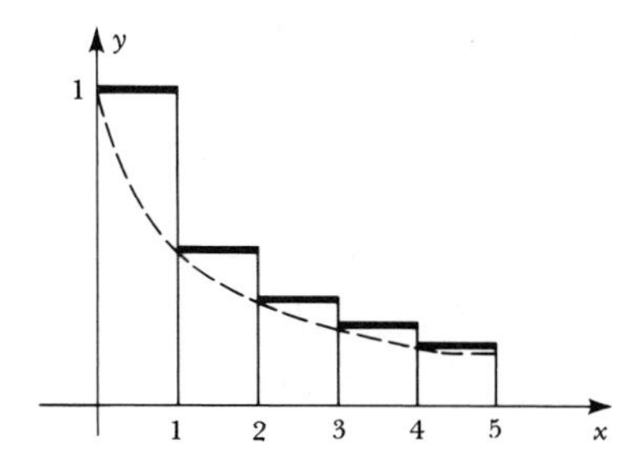

FIGURE 5-1 Areas for integral comparison test

than the area under the smooth curve that represents the function $1/(x+1) = y$. Use this to prove that $\sum_{n=1}^{\infty} 1/n$ diverges logarithmically.

Do the following series converge or not?

5-2
$$\frac{\frac{1}{2}\cdot\frac{1}{2}}{9\cdot 7\cdot 25\cdot 1!} + \frac{\frac{1}{2}\cdot\frac{3}{2}\cdot\frac{3}{2}}{11\cdot 9\cdot 49\cdot 2!} + \frac{\frac{1}{2}\cdot\frac{3}{2}\cdot\frac{5}{2}\cdot\frac{5}{2}}{13\cdot 11\cdot 81\cdot 3!} + \cdots$$

5-3
$$\frac{(1\cdot 3)^2}{1\cdot 1\cdot 1^2} + \frac{(1\cdot 3\cdot 5)^2}{4\cdot 2\cdot(1\cdot 2)^2} + \frac{(1\cdot 3\cdot 5\cdot 7)^2}{16\cdot 3\cdot(1\cdot 2\cdot 3)^2} + \frac{(1\cdot 3\cdot 5\cdot 7\cdot 9)^2}{64\cdot 4\cdot(1\cdot 2\cdot 3\cdot 4)^2} + \cdots$$

5-2 FAMILIAR SERIES

The reader should be familiar with at least the following simple series:

$$(1+x)^n = 1 + nx + n(n-1)\frac{x^2}{2!} + \cdots$$

$$= \sum_{\alpha=0}^{\infty} \frac{n!}{(n-\alpha)!} \frac{x^\alpha}{\alpha!} \qquad \text{(binomial series)} \tag{5-9}$$

$$e^x = 1 + x + \frac{x^2}{2!} + \frac{x^3}{3!} + \cdots = \sum_{\alpha=0}^{\infty} \frac{x^\alpha}{\alpha!} \tag{5-10}$$

Using the Euler relation $e^{ix} = \cos x + i \sin x$, we deduce from (5-10)

$$\sin x = x - \frac{x^3}{3!} + \frac{x^5}{5!} - + \cdots \tag{5-11}$$

$$\cos x = 1 - \frac{x^2}{2!} + \frac{x^4}{4!} - + \cdots \tag{5-12}$$

Term-by-term integration of the series for $(1 + x)^{-1}$ and $(1 + x^2)^{-1}$ yields

$$\ln(1 + x) = x - \frac{x^2}{2} + \frac{x^3}{3} - \frac{x^4}{4} + - \cdots \tag{5-13}$$

$$\tan^{-1} x = x - \frac{x^3}{3} + \frac{x^5}{5} - + \cdots \tag{5-14}$$

5-3 TRANSFORMATION OF SERIES

Various devices may be used to reduce a given unfamiliar series to a known one. Differentiation and integration are often useful.

Example

$$f(x) = 1 + 2x + 3x^2 + 4x^3 + \cdots$$

Integrating term-by-term,

$$\int_0^x f(x')\, dx' = x + x^2 + x^3 + \cdots = \frac{x}{1-x}$$

Then, differentiating

$$f(x) = \frac{d}{dx}\left(\frac{x}{1-x}\right) = \frac{1}{(1-x)^2}$$

[We might have recognized this immediately from the binomial series (5-9)].

Example

$$f(x) = \frac{1}{1\cdot 2} + \frac{x}{2\cdot 3} + \frac{x^2}{3\cdot 4} + \frac{x^3}{4\cdot 5} + \cdots \tag{5-15}$$

$$x^2 f(x) = \frac{x^2}{1 \cdot 2} + \frac{x^3}{2 \cdot 3} + \frac{x^4}{3 \cdot 4} + \cdots$$

Differentiating twice,

$$(x^2 f)'' = 1 + x + x^2 + x^3 + \cdots = \frac{1}{1 - x}$$

From this, two integrations give

$$f(x) = \frac{1}{x} + \frac{(1 - x)}{x^2} \ln(1 - x)$$

The constants of integration that arise in this procedure must be evaluated by knowing the series at certain values of x; for example it is obvious from (5-15) that $f(0) = \frac{1}{2}$.

Complex variables sometimes provide useful transformations.

Example

$$f(\theta) = 1 + a \cos \theta + a^2 \cos 2\theta + \cdots$$

$$= \text{Re}(1 + ae^{i\theta} + a^2 e^{2i\theta} + \cdots)$$

This is just a simple geometric series. Therefore,

$$f(\theta) = \text{Re}\left(\frac{1}{1 - ae^{i\theta}}\right) = \frac{1 - a \cos \theta}{1 - 2a \cos \theta + a^2}$$

The trick of differentiation or integration may be employed even if the series does not contain a variable.

Example

$$S = \frac{1}{2!} + \frac{2}{3!} + \frac{3}{4!} + \cdots$$

Define

$$f(x) = \frac{x^2}{2!} + \frac{2x^3}{3!} + \frac{3x^4}{4!} + \cdots$$

Then $S = f(1)$

$$f'(x) = x + x^2 + \frac{x^3}{2!} + \frac{x^4}{3!} + \cdots = xe^x$$

$$f(x) = \int_0^x xe^x\,dx = xe^x - e^x + 1$$

$$S = f(1) = 1$$

Example

$$S = 1 + m + \frac{m(m-1)}{2!} + \frac{m(m-1)(m-2)}{3!} + \cdots = \sum_n \frac{m!}{n!(m-n)!}$$

Let

$$f(x) = \sum_n \frac{m!x^n}{n!(m-n)!} \qquad S = f(1)$$

But

$$f(x) = (1 + x)^m$$

Therefore,

$$S = 2^m$$

Shortcuts such as the ones discussed here can be justified mathematically in many cases; however, there are cases in which they cannot be rigorously justified and may even lead to an incorrect result. Physicists, in general, tend to assume that the cases arising in physical problems are well behaved, and that tricks can be used without detailed rigorous justification. One should always check the reasonableness of an answer derived by shortcut methods; if peculiarities are evident, a more careful mathematical analysis is warranted.

PROBLEMS

Find the sums of the series in 5-4 through 5-7

5-4 $1 + \frac{1}{4} - \frac{1}{16} - \frac{1}{64} + \frac{1}{256} + \frac{1}{1024} - - + + \cdots$

5-5 $\frac{1}{1 \cdot 3} + \frac{1}{2 \cdot 4} + \frac{1}{3 \cdot 5} + \frac{1}{4 \cdot 6} + \cdots$

5-6 $1 - \frac{1}{5 \cdot 3^2} - \frac{1}{7 \cdot 3^3} + \frac{1}{11 \cdot 3^5} + \frac{1}{13 \cdot 3^6} - - + + \cdots$

5-7 $\frac{1}{0!} + \frac{2}{1!} + \frac{3}{2!} + \cdots$

5-8 Evaluate in closed form the sum

$$f(\theta) = \sin\theta + \tfrac{1}{3}\sin 2\theta + \tfrac{1}{5}\sin 3\theta + \tfrac{1}{7}\sin 4\theta + \cdots$$

(you may assume $0 < \theta < \pi$ for definiteness)

5-9 Evaluate the series

$$f(x) = \sum_{n=0}^{\infty} \frac{(-1)^{n+1} n^2 x^{2n-1}}{(2n-1)!}$$

$$= x - \frac{4x^3}{3!} + \frac{9x^5}{5!} - + \cdots$$

in closed form, by comparing with

$$\sin x = x - \frac{x^3}{3!} + \frac{x^5}{5!} - + \cdots$$

5-10 Of what function is

$$\cos\theta + \frac{\cos 3\theta}{9} + \frac{\cos 5\theta}{25} + \cdots$$

the Fourier series? Work it out; don't look it up!

Evaluate the following series:

5-11 $s = 1 - \dfrac{1}{4} + \dfrac{1}{6 \cdot 2!} - \dfrac{1}{8 \cdot 3!} + \dfrac{1}{10 \cdot 4!} - + \cdots$

5-12 $s(x) = 1 + \dfrac{x^2}{2} - \dfrac{x^4}{4} - \dfrac{x^6}{6} + \dfrac{x^8}{8} + \dfrac{x^{10}}{10} - - + + \cdots$

5-13 $s(x) = 1 + 3x + 5x^2 + 7x^3 + \cdots$

SUMMARY

A series $\sum_{n=1}^{\infty} a_n$ is said to converge absolutely if the related series $\sum_{n=1}^{\infty} |a_n|$ converges. Absolute convergence implies convergence, but a series that does not converge absolutely may still converge. In fact if the signs of the a_n alternate and a_n approaches zero monotonically, then the series $\sum_{n=1}^{\infty} a_n$ will converge.

One method for deciding whether a series $\sum a_n$ will converge absolutely is to examine the convergence of $\int |a_n|\, dn$ at its upper limit. If the integral diverges, then the sum $\sum |a_n|$ does also, and conversely. Another method is to compare the ratio of successive terms a_{n+1}/a_n with this ratio for a series of known convergence. If the terms in the series to be tested drop off faster than those in a series that is known to converge, then the new series will also converge. By applying these methods we were able to discover the following fact: for $a_{n+1}/a_n = 1 - s_1/n - s_2/n \ln n +$ (terms which drop off faster in n) the series will converge if s_1 is greater than 1, or if s_1 is 1 and s_2 is greater than 1.

Series for binomial expansions, exponentials, logarithms, and trigonometric functions should be part of the conscious memory of every physicist. Many other series can be related to these by simple means such as term-by-term differentiation and integration.

CHAPTER 6 Evaluation of Integrals

In practical work, it is often necessary to evaluate integrals explicitly. The many techniques that can be useful in such situations appear in various places in this text: contour integration is discussed in Section 8-1, and certain tabulated functions that are defined in terms of integrals are discussed in Section 10-2 as well as in Section 6-1. The bulk of this Chapter is devoted to some more elementary considerations.

Section 6-1 is a review of commonly used devices such as change of variables, use of complex numbers, and differentiation with respect to a parameter, which may be used to relate the given integral to one already known. In Section 6-2 the power of symmetry arguments in evaluating integrals containing vectors is illustrated by a number of examples.

6-1 ELEMENTARY METHODS

We first review some useful elementary techniques for doing integrals. The simplest device is changing variables. For example, there is a standard trick for evaluating

$$I = \int_0^\infty e^{-t^2}\, dt$$

by using polar coordinates ($x = r\cos\phi$, $y = r\sin\phi$, $dx\,dy = r\,dr\,d\phi$). Consider

$$I^2 = \int_0^\infty e^{-x^2}\, dx \int_0^\infty e^{-y^2}\, dy = \frac{1}{4}\int_{-\infty}^\infty e^{-x^2}\, dx \int_{-\infty}^\infty e^{-y^2}\, dy$$

Change to polar coordinates

$$I^2 = \frac{1}{4}\int_0^\infty e^{-r^2}\, r\, dr \int_0^{2\pi} d\phi$$

$$= \frac{\pi}{2} \int_0^\infty e^{-r^2} r \, dr = -\frac{\pi}{4} \int_0^\infty \frac{d}{dr} [e^{-r^2}] \, dr$$

$$= \frac{\pi}{4}$$

Hence

$$I = \int_0^\infty e^{-t^2} \, dt = \frac{\sqrt{\pi}}{2} \tag{6-1}$$

From this result, by setting $t = u\sqrt{a}$ we find

$$\int_0^\infty e^{-au^2} \, du = \tfrac{1}{2} \sqrt{\frac{\pi}{a}} \tag{6-2}$$

or by setting $t = u^2$, we may deduce

$$\int_0^\infty e^{-u^4} u \, du = \tfrac{1}{4} \sqrt{\pi} \tag{6-3}$$

How about $\int_0^\infty e^{-t^4} \, dt$? This is not so easy. Let

$$I_\alpha = \int_0^\infty e^{-t^\alpha} \, dt$$

Make the change of variable

$$t^\alpha = u \qquad t = u^{1/\alpha} \qquad dt = \frac{1}{\alpha} u^{(1/\alpha)-1} \, du$$

Then

$$I_\alpha = \frac{1}{\alpha} \int_0^\infty e^{-u} u^{(1/\alpha)-1} \, du$$

This does not look any easier, but one defines the *gamma function* $\Gamma(z)$ by

$$\Gamma(z) = \int_0^\infty e^{-u} u^{z-1} \, du \tag{6-4}$$

so that

$$I_\alpha = \int_0^\infty e^{-t^\alpha} \, dt = \frac{1}{\alpha} \Gamma\left(\frac{1}{\alpha}\right) \tag{6-5}$$

We shall say more about the gamma function in Section 10-2; note that (6-1) gives $\Gamma(\frac{1}{2}) = \sqrt{\pi}$.

Another useful technique is to introduce complex variables.

Example

$$I = \int_0^\infty e^{-\alpha x} \cos \lambda x \, dx$$

$$= \text{Re} \int_0^\infty e^{-\alpha x} e^{i\lambda x} \, dx$$

$$= \text{Re} \frac{1}{\alpha - i\lambda}$$

Therefore,

$$I = \frac{\alpha}{\alpha^2 + \lambda^2}$$

This method gives us another integral at the same time, from the imaginary part,

$$\int_0^\infty e^{-\alpha x} \sin \lambda x \, dx = \frac{\lambda}{\alpha^2 + \lambda^2} \tag{6-6}$$

The method of integration by parts is very useful, and is presumably familiar by now.

Another useful trick is differentiation or integration with respect to a parameter.

Example

$$I = \int_0^\infty e^{-\alpha x} \cos \lambda x \; x \, dx$$

Let

$$I(\alpha) = \int_0^\infty e^{-\alpha x} \cos \lambda x \, dx = \frac{\alpha}{\alpha^2 + \lambda^2}$$

Then

$$I = -\frac{d}{d\alpha} I(\alpha) = \frac{\alpha^2 - \lambda^2}{(\alpha^2 + \lambda^2)^2}$$

We have assumed that the order of differentiation and integration can be reversed. For necessary and sufficient conditions see mathematics books, such as Whittaker and Watson (34), Chapter IV, or Apostol (3), Chapter 9. In physical applications it will nearly always work.

Example

$$I = \int_0^\infty \frac{\sin x}{x}\, dx$$

Let

$$I(\alpha) = \int_0^\infty \frac{e^{-\alpha x}\sin x}{x}\, dx \qquad \text{so that } I = I(0)$$

$$\frac{dI(\alpha)}{d\alpha} = -\int_0^\infty e^{-\alpha x}\sin x\, dx = \frac{-1}{\alpha^2+1}$$

$$I(\alpha) = -\int \frac{d\alpha}{\alpha^2+1} = C - \tan^{-1}\alpha$$

But $I(\infty) = 0$. Therefore $C = \pi/2$.

$$I(\alpha) = \frac{\pi}{2} - \tan^{-1}\alpha \qquad \text{and} \qquad I = I(0) = \frac{\pi}{2}$$

Sometimes one can combine several derivatives of an integral to form a differential equation.

Example

$$I(\alpha) = \int_0^\infty \frac{e^{-\alpha x}}{1+x^2}\, dx$$

$$I''(\alpha) + I(\alpha) = \int_0^\infty e^{-\alpha x}\, dx = \frac{1}{\alpha}$$

This equation is easily solved by the variation of parameters method discussed in Section 3-1. The result is

$$I(\alpha) = -\cos\alpha \int^\alpha \frac{\sin t}{t}\, dt + \sin\alpha \int^\alpha \frac{\cos t}{t}\, dt$$

But $I(\alpha)$ and all its derivatives vanish at $\alpha = \infty$. Thus

$$I(\alpha) = \sin\alpha \int_\infty^\alpha \frac{\cos t}{t}\, dt - \cos\alpha \int_\infty^\alpha \frac{\sin t}{t}\, dt$$

The *cosine-integral* and *sine-integral* functions are defined by

$$\text{Ci}\, x = \int_\infty^x \frac{\cos t}{t}\, dt \qquad \text{Si}\, x = \int_0^x \frac{\sin t}{t}\, dt \tag{6-7}$$

Thus $I(\alpha) = (\text{Ci}\,\alpha)\sin\alpha + (\pi/2 - \text{Si}\,\alpha)\cos\alpha$. The functions $\text{Si}\,\alpha$ and $\text{Ci}\,\alpha$ are tabulated in Abramowitz and Stegun (1).

Finally, the useful integrals

$$\int_0^\infty e^{-\alpha x^2} x^n \, dx (n = 0, 1, 2, \ldots)$$

can be obtained from the first two,

$$\int_0^\infty e^{-\alpha x^2} dx = \frac{1}{2}\sqrt{\frac{\pi}{\alpha}} \qquad \int_0^\infty e^{-\alpha x^2} x \, dx = \frac{1}{2\alpha} \tag{6-8}$$

by repeated differentiation.

We observe in passing that if a parameter in an integral also appears in the limit(s), the differentiation with respect to that parameter proceeds according to the following rule:

$$\frac{d}{d\alpha}\left[\int_{a(\alpha)}^{b(\alpha)} f(x, \alpha) \, dx\right] = \int_{a(\alpha)}^{b(\alpha)} \frac{\partial f(x, \alpha)}{\partial \alpha} dx + \frac{db}{d\alpha} f[b(\alpha), \alpha] - \frac{da}{d\alpha} f[a(\alpha), \alpha] \tag{6-9}$$

PROBLEMS

Evaluate the following integrals:

6-1 $\displaystyle\int_0^\infty \frac{e^{-ay} - e^{-by}}{y} dy$

6-2 $\displaystyle\int_0^\infty \sin bx \, dx$ (multiply the integrand by $e^{-\lambda x}$ to apply a convergence factor; do integral; remove the convergence factor by setting $\lambda = 0$)

6-3 $\displaystyle\int_0^\infty \frac{\cos ax}{1 + x^2} dx$

6-4 $\displaystyle\int_0^\infty \frac{\cos ax}{(1 + x^2)^2} dx$

6-5 $\displaystyle\int_0^1 \frac{dx}{x} \ln\left[\frac{1 + x}{1 - x}\right]$

Hint: expand the integrand in a power series.

6-6 $\displaystyle\int_0^\infty \frac{dx}{\cosh x}$

Hint: expand the integrand in a series that is useful near $x = \infty$.

6-7 $\displaystyle\int_{-\infty}^\infty \frac{dx}{(1 + x^2)^2}$

6-8 Show that

$$\int_0^\infty \frac{\ln t\, dt}{t^2 - 1} = \int_{-\infty}^{\infty} \frac{u\, du}{e^u - e^{-u}}$$

and evaluate the integral in terms of a series. Can you sum the series using elementary techniques? Integrals of this form are best handled with the complex variable techniques of Chapter 8.

6-2 USE OF SYMMETRY ARGUMENTS

The evaluation of some integrals may be greatly simplified by exploiting the symmetries present in the problem. We shall illustrate the principles involved by means of some integrals over solid angle in three dimensions.

Example

Consider the integral

$$I_1(\mathbf{k}) = \int \frac{d\Omega}{1 + \mathbf{k}\cdot\hat{\mathbf{r}}} = \int_0^{2\pi} d\phi \int_{-1}^{+1} d(\cos\theta) \frac{1}{1 + \mathbf{k}\cdot\hat{\mathbf{r}}}$$

where $\hat{\mathbf{r}}$ is the unit radius vector and (θ, ϕ) are conventional spherical polar coordinates:

$$\hat{\mathbf{r}}_x = \sin\theta\cos\phi \qquad \hat{\mathbf{r}}_y = \sin\theta\sin\phi \qquad \hat{\mathbf{r}}_z = \cos\theta$$

Since the orientation of our coordinate system is arbitrary, we may choose the z-axis along $\mathbf{k}$ and obtain (we assume $k < 1$)

$$I_1(\mathbf{k}) = \int_0^{2\pi} d\phi \int_{-1}^{+1} \frac{d(\cos\theta)}{1 + k\cos\theta} = \frac{2\pi}{k} \ln\left(\frac{1+k}{1-k}\right) \tag{6-10}$$

The integrals

$$I_m(\mathbf{k}) = \int \frac{d\Omega}{(1 + \mathbf{k}\cdot\hat{\mathbf{r}})^m}$$

may be obtained from I_1 by differentiation (replace the 1 in the denominator by α, differentiate with respect to α, and set α equal to 1 again). For example,

$$I_2(\mathbf{k}) = \int \frac{d\Omega}{(1 + \mathbf{k}\cdot\hat{\mathbf{r}})^2} = \frac{4\pi}{1 - k^2} \tag{6-11}$$

Another example is

$$I_1(\mathbf{k}, \mathbf{a}) = \int \frac{\mathbf{a}\cdot\hat{\mathbf{r}}\, d\Omega}{1 + \mathbf{k}\cdot\hat{\mathbf{r}}} \tag{6-12}$$

This integral is complicated by the fact that *two* directions are given by the two vectors **a** and **k**, and we cannot choose our polar axis along both of them. However, the direction **a** is trivial, in that it may be factored out. For consider

$$\mathbf{J}(\mathbf{k}) = \int \frac{\hat{\mathbf{r}}\, d\Omega}{1 + \mathbf{k} \cdot \hat{\mathbf{r}}} \tag{6-13}$$

Clearly

$$I_1(\mathbf{k}, \mathbf{a}) = \mathbf{a} \cdot \mathbf{J}$$

Now $\mathbf{J}(\mathbf{k})$ is a vector and must point in the direction **k**, since no other direction is specified in the definition (6-13) of **J**. Therefore,

$$\mathbf{J}(\mathbf{k}) = A\mathbf{k} \tag{6-14}$$

To evaluate the scalar A, we dot **k** into both sides of (6-14) and obtain

$$\begin{aligned} A &= \frac{1}{k^2}\, \mathbf{k} \cdot \mathbf{J}(\mathbf{k}) = \frac{1}{k^2} \int \frac{\mathbf{k} \cdot \hat{\mathbf{r}}}{1 + \mathbf{k} \cdot \hat{\mathbf{r}}}\, d\Omega \\ &= \frac{1}{k^2} \int d\Omega \left(1 - \frac{1}{1 + \mathbf{k} \cdot \hat{\mathbf{r}}}\right) \\ &= \frac{4\pi}{k^2} \left(1 - \frac{1}{2k} \ln \frac{1 + k}{1 - k}\right) \end{aligned}$$

Thus our original integral (6-12) is

$$I_1(\mathbf{k}, \mathbf{a}) = A\mathbf{a} \cdot \mathbf{k} = \frac{4\pi}{k^2}\, \mathbf{a} \cdot \mathbf{k} \left[1 - \frac{1}{2k} \ln \frac{1 + k}{1 - k}\right]$$

What about the integral

$$I_2(\mathbf{k}, \mathbf{a}) = \int \frac{\mathbf{a} \cdot \hat{\mathbf{r}}\, d\Omega}{(1 + \mathbf{k} \cdot \hat{\mathbf{r}})^2}\ ?$$

We could obtain $I_2(\mathbf{k}, \mathbf{a})$ from $I_1(\mathbf{k}, \mathbf{a})$ by replacing the 1 by α and differentiating as before, but a simpler method is to differentiate $I_1(\mathbf{k})$ with respect to **k**. On the one hand,

$$\frac{\partial I_1(\mathbf{k})}{\partial \mathbf{k}} = \frac{\partial}{\partial \mathbf{k}} \int \frac{d\Omega}{1 + \mathbf{k} \cdot \hat{\mathbf{r}}} = -\int \frac{d\Omega\, \hat{\mathbf{r}}}{(1 + \mathbf{k} \cdot \hat{\mathbf{r}})^2} \tag{6-15}$$

On the other hand, using (6-10)

$$\frac{\partial I_1(\mathbf{k})}{\partial \mathbf{k}} = \frac{\mathbf{k}}{k} \frac{\partial I_1(\mathbf{k})}{\partial k} = \frac{2\pi \mathbf{k}}{k^2} \left[\frac{2}{1 - k^2} - \frac{1}{k} \ln\left(\frac{1 + k}{1 - k}\right)\right] \tag{6-16}$$

Comparing (6-15) and (6-16) gives

$$\int \frac{d\Omega\, \hat{\mathbf{r}}}{(1 + \mathbf{k} \cdot \hat{\mathbf{r}})^2} = \frac{2\pi \mathbf{k}}{k^2} \left(\frac{-2}{1 - k^2} + \frac{1}{k} \ln \frac{1 + k}{1 - k}\right)$$

and therefore

$$\int \frac{\mathbf{a}\cdot\hat{\mathbf{r}}\,d\Omega}{(1+\mathbf{k}\cdot\hat{\mathbf{r}})^2} = \mathbf{a}\cdot\int \frac{\hat{\mathbf{r}}\,d\Omega}{(1+\mathbf{k}\cdot\hat{\mathbf{r}})^2} = \frac{2\pi\mathbf{k}\cdot\mathbf{a}}{k^2}\left[\frac{-2}{1-k^2} + \frac{1}{k}\ln\left(\frac{1+k}{1-k}\right)\right]$$

Other examples of the use of symmetry arguments are the evaluation of integrals such as

$$\phi_1(\mathbf{a},\mathbf{b}) = \int d\Omega\,\hat{\mathbf{r}}\cdot\mathbf{a}\hat{\mathbf{r}}\cdot\mathbf{b}$$

$$\phi_2(\mathbf{a},\mathbf{b},\mathbf{c},\mathbf{d}) = \int d\Omega\,(\hat{\mathbf{r}}\cdot\mathbf{a})(\hat{\mathbf{r}}\cdot\mathbf{b})(\hat{\mathbf{r}}\cdot\mathbf{c})(\hat{\mathbf{r}}\cdot\mathbf{d})$$

etc.

To evaluate ϕ_1, we observe that it is a scalar that is linear in both **a** and **b**. The only possibility is that $\phi_1 = A\mathbf{a}\cdot\mathbf{b}$, where A is a number. To find A, let **a** and **b** both equal $\hat{\mathbf{z}}$. Then

$$\phi_1(\hat{\mathbf{z}},\hat{\mathbf{z}}) = A = \int d\Omega\,(\hat{\mathbf{r}}\cdot\hat{\mathbf{z}})^2 = \int d\Omega\cos^2\theta = \frac{4\pi}{3}$$

Therefore $\phi_1(\mathbf{a},\mathbf{b}) = (4\pi/3)\mathbf{a}\cdot\mathbf{b}$.

The evaluation of ϕ_2 proceeds similarly. Since ϕ_2 is a scalar, linear in the four vectors **a**, **b**, **c**, **d**, as well as being invariant under any interchange of these vectors, it must have the form

$$\phi_2 = B(\mathbf{a}\cdot\mathbf{b}\mathbf{c}\cdot\mathbf{d} + \mathbf{a}\cdot\mathbf{c}\mathbf{b}\cdot\mathbf{d} + \mathbf{a}\cdot\mathbf{d}\mathbf{b}\cdot\mathbf{c})$$

where B is a number. B is found by setting $\mathbf{a} = \mathbf{b} = \mathbf{c} = \mathbf{d} = \hat{\mathbf{z}}$, so that

$$\phi_2(\hat{\mathbf{z}},\hat{\mathbf{z}},\hat{\mathbf{z}},\hat{\mathbf{z}}) = 3B = \int d\Omega\,(\hat{\mathbf{r}}\cdot\hat{\mathbf{z}})^4 = \int d\Omega\cos^4\theta = \frac{4\pi}{5}$$

Therefore $B = 4\pi/15$, and

$$\phi_2(\mathbf{a},\mathbf{b},\mathbf{c},\mathbf{d}) = \frac{4\pi}{15}(\mathbf{a}\cdot\mathbf{b}\mathbf{c}\cdot\mathbf{d} + \mathbf{a}\cdot\mathbf{c}\mathbf{b}\cdot\mathbf{d} + \mathbf{a}\cdot\mathbf{d}\mathbf{b}\cdot\mathbf{c})$$

As a final example of the sort of device one can use to simplify integrals, we mention the identity

$$\frac{1}{ab} = \int_0^1 \frac{du}{[au + b(1-u)]^2} \tag{6-17}$$

which Feynman has used to simplify the evaluation of integrals arising in quantum field theory. As an application of (6-17), we evaluate the integral

$$\psi(\mathbf{k},\mathbf{l}) = \int \frac{d\Omega}{(1+\mathbf{k}\cdot\hat{\mathbf{r}})(1+\mathbf{l}\cdot\hat{\mathbf{r}})} \tag{6-18}$$

Use of (6-17) converts (6-18) to

$$\psi(\mathbf{k}, \mathbf{l}) = \int_0^1 du \int \frac{d\Omega}{\{1 + \hat{\mathbf{r}} \cdot [\mathbf{k}u + \mathbf{l}(1-u)]\}^2} \tag{6-19}$$

The solid angle integral in (6-19) is just $I_2[\mathbf{k}u + \mathbf{l}(1-u)]$, as given by (6-11). Thus

$$\psi(\mathbf{k}, \mathbf{l}) = 4\pi \int_0^1 \frac{du}{1 - [\mathbf{k}u + \mathbf{l}(1-u)]^2}$$

This is an elementary integral, although rather tedious; the answer has the interesting form

$$\psi(\mathbf{k}, \mathbf{l}) = \frac{4\pi}{\sqrt{A^2 - B^2}} \cosh^{-1} \frac{A}{B} \tag{6-20}$$

where

$$A = 1 - \mathbf{k} \cdot \mathbf{l}$$
$$B = \sqrt{(1-k^2)(1-l^2)}$$

The proof of (6-20) is left as an exercise.

PROBLEMS

Evaluate the following integrals:

6-9 $\int d^3 x e^{i\mathbf{a} \cdot \mathbf{x}} e^{-br^2}$ (the symbol d^3x stands for $dx\, dy\, dz$, or in general a volume element in three dimensions)

6-10 $\int d^3 x \mathbf{x} e^{i\mathbf{a} \cdot \mathbf{x}} e^{-br^2}$

6-11 $\displaystyle\iiint \frac{\mathbf{k}_1 \cdot \mathbf{r} \mathbf{k}_2 \cdot \mathbf{r}\, d^3r}{(1+r^2)^3}$

6-12 $\mathbf{J}(\mathbf{a}, \mathbf{b}) = \int d\Omega\, \hat{\mathbf{r}}\, \hat{\mathbf{r}} \cdot \hat{\mathbf{a}}\, \hat{\mathbf{r}} \cdot \hat{\mathbf{b}}$

6-13 $\mathbf{J} = \int d\Omega\, \hat{\mathbf{r}} \hat{\mathbf{r}} \cdot (\hat{\mathbf{a}} \times \hat{\mathbf{b}})$

6-14 Derive equation (4-21).

6-15 Prove equation (4-18).

6-16 Evaluate

$$\int d\Omega_x \int d\Omega_y\, \hat{\mathbf{a}} \cdot \hat{\mathbf{x}}\, \hat{\mathbf{x}} \cdot \hat{\mathbf{y}}\, \hat{\mathbf{y}} \cdot \hat{\mathbf{b}}\, \hat{\mathbf{b}} \cdot \hat{\mathbf{x}}\, \hat{\mathbf{x}} \cdot \hat{\mathbf{y}}\, \hat{\mathbf{y}} \cdot \hat{\mathbf{a}}$$

SUMMARY

The following possibilities should be considered in evaluation of an unfamiliar integral:

1. Change of variables may reduce it to a familiar form.

2. It may be related to a tabulated integral.

3. It might be the real or imaginary part of a complex expression that is easier to evaluate.

4. It may be necessary to introduce a convergence factor in order to evaluate the integral in a well-defined way.

5. Differentiation with respect to a parameter may reduce the integrand to a simpler form. In some cases a parameter can be introduced just for this purpose.

When the integrand contains external vectors, the value of the integral must be made up of these vectors in an appropriate way: if the integrand is a vector, the integral must be a vector, and so on. These symmetry considerations will often completely determine the form of the answer. If only one "outside" vector occurs inside an integration, choosing it as the z-axis for the integration variables usually simplifies evaluation.

CHAPTER 7 Some Properties of Functions of a Complex Variable

In the preceding chapters we have occasionally used complex numbers to simplify a calculation; however, these applications have all been relatively trivial extensions of properties of real functions. Many more useful methods can be obtained by exploiting the properties of functions of a complex variable. This chapter is devoted to an elementary discussion of some of the properties most useful in physical applications; Chapters 8 and 9 discuss their application to problem solving.

Most beginning students have some trouble "visualizing" complex functions with complex arguments. In Section 7-1 we discuss these functions as mappings from one complex plane to another, and illustrate some methods for discovering and displaying their properties. Branch points and cuts, in particular, are studied.

Section 7-2 concentrates on the property of analyticity. This is a generalization to complex variables of the notion of differentiability for real variables, but in practice it is much more restrictive than mere differentiability. Because analyticity is so restrictive, a large number of theorems can be proved about analytic functions, and these provide us with a battery of tools for application. Cauchy's integral theorem, which applies to line integrals in the complex plane, is perhaps the most important of these tools for the applications in this book, but the property of conformal mapping allows us to solve many boundary value problems (see Chapter 9).

Power series expansions of complex functions in their regions of analyticity are discussed in Section 7-3. We then go on, in the next section, to examine a problem that occurs frequently in evaluating path integrals generated by physics problems: what to do when a pole appears on the contour of integration. In Chapter 12 we will see that the choice of treatment of such poles is frequently related to boundary conditions imposed by the physics of the problem.

Another useful property of analytic functions is discussed in Section 7-5. This is the notion of analytic continuation: if you know the values of a function on one region of the complex plane, you can discover its values on any other region where it is analytic (provided there are no barriers of singularities in the way). We do not explicitly apply this method in future chapters, but the knowledge that it can be done is essential to some of the solutions. Many calculations in modern physics also depend on this property.

7-1 FUNCTIONS OF A COMPLEX VARIABLE AS MAPPINGS

A complex number has the form

$$z = x + iy = re^{i\theta} \tag{7-1}$$

where x, y, r, and θ are real, $i^2 = -1$, and $e^{i\theta} = \cos\theta + i\sin\theta$. x and y are the *real* (Re z) and *imaginary* (Im z) parts of z, respectively, $r = |z|$ is the *magnitude*, and $\theta = \arg z$ is the *phase* or *argument*. Such a number may be represented geometrically by a point on the *complex z-plane*, or *xy*-plane, as shown in Figure 7-1. The *complex conjugate* of z will be denoted by z^*; $z^* = x - iy$.

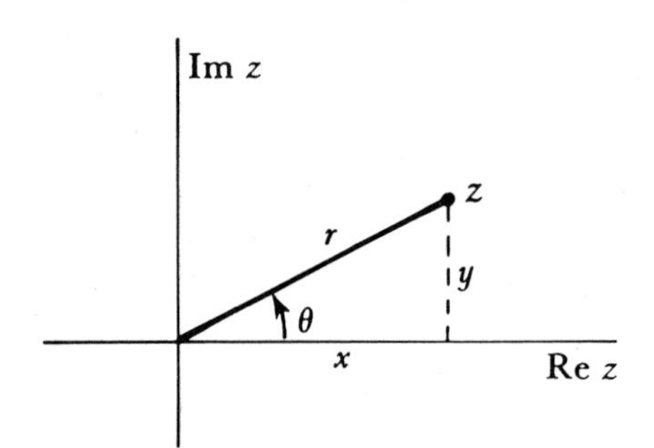

FIGURE 7-1 A point in the complex plane

A function $W(z)$ of the complex variable z is itself a complex number whose real and imaginary parts U and V depend on the position of z in the *xy*-plane.

$$W(z) = U(x, y) + iV(x, y) \tag{7-2}$$

Two different graphical representations of the function $W(z)$ are useful. One is simply to plot the real and (or) imaginary parts $U(x, y)$ and $V(x, y)$ as surfaces above the *xy*-plane. The other is to represent the complex number $W(z)$ by a point in the complex "W-plane," or U, V plane, so that each point in the z-plane corresponds to one (or more) points in the W-plane. In this way, the function $W(z)$ produces a *mapping* of the *xy*-plane onto the U, V plane.

Example

$$W(z) = z^2 = (x + iy)^2 = x^2 - y^2 + 2ixy$$

$$U = x^2 - y^2 \qquad V = 2xy$$

Alternatively,

$$W = z^2 = r^2 e^{2i\theta}$$

The mapping of a number of points and two curves from the z-plane onto the W-plane is shown in Figure 7-2. For example, the line $x = 1$ becomes the parabola $4U = 4 - V^2$.

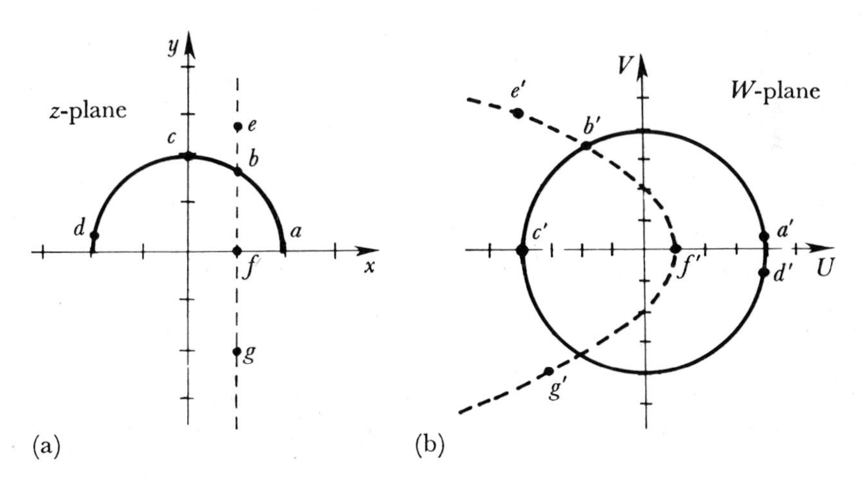

FIGURE 7-2 Illustration by some points and curves of the mapping produced by the function $W(z) = z^2$. The points $a, b, \ldots$ in the z-plane are mapped into $a', b', \ldots$ in the W-plane.

In the above example, two points z and $-z$ go into the same point $W(z)$. The upper half of the z-plane maps onto the entire W-plane, and so does the lower half z-plane. Clearly, this situation presents difficulties for the inverse mapping, which is represented by the square root $z(W) = W^{1/2}$.

In order to have a convention for mappings where the z-plane is always the original plane, and the W-plane is the image plane, we discuss instead the map

$$W(z) = z^{1/2}$$

On the z-plane, the argument θ is specified only up to a multiple of 2π; that is, $\theta = \theta_1$ and $\theta = \theta_1 + 2\pi$ both produce the same value of $z = re^{i\theta}$. However, if we define

$$z^{1/2} = \sqrt{r}\, e^{i\theta/2}$$

these two angles will give rise to different values of W:

$$W(re^{i\theta_1}) = \sqrt{r}\, e^{1/2(i\theta_1)}$$

$$W(re^{i\theta_1 + i2\pi}) = -\sqrt{r}\, e^{1/2(i\theta_1)}$$

This situation is illustrated in Figure 7-3.

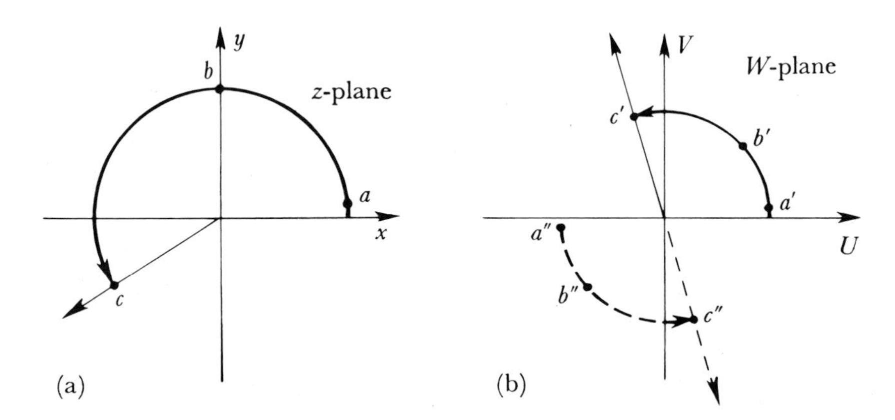

FIGURE 7-3 Illustration of the mapping produced by $W(z) = z^{1/2}$

Suppose we try to make the mapping single valued by agreeing that a point in the z-plane corresponds to one of these points, p', and not the other, p''. We must make sure that if we start at p and trace a closed curve in the z-plane, the mapping will produce a closed curve in the W-plane starting at p' and returning to p', not p''. This is true provided the closed curve in the xy-plane does not encircle the origin. However, if the curve encircles the origin once, θ changes by 2π and the mapped curve in the W-plane will not return to its starting point.

Thus the multivalued feature can be avoided only if we agree never to encircle the origin $z = 0$. To ensure this, we draw a so-called *branch line* or *branch cut* from $z = 0$ to infinity and agree not to cross it. The singular point $z = 0$ is called a *branch point.* The branch line may be drawn from $z = 0$ to infinity in any way but it is usually convenient to take it along the positive or negative real axis.

The z-plane, when cut in this way, is called a *sheet*, or *Riemann sheet*, of the function $W(z)$. This sheet maps in a single-valued manner onto a portion (in our example, half) of the W-plane, this portion being called a *branch* of the function. A second sheet, similarly cut, is needed to map onto the other half of the W-plane. We may now cross the branch line without getting into multivalue troubles if we transfer from one sheet to the other, when crossing the cut. To picture this, imagine that the edges of the sheets along the cut are joined to each other in the manner indicated in Figure 7-4. The sheets so joined form a

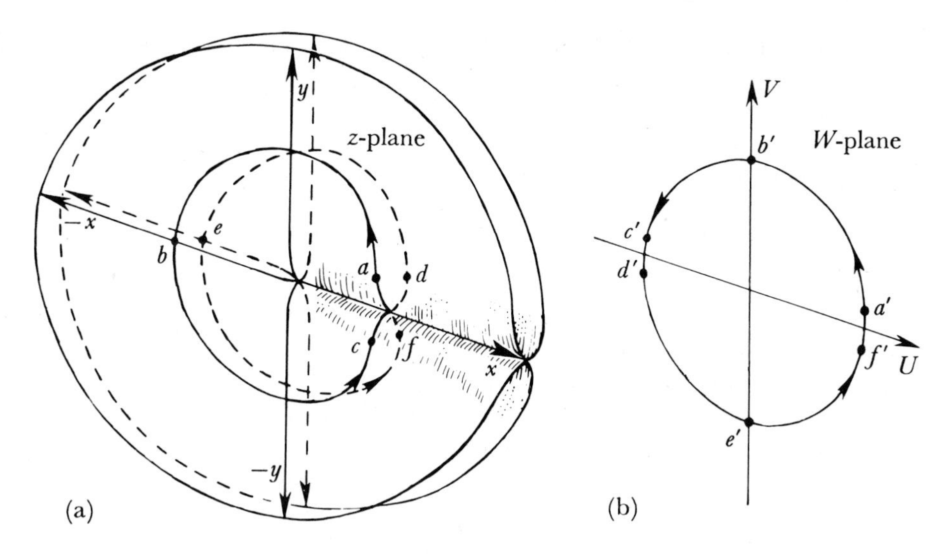

FIGURE 7-4 Riemann surface and mapping for the function $W(z) = z^{1/2}$. The dashed part of the curve in the z-plane lies on the lower sheet.

Riemann surface, which maps in a single-valued manner onto the entire W-plane. If we now go around the branch point $z = 0$ twice, once on each sheet, we come back to the starting point in the W-plane, as indicated in Figure 7-4.

Other roots may be described in the same way.

Example

$$W = z^{1/3}$$

The mapping produced by this function is indicated in Figure 7-5. The

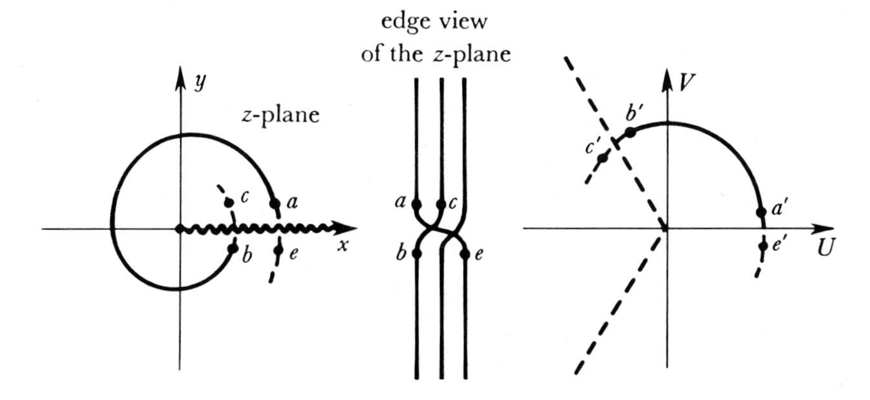

FIGURE 7-5 Riemann surface and mapping for $W = z^{1/3}$. This sketch is a less pictorial way of conveying information similar to that in Figure 7-4.

origin is again a branch point, said to be of order 2 because the Riemann surface contains 3 (= 2 + 1) sheets.

The function $W(z) = \ln z$ is a more drastic case:

$$z = re^{i\theta}$$

$$\ln z = \ln r + i\theta$$

The origin is again a branch point, this time of infinite order because the Riemann surface has an infinite number of sheets. Each sheet maps onto a horizontal strip in the W-plane of width $\Delta V = 2\pi$ in the "imaginary" direction. Continued circling of the origin $z = 0$ in the same sense will take us farther and farther from the starting point on the map.

Another important type of function is one containing two branch points arising from square roots. Consider

$$W(z) = \sqrt{(z-a)(z-b)}$$

with branch points at $z = a$ and $z = b$. In order to sketch the mapping, it is convenient to introduce polar coordinates of z centered at each branch point, that is

$$z - a = r_1 e^{i\theta_1} \qquad z - b = r_2 e^{i\theta_2}$$

$$W(z) = \sqrt{(z-a)(z-b)} = (r_1 r_2)^{1/2} e^{i(\theta_1 + \theta_2)/2}$$

The Riemann surface of this function may be formed by drawing branch cuts from each of the two branch points to infinity in arbitrary directions, or by making a single cut connecting the two points. The resulting Riemann sheets and the branch of the function corresponding to a given sheet depend on the choice of cuts, as shown for example in Figure 7-6.

Similar methods may be used to study more complicated functions.

Example

$$f(z) = \sqrt[3]{(z-a)(z-b)}$$

We introduce polar coordinates $z - a = r_1 e^{i\theta_1}$; $z - b = r_2 e^{i\theta_2}$ so that

$$f(z) = (r_1 r_2)^{1/3} e^{i(\theta_1 + \theta_2)/3}$$

The branch cuts must be drawn out from a and b in such a way that $f(z)$ remains single valued on each cut Riemann sheet. Begin on one sheet with a value of z corresponding to r_1, θ_1, r_2, θ_2. Consider a point z' corresponding to r_1, $\theta_1 + 2\pi$, r_2, θ_2. (This point would be obtained from z by going around a once in a counterclockwise direction.) The point z' cannot be on the same sheet as z, because $f(z') = e^{i2\pi/3} f(z) \neq f(z)$. Hence one branch cut must prevent us from going around a.

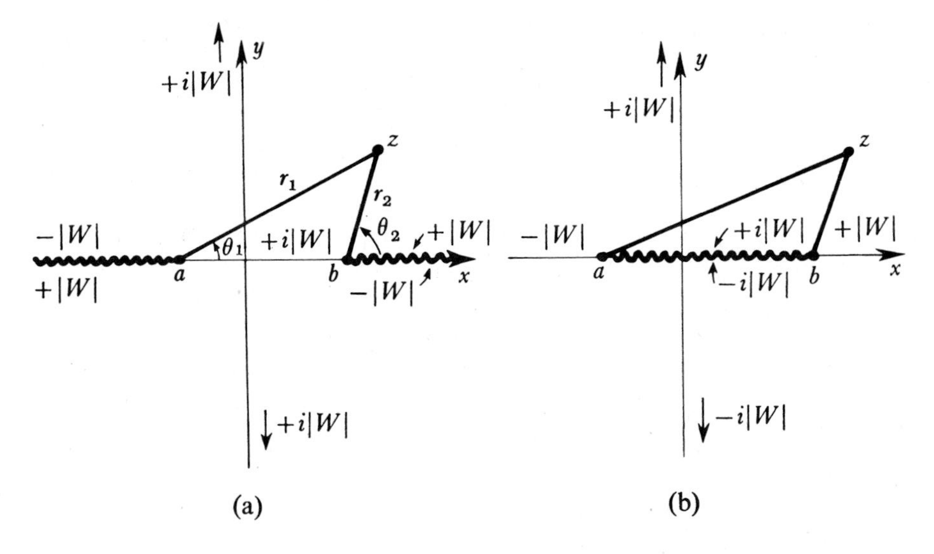

FIGURE 7-6 Two of the ways of drawing branch lines for the function $W(z) = \sqrt{(z-a)(z-b)}$, with the behavior of W indicated in various regions of one Riemann sheet covering the z-plane. In both drawings, the sheet is that one for which $W(z)$ is positive along the upper side of the real axis to the right of b. The symbol $-i\,|W|$, for example, means that $W(z)$ is pure negative imaginary at the place indicated.

Similar arguments show that $z \leftrightarrow r_1, \theta_1, r_2, \theta_2$ cannot be on the same sheet as $z'' \leftrightarrow r_1, \theta_1, r_2, \theta_2 + 2\pi$ or $z''' \leftrightarrow r_1, \theta_1 + 2\pi, r_2, \theta_2 + 2\pi$. Thus the plane must be cut in such a way that we cannot go around a, or b, or a and b, and remain on the top Riemann sheet. The pattern of branch cuts could therefore resemble Figure 7-6(a), but not 7-6(b). How would the Riemann sheets be connected?

Finally, consider the function $Q_n(z)$, defined in

$$Q_n(z) = \frac{1}{2} P_n(z) \ln\left(\frac{z+1}{z-1}\right) + f_{n-1}(z)$$

[$P_n(z)$ is the Legendre polynomial, and $f_{n-1}(z)$ is a polynomial in z of order $n-1$]. This is a many-valued function; to make it single valued, it is conventional to cut the z-plane between -1 and $+1$ along the real axis and to define $Q_n(z)$ real for real $z > 1$. (What other positions of the branch cuts are possible?) Then, if $-1 < x < +1$

$$Q_n(x \pm i\varepsilon) = \frac{1}{2} P_n(x)\left[\ln\left(\frac{1+x}{1-x}\right) \mp i\pi\right] + f_{n-1}(x) \tag{7-3}$$

If $-1 < x < 1$, one usually means by $Q_n(x)$ the arithmetic mean

$$Q_n(x) = \frac{1}{2}[Q_n(x + i\varepsilon) + Q_n(x - i\varepsilon)] = \frac{1}{2}P_n(x)\ln\left(\frac{1+x}{1-x}\right) + f_{n-1}(x) \tag{7-4}$$

PROBLEMS

7-1 Properties of $w = 1/z$

(a) For the shaded region in the z-plane in Figure 7-7(a), find the corresponding region in the w-plane.

(b) For the shaded region in the w-plane in Figure 7-7(b), find the corresponding region in the z-plane.

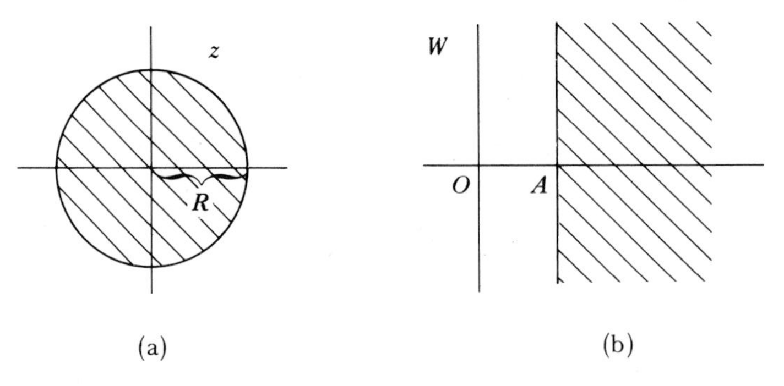

FIGURE 7-7

7-2 Properties of $w = z^2$

(a) For the shaded region in the w-plane in Figure 7-8(a), find the corresponding region in the z-plane.

(b) For the shaded region in the z-plane in Figure 7-8(b), find the corresponding region in the w-plane.

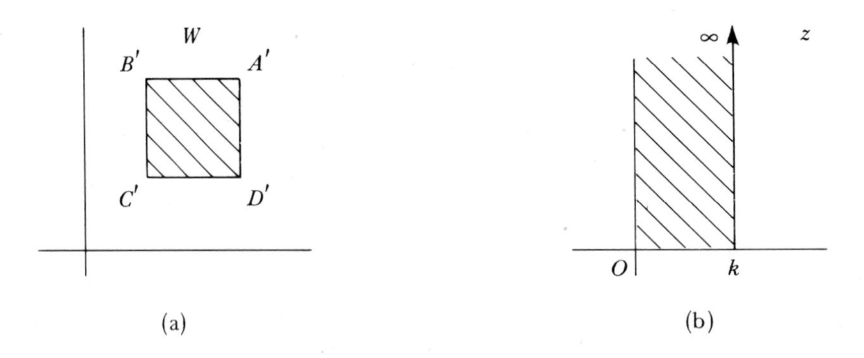

FIGURE 7-8

7-3 Properties of $w = e^z$

For shaded areas a, b, and c in the z-plane of Figure 7-9, find the corresponding region in the w-plane.

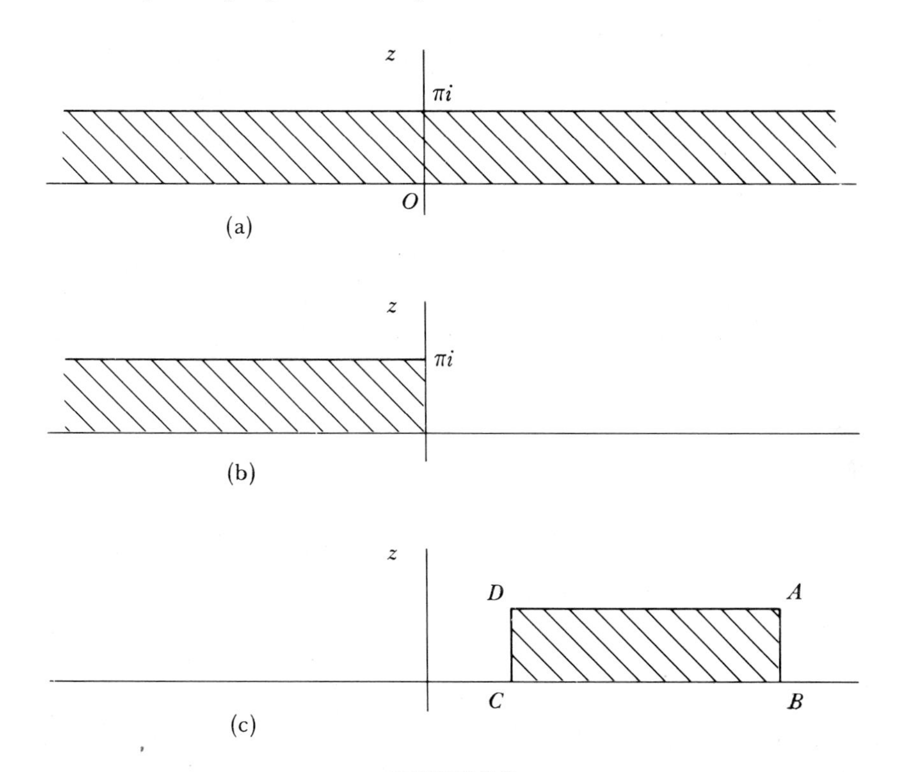

FIGURE 7-9

7-4 For the function

$$W(z) = \frac{1}{\sqrt{(z^2+1)(z-2)}}$$

(a) locate the branch points.
(b) indicate two different methods of drawing branch cuts such that the function $W(z)$ is single valued in the cut plane.
(c) for each of the choices of branch cuts in part (b), describe the mapping produced by the function.

7-5 Consider the function

$$W(z) = \frac{1}{\sqrt{z-1} - i\sqrt{2}}$$

(a) Explain how to cut the z-plane to make the mapping single valued.
(b) Discuss the behavior of $W(z)$ near the point $z = -1$ on each of the sheets of the cut z-plane.

7-2 BASIC PROPERTIES OF ANALYTIC FUNCTIONS

CAUCHY-RIEMANN EQUATIONS

A function is *analytic* at a point z if it has a derivative there and at each point in the neighborhood of z. We define the derivative of a complex function by analogy to the real case: the derivative exists if

$$f'(z) = \lim_{h \to 0} \frac{f(z+h) - f(z)}{h} \tag{7-5}$$

exists and is independent of the path (in the complex plane) by which $h \to 0$. If a function is both analytic and single valued throughout a region R, we shall call it *regular* in R. A region of regularity of a multivalued function should be specified on a cut Riemann sheet. If $f(z)$ is not analytic at $z = z_0$, it is said to have a *singularity* there.

If $W(z) = U(x, y) + iV(x, y)$ is an analytic function and we write

$$h = h_x + ih_y$$

then two paths for $h \to 0$ are along the horizontal and vertical directions, for which $h_y = 0$ and $h_x = 0$, respectively. The limits (7-5) obtained for these paths must be equal:

$$\frac{\partial U}{\partial x} + i\frac{\partial V}{\partial x} = \frac{1}{i}\frac{\partial U}{\partial y} + \frac{\partial V}{\partial y}$$

Equating real and imaginary parts of this equation gives the *Cauchy-Riemann differential equations*:

$$\frac{\partial U}{\partial x} = \frac{\partial V}{\partial y} \qquad \frac{\partial V}{\partial x} = -\frac{\partial U}{\partial y} \tag{7-6}$$

These equations are necessary for regularity but not sufficient. Sufficient conditions can be shown to be the Cauchy-Riemann equations together with the *continuity* of the first partial derivatives.

Example

$$W = z^2$$

$$U = x^2 - y^2 \qquad V = 2xy$$

$$\frac{\partial U}{\partial x} = 2x = \frac{\partial V}{\partial y} \qquad \frac{\partial V}{\partial x} = 2y = -\frac{\partial U}{\partial y}$$

It is useful to keep in mind that some simple functions are not analytic anywhere. The classic example is

$$W = z^* \qquad \text{(complex conjugate of } z\text{)}$$

Then

$$U = x \qquad V = -y$$

$$\frac{\partial U}{\partial x} = +1 \qquad \frac{\partial V}{\partial y} = -1$$

and the Cauchy-Riemann equations are not satisfied.

LIOUVILLE'S THEOREM

Notice that equations (7-6) have as a consequence

$$\frac{\partial^2 U}{\partial x^2} + \frac{\partial^2 U}{\partial y^2} = 0 \tag{7-7a}$$

and

$$\frac{\partial^2 V}{\partial x^2} + \frac{\partial^2 V}{\partial y^2} = 0 \tag{7-7b}$$

Because U and V obey these equations, they never have absolute maxima or minima within regions of regularity. For suppose U is to have a maximum for variations in x at some point. Then

$$\frac{\partial U}{\partial x} = 0 \qquad \frac{\partial^2 U}{\partial x^2} < 0$$

From (7-7a), we find

$$\frac{\partial^2 U}{\partial y^2} > 0$$

and the function cannot have a maximum for variations in y at this point also.

For example, in the function $W = z^2$ discussed above, $U = x^2 - y^2$ gives

$$\frac{\partial^2 U}{\partial x^2} = +2 = -\frac{\partial^2 U}{\partial y^2}$$

that is, the curvature in the x direction has the opposite sign to curvature in the y direction. What is the corresponding statement for $V = 2xy$?

More generally, the curvature of U (or V) in *any* two perpendicular directions in the xy-plane takes opposite signs.

Suppose that a function $f(z)$ is regular everywhere in the z-plane, including the point at infinity. Then this argument tells us that Re f and Im f can never have maxima; thus they must be constant over the entire z-plane. This is known as *Liouville's theorem*.

CONFORMAL MAPPING

The mapping from the z-plane to the *W-plane* for an analytic function $W(z)$ is *conformal,* that is it preserves the angle between two curves. Consider for example the curves shown in Figure 7-10. If we parametrize the tangent to

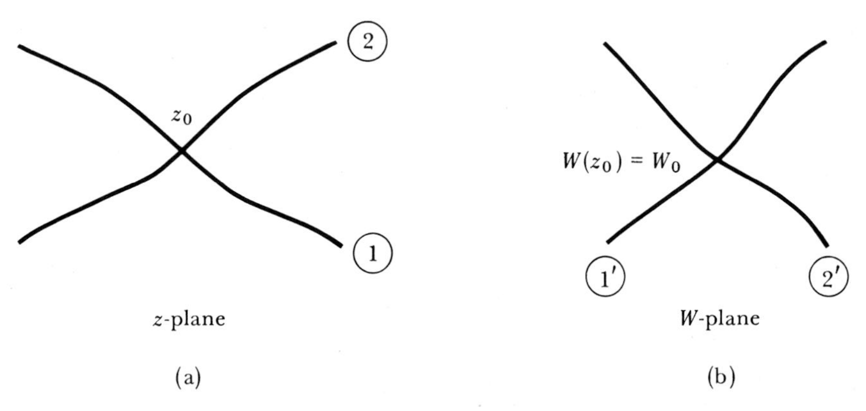

FIGURE 7-10 The angle between two curves is preserved by a change of variables given by the analytic function $W(z)$.

curve 1 at the point of intersection by $z - z_0 = re^{i\theta_1}$, and we parametrize the tangent to curve 2 by $z - z_0 = \rho e^{i\theta_2}$, then the angle between the two curves is defined to be $(\theta_2 - \theta_1)$. These curves are mapped into curve 1′ and curve 2′ in the W-plane, respectively. The tangent to curve 1′ near the intersection can then be parametrized by $W - W_0 = r'\, e^{i\theta_1'}$ and the tangent to curve 2′ by $W - W_0 = \rho'\, e^{i\theta_2'}$.

Since W is an analytic function of z,

$$W(z) - W(z_0) \approx W'(z_0)(z - z_0) \tag{7-8}$$

near the point z_0. Hence if $W'(z_0) = \alpha e^{i\beta}$, we have

$$r'e^{i\theta_1'} = \alpha e^{i\beta} re^{i\theta_1}$$

$$\rho' e^{i\theta_2'} = \alpha e^{i\beta} \rho e^{i\theta_2}$$

Equating the phases on both sides gives

$$\theta_1' = \beta + \theta_1$$

and

$$\theta_2' = \beta + \theta_2$$

hence

$$\theta_2' - \theta_1' = \theta_2 - \theta_1 \tag{7-9}$$

and the angle between the two curves is preserved under our transformation.

Another useful property of analytic functions that can be proved from the Cauchy-Riemann equations is that U changes most rapidly along those lines in the xy-plane where V is constant, and vice-versa. The direction along which U changes most rapidly is given by

$$\mathbf{grad\ U} = \frac{\partial U}{\partial x}\hat{\mathbf{x}} + \frac{\partial U}{\partial y}\hat{\mathbf{y}}$$

Likewise the direction along which V changes most rapidly is

$$\begin{aligned}\mathbf{grad\ V} &= \frac{\partial V}{\partial x}\hat{\mathbf{x}} + \frac{\partial V}{\partial y}\hat{\mathbf{y}} \\ &= -\frac{\partial U}{\partial y}\hat{\mathbf{x}} + \frac{\partial U}{\partial x}\hat{\mathbf{y}}\end{aligned}$$

Hence

$$\mathbf{grad\ U} \cdot \mathbf{grad\ V} = 0 \tag{7-10}$$

and at any point where the gradient is non-zero, the lines along which V changes most rapidly are perpendicular to the ones along which U changes most rapidly. Since the lines along which U changes most rapidly are perpendicular to lines along which U is a constant, the contours U = constant are parallel to the lines along which V changes most rapidly, as claimed.

PATH INTEGRALS

Because the functions $f(z)$ depend on two real parameters x and y, it is possible to integrate $f(z)$ along any line in the complex plane, rather than just along the real axis. The integral

$$\int_{z_1}^{z_2} f(z)\,dz$$

should be thought of as a sum of complex values along the points of some path γ connecting z_2 and z_1. A convenient way to do this is to parametrize the path $z(t)$ by a real parameter t; then the definition

$$\underset{\text{along } \gamma}{\int_{z_1}^{z_2}} f(z)\,dz \equiv \int_{t_1}^{t_2} f[z(t)]z'(t)\,dt$$

enables us to evaluate our new *path integral* in terms of a standard integration along the real axis. Clearly the result obtained will depend in general upon the path followed from z_1 to z_2 (Figure 7-11).

CAUCHY'S THEOREM

In certain cases, however, the value of the integral will depend only on the endpoints and not on the particular path taken between them. When this happens, the integral around a closed contour such as $\gamma_1 - \gamma_2$ in Figure 7-11 will vanish. Let us explore the conditions under which this is true.

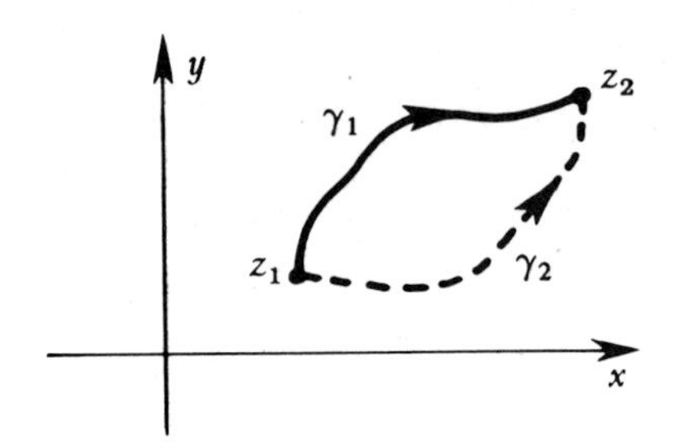

FIGURE 7-11 Paths of integration in the complex plane

For $f = U + iV$, and $dz = dx + i\,dy$, we obtain

$$\oint f\,dz = \oint (U\,dx - V\,dy) + i \oint (V\,dx + U\,dy)$$

(The symbol $\oint$ stands for "integral around a closed contour.") Each of the two integrals on the right side of this equation may be thought of as a "surface" integral in the xy-plane. By Green's theorem, these "surface" integrals may be rewritten as "volume" integrals

$$\oint f\,dz = -\iint \left[\frac{\partial U}{\partial y} + \frac{\partial V}{\partial x}\right] dx\,dy + i \iint \left[-\frac{\partial V}{\partial y} + \frac{\partial U}{\partial x}\right] dx\,dy$$

provided all the first partial derivatives of U and V exist and are continuous throughout the region. If, in addition, U and V obey the Cauchy-Riemann equations, the right-hand side will vanish.

Thus, if f is analytic, and if $f'(z)$ is continuous throughout the region enclosed by the contour,

$$\oint_C f(z)\,dz = 0$$

The requirement that $f'(z)$ be continuous can be removed, but in this case the proof becomes more complicated [see, for example, Churchill (7), p. 115]. We will assume this result:

$$\oint_C f(z)\,dz = 0 \tag{7-11}$$

if $f(z)$ is analytic at all points within the closed contour C. This is known as *Cauchy's theorem.*

Example

Compute $I = \oint_C z^2\,dz$, where C is a circle of unit radius about the origin. The function $z^2 = f(z)$ has no singularities within C, hence by Cauchy's theorem, $I = 0$.

We can check this by explicit computation. Along C, $z = e^{i\theta}$. Hence $dz = ie^{i\theta}\,d\theta$ and

$$I = \int_{\theta=0}^{2\pi} e^{2i\theta}\,ie^{i\theta}\,d\theta = \left.\frac{ie^{3i\theta}}{3i}\right|_0^{2\pi} \equiv 0$$

as expected.

Suppose that $f(z)$ has a singularity at some point z_0. Then the integral $\oint f(z)\,dz$ around some closed contour containing z_0 may be different from zero. This is illustrated by the following example.

Example

Compute $I = \oint_C 1/z\,dz$, where C is again the circle of unit radius about the origin.

$$\frac{dz}{z} = i\,d\theta$$

hence

$$I = \int_0^{2\pi} i\,d\theta = 2\pi i$$

Cauchy's theorem does not apply in this case because $f(z) = 1/z$ is not analytic at $z = 0$, a point inside the contour of integration.

These results for integrals around closed contours can be restated in terms of integrations over open arcs. Suppose we have two points a and b connected by various arcs $\gamma_1, \gamma_2, \gamma_3 \ldots$. Then:

1. If γ_1 can be continuously deformed into γ_2 without crossing any singularities of $f(z)$, the closed contour $\gamma_1 - \gamma_2$ contains no singularities. Hence

$$\oint_{\gamma_1 - \gamma_2} f(z)\,dz = 0 \qquad \text{by Cauchy's theorem}$$

and

$$\underset{\text{along } \gamma_1}{\int_a^b f(z)\,dz} = \underset{\text{along } \gamma_2}{\int_a^b f(z)\,dz}$$

2. Suppose that in order to deform arc γ_1 into arc γ_3, one must cross a singularity of $f(z)$. Then the arc $\gamma_1 - \gamma_3$ will contain this singularity and we have a priori no knowledge of $\oint_{\gamma_1 - \gamma_3} f(z)\,dz$. In this case

$$\underset{\text{along } \gamma_1}{\int_a^b f(z)\,dz} \quad \text{may be different from} \quad \underset{\text{along } \gamma_3}{\int_a^b f(z)\,dz}$$

Example

Consider the integral of $f(z) = 1/z$ along the arcs shown in Figure 7-12.

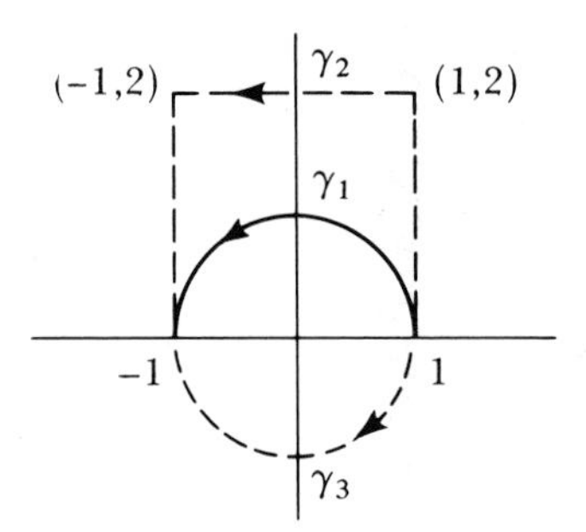

FIGURE 7-12 Various paths for evaluation of $\int \frac{1}{z}\,dz$

The various integrals are:

$$I_1 = \int_{\gamma_1} \frac{1}{z}\,dz = \int_0^{\pi} i\,d\theta = i\pi$$

$$I_2 = \int_{\gamma_2} \frac{1}{z}\,dz = \log z \Big|_{1+i0}^{1+i2} + \log z \Big|_{1+i2}^{-1+i2} + \log z \Big|_{-1+i2}^{-1}$$

$$= \log z \Big|_{1}^{-1} = \log(-1) - \log(1)$$

$$= i\pi$$

$$I_3 = \int_{\gamma_3} \frac{1}{z}\,dz = \int_0^{-\pi} i\,d\theta = -i\pi$$

We notice that $I_1 = I_2$, as expected (since γ_1 can be deformed into γ_2 without crossing the singular point $z = 0$); but $I_1 \neq I_3$, since the closed contour $\gamma_1 - \gamma_3$ encloses a singularity. $I_1 - I_3 = I = 2\pi i$ as calculated in the previous example.

THEOREM OF RESIDUES

These results obtained by explicit calculation with the function $f(z) = 1/z$ can be generalized. Consider the integral

$$\oint_C \frac{R(z)\,dz}{z-b}$$

where C is a closed contour around b containing no singularities of $R(z)$. Then by Cauchy's theorem, we can shrink C to a tiny circle about b without changing the value of the integral (the contour will cross no singularities of the integrand in this shrinking process). For a sufficiently small circle, $R(z)$ on the edge of the circle can be approximated by $R(b)$ and we have

$$\oint_C \frac{R(z)\,dz}{z-b} = R(b)\oint_C \frac{d(z-b)}{z-b} = 2\pi i\,R(b) \tag{7-12}$$

In general, if a function $f(z)$ goes to infinity at $z = b$ like $R(b)/(z-b)$, it is said to have a *simple pole* at $z = b$. The number $R(b)$ is called the residue of the pole. For a function $f(z)$ with many poles and a contour C that encloses several of them, the arguments which led to (7-12) then give

$$\oint_C f(z)\,dz = 2\pi i \sum \text{(residues of poles enclosed)} \tag{7-13}$$

These formulae have many uses besides evaluation of integrals around simple poles. In (7-12), b may be any point in a region where $R(z)$ is regular. Hence we can rewrite this as

$$R(z) = \frac{1}{2\pi i}\oint_C \frac{R(t)\,dt}{t-z} \tag{7-14}$$

which gives a representation of $R(z)$ at a point in terms of values surrounding this point. Equation (7-14) is referred to as *Cauchy's integral formula.* Cauchy's formula may be differentiated any number of times to obtain representations for the various derivatives:

$$\begin{aligned} R'(z) &= \frac{1}{2\pi i}\oint_C \frac{R(t)\,dt}{(t-z)^2} \\ R^{(n)}(z) &= \frac{n!}{2\pi i}\oint_C \frac{R(t)\,dt}{(t-z)^{n+1}} \end{aligned} \tag{7-15}$$

Many more integrals may be evaluated by use of (7-15). A function $f(z)$ is said to have a pole of order n at b if it behaves like $g/(z-b)^n$ as z approaches b. If we set $f(t) = R(t)/(t-b)^n$ in equation (7-15), we find (for a contour C enclosing b)

$$R^{(n-1)}(b) = \frac{(n-1)!}{2\pi i} \oint_C f(t)\, dt$$

that is

$$\frac{1}{2\pi i} \oint_C f(t)\, dt = \frac{1}{(n-1)!} \left\{ \left(\frac{d}{dz} \right)^{n-1} [(z-b)^n f(z)] \right\}_{z=b} \tag{7-16}$$

We call the right-hand side of equation (7-16) the residue of the nth order pole. Then equation (7-13) is true, with residues defined in this way, regardless of the orders of the poles enclosed by C. With this interpretation (7-13) is sometimes referred to as the *theorem of residues.*

PROBLEMS

7-6 Which of the following are analytic functions of the complex variable z?
(a) $|z|$
(b) $\text{Re}\, z$
(c) $e^{\sin z}$

7-7 Show that a function such as $(z - z_0)^{5/3}$ is not analytic at the branch point z_0.

7-8 Suppose that the function $f(z)$ has a pole of order n at $z = z_0$. Show that the function $f'(z)/f(z)$ has a simple pole at z_0. What is the residue?

7-9 Prove by explicit calculation that

$$\int_{C_1} z\, dz = \int_{C_2} z\, dz$$

for the contours C_1 and C_2 indicated in Figure 7-13.

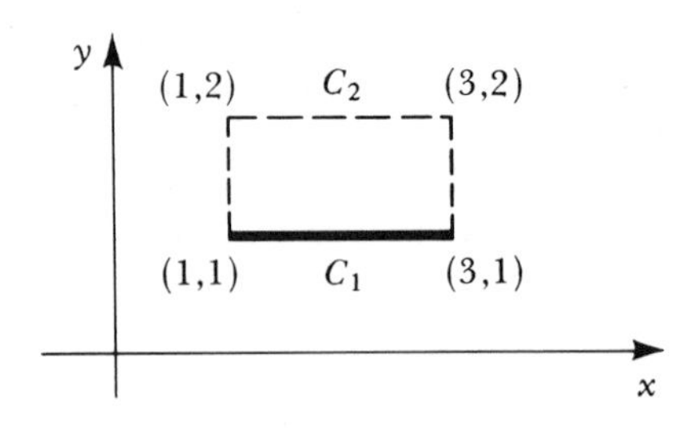

FIGURE 7-13

7-10 One may define the function $y = \sin x$ as the inverse of

$$x = \sin^{-1} y = \int_0^y \frac{dt}{\sqrt{1-t^2}}$$

(a) Use the two contours shown in Figure 7-14 to demonstrate that one y may correspond to more than one x.

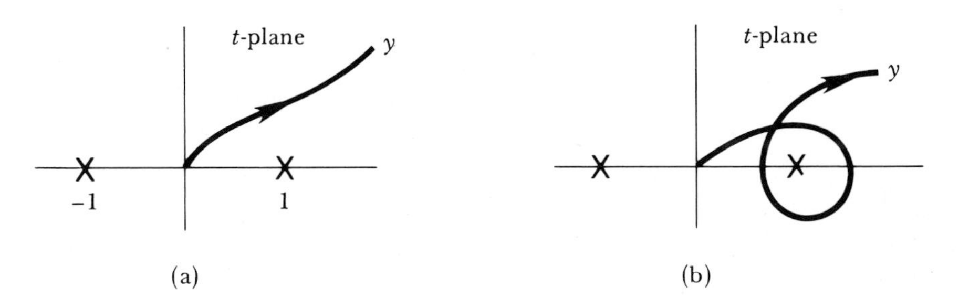

FIGURE 7-14

(b) Do the same problem for Figure 7-15.

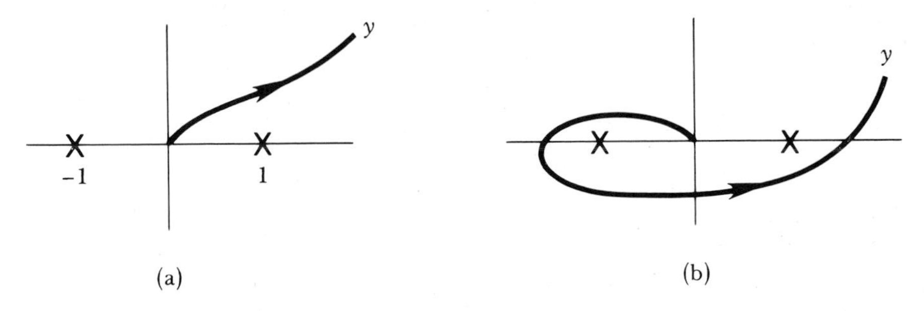

FIGURE 7-15

7-11 Suppose you knew only the following definition: $f(z)$ is analytic if

$$\int_a^b f(z)\,dz = \int_a^b (u\,dx - v\,dy) + i\int_a^b (v\,dx + u\,dy)$$

depends only on the endpoints in the xy-plane and not on the path taken between them. Derive the Cauchy-Riemann equations.

7-12 Suppose you knew only the integral representation

$$R(z) = \frac{1}{2\pi i}\int_C \frac{R(t)\,dt}{t - z}$$

Use this to prove $R(z)$ is analytic in the region contained in C. (This is the inverse of the proof in the text.)

7-13 Compute the integral

$$I = \oint_C \frac{dz}{(z - 2i)^2 (z + 5)z}$$

where the contour C is a circle of radius 3 about the origin of the z-plane.

7-3 SERIES EXPANSIONS OF COMPLEX FUNCTIONS

TAYLOR'S SERIES EXPANSIONS

Recall that we define a function to be regular in a region R if it is analytic and single valued throughout the region. If $f(z)$ is regular in a region, its derivatives of all orders exist and are regular there. This allows us to write a Taylor's series expansion about any point z_0 within the region where $f(z)$ is regular:

$$f(z) = a_0 + a_1(z - z_0) + a_2(z - z_0)^2 + \cdots$$
$$a_0 = f(z_0) \qquad a_n = \frac{1}{n!} f^{(n)}(z_0) \tag{7-17}$$

The region of the z-plane in which the series converges is a circle. This *circle of convergence* extends to the nearest singularity of $f(z)$, that is, to the nearest point where $f(z)$ is not analytic.

These statements can be understood quite simply by use of the Cauchy integral representation for $f(z)$. Because $f(z)$ is regular in R, we can write

$$f(z) = \frac{1}{2\pi i} \oint_C \frac{f(z')}{z' - z}\, dz' \tag{7-18}$$

with C any closed contour entirely in R. For a particular point z_0 in R,

$$z' - z = z' - z_0 - (z - z_0) = (z' - z_0)\left[1 - \left(\frac{z - z_0}{z' - z_0}\right)\right]$$

Let us now choose C to be a circle centered on z_0; then $|(z - z_0)/(z' - z_0)| < 1$ for points z within C, and the expansion

$$\frac{1}{1 - \left(\dfrac{z - z_0}{z' - z_0}\right)} = 1 + \left(\frac{z - z_0}{z' - z_0}\right) + \left(\frac{z - z_0}{z' - z_0}\right)^2 + \cdots$$

converges absolutely. Thus our integral representation (7-18) becomes

$$f(z) = \frac{1}{2\pi i} \oint_C \frac{f(z')}{z' - z_0}\left[1 + \left(\frac{z - z_0}{z' - z_0}\right) + \left(\frac{z - z_0}{z' - z_0}\right)^2 + \cdots\right]$$
$$= f(z_0) + f'(z_0)(z - z_0) + \frac{f''(z_0)}{2!}(z - z_0)^2 + \cdots$$

The contour C can be any circle centered on z_0 that is entirely contained in R; therefore it can have a radius as large as the distance d from z_0 to the nearest singularity. Hence we see that the series should converge for $|z - z_0|$ less than d. [We have assumed that interchanging the order of integration and summation does not ruin the convergence; for a proof see Churchill (7), p. 147.]

What about points outside of this circle of radius d? The derivation above is not valid in this case; hence other considerations must be used. We first note that the power series (7-17) cannot converge to the function $f(z)$ right at the singularity, for if it converged, the power series would have a well-defined derivative at this point. (But lack of such a derivative is what makes f singular!)

The rest of the proof depends on a simple theorem: *If* $\sum_n a_n(z_1 - z_0)^n$ *converges for some* z_1, *then* $\sum_n a_n(z_2 - z_0)^n$ *converges absolutely for all* z_2 *such that* $|z_2 - z_0| < |z_1 - z_0|$. This theorem is so easy to prove that we leave it as an exercise for the reader (Problem 7-14). If the series (7-17) did converge for some point outside our circle of radius d, then by this theorem it would converge absolutely for all points closer to z_0. In particular it would converge at the singular point. We know this does not happen; hence the series diverges at all points outside the circle.

Although the power series representation diverges at the singular point of $f(z)$, it may converge at other points along the circle of convergence.

Example

$$f(z) = \sum_{n=1}^{\infty} \frac{z^n(-1)^{n+1}}{n}$$

the ratio of two successive terms is

$$\left|\frac{a_{n+1}}{a_n}\right| = \left|\frac{z^{n+1}\, n}{(n+1)z^n}\right| = r\left(1 - \frac{1}{n} + \frac{1}{n^2} \cdots\right)$$

Hence we have absolute convergence if $r < 1$, but the series will not converge absolutely for $r \geq 1$. This radius of convergence is clearly determined by the singularity at $z = -1$ [see equation (5-13)].

We know from Chapter 5 that the series diverges for $z = -1$ but converges for $z = +1$. What happens at other points on the circle of convergence $z = e^{i\theta}$? Put in some values of θ and see.

The converse of the above remarks is also true: any power series convergent within a circle R represents a regular function there.

EXPANSIONS ABOUT $z = \infty$

If a function f is regular at $z \to \infty$, then it is regular near $1/z \to 0$ and we can write

$$f(z) = a_0 + \frac{a_1}{z} + \frac{a_2}{z^2} + \frac{a_3}{z^3} + \cdots \tag{7-19}$$

The power series $g(x) = a_0 + a_1 x + a_2 x^2 + a_3 x^3 + \cdots$ will converge within some radius $|x| < d$, where d is the distance to the nearest singularity from the origin. Hence

$$a_n = \frac{1}{2\pi i} \oint_C \frac{g(x)\, dx}{x^{n+1}}$$

for a path C entirely within the circle $|x| = d$. Under the change of variables $x = 1/z$, C becomes a contour C' outside the circle $|z| = 1/d$, and our expression is transformed into

$$a_n = -\frac{1}{2\pi i} \oint_{C'} f(z) z^{n-1}\, dz$$

where the path C' is traversed in the clockwise direction.

In terms of a normal, anticlockwise, contour $C'' = -C'$, we have

$$a_n = \frac{1}{2\pi i} \oint_{C''} f(z) z^{n-1}\, dz \tag{7-20}$$

for any contour C'' enclosing the origin and entirely outside any singularities of $f(z)$.

LAURENT SERIES

Suppose we are given a function that is regular at neither infinity nor the origin, but that is regular for some range of values of $|z|$ (see Figure 7-16).

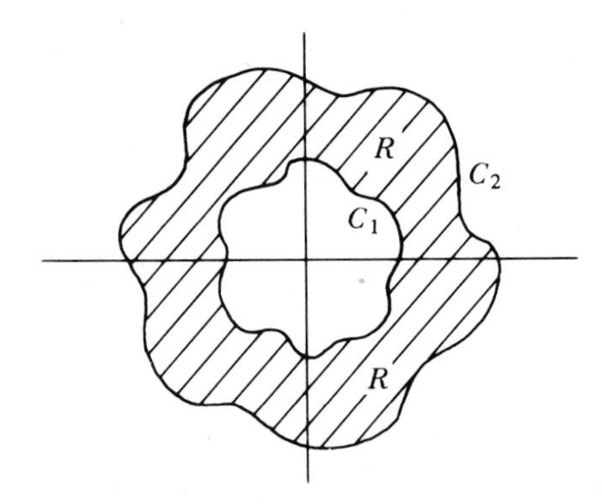

FIGURE 7-16 Region within which f is known to be regular

Consider the problem of finding a power series expansion for f that will converge in R. We could write f as a Taylor series expansion about some point in R, but this would converge only within a small circle about this point, and would not be good for all the points in R. Clearly an expansion about the origin is the most natural way to obtain a series that converges for a given range of $|z|$. A Taylor's series cannot be used, however, since the origin is not included in the region of regularity. Thus we must devise a new procedure.

Divide the function into two pieces

$$f(z) = f_1(z) + f_2(z) \tag{7-21}$$

where $f_1(z)$ is analytic outside C_1, and $f_2(z)$ is analytic inside C_2. (See Problem 7-15.) Because f_1 is analytic at infinity, it can be written

$$f_1(z) = \sum_{n=0}^{\infty} \frac{a_n^1}{z^n} \tag{7-22}$$

Likewise, because f_2 is analytic at the origin, it can be written as

$$f_2(z) = \sum_{n=0}^{\infty} a_n^2 z^n \tag{7-23}$$

Expansion (7-22) converges for $|z|$ greater than the maximum $|z|$ on C_1; expansion (7-23) converges for $|z|$ smaller than the minimum $|z|$ on C_2. Hence in the ring-shaped region between the largest radius on C_1 and the smallest radius on C_2, the expansion

$$f(z) = \sum_{n=-\infty}^{\infty} a_n z^n \tag{7-24}$$

with

$$a_n = \frac{1}{2\pi i} \int_C \frac{f(z)}{z^{n+1}}\, dz \tag{7-25}$$

converges (the contour C is taken about the origin but within the annular region of convergence).

This representation of the function is known as a *Laurent* series. More generally, if the ring is centered on point z_0, the formulae become

$$f(z) = \sum_{n=-\infty}^{\infty} a_n (z - z_0)^n$$

$$a_n = \frac{1}{2\pi i} \int_C \frac{f(z)\, dz}{(z - z_0)^{n+1}}$$

If the inner circle can be shrunk to an infinitesimal size without crossing other singularities, then the point $z = z_0$ is called an isolated singularity of the function. It is a pole of order m if the series contains no terms more singular than $(z - z_0)^{-m}$. However, if the series contains an infinite number of powers of $(z - z_0)^{-1}$, then $z = z_0$ is said to be an essential singularity of the function.

For example, $e^{1/z}$ has an essential singularity at $z = 0$. Likewise $e^{1/(z-5)}$ has an essential singularity at $z = 5$. But for any region around $z = 5$ that excludes this point, we have a simple expansion

$$e^{1/(z-5)} = \sum_{n=0}^{\infty} \frac{1}{n!} \frac{1}{(z - 5)^n}$$

Example

The function $\sqrt{1-z^2}$ has branch points at $z = \pm 1$. If we draw the branch cut between them, the region $|z| > 1$ has no singularities except at ∞. We can therefore write the Laurent expansion

$$\sqrt{1-z^2} = i\sqrt{z^2-1} = iz\sqrt{1-\frac{1}{z^2}}$$

$$= iz\left\{1 - \frac{1}{2z^2} + \cdots\right\}$$

This series does not converge for points inside the circle $|z| < 1$, even those points not on the branch cut.

Example

The function $f = \sqrt{(1-z^2)(25-z^2)}$ has branch points at $z = \pm 1$ and $z = \pm 5$. If the branch cuts are drawn as in Figure 7-17, our function is

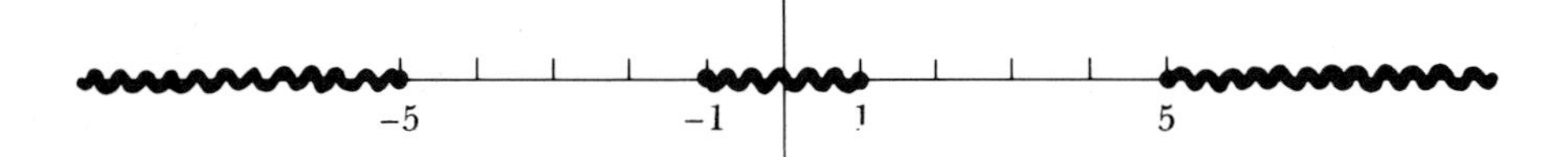

FIGURE 7-17 Branch cuts for the function $\sqrt{(1-z^2)(25-z^2)}$

regular within the ring $1 < |z| < 5$. This allows us to write a Laurent expansion for it about $z = 0$. It can be obtained by the expansion

$$f = iz\sqrt{1-\frac{1}{z^2}} \quad 5\sqrt{1-\frac{z^2}{25}}$$

$$= 5iz\left\{1 - \frac{1}{2z^2}\cdots\right\}\left\{1 - \frac{1}{2}\frac{z^2}{25}\cdots\right\}$$

which gives both positive and negative powers of z, as expected.

Consider a function $f(z)$ that is analytic in a region R of the complex plane, and assume that a finite part of the real axis is included in R. We can expand the function in a power series about one of these points on the real axis. If $f(z)$ assumes only *real* values on that part of the real axis in R, then all the coefficients in our power series expansion must be real. Hence $f(z^*) = [f(z)]^*$

throughout the circle of convergence of the power series. This equality can be extended to all of R by analytic continuation (see Section 7-5). It means that in going from a point z to its "image" in the real axis, namely z^*, we carry the *value* f of the function over into *its* image f^*. We will refer to this property as the *Schwartz reflection principle.*

A power series that converges everywhere defines a single-valued analytic function with no singularities in the entire plane (excluding ∞). Such a function is called an *entire function.* Examples are polynomials, e^z, and sin z. A single-valued function that has no singularities other than poles in the entire plane (excluding ∞) is called a *meromorphic* function. Examples are rational functions, that is, ratios of polynomials.

PROBLEMS

7-14 If $\sum a_n(z_1 - z_0)^n$ converges, then we know the terms are bounded: $|a_n(z_1 - z_0)^n| < M$ for some number M. Hence

$$\sum \left| a_n(z_2 - z_0)^n \right| = \sum \left| a_n(z_1 - z_0)^n \left(\frac{z_2 - z_0}{z_1 - z_0} \right)^n \right| < M \sum_{n=0}^{\infty} \left| \frac{z_2 - z_0}{z_1 - z_0} \right|^n$$

Use this to prove that $\sum a_n(z_2 - z_0)^n$ converges absolutely for

$$|z_2 - z_0| < |z_1 - z_0|.$$

7-15 Given a function $f(z)$, which is analytic in the ring $r_1 < |z| < r_2$, we can represent it in the ring by

$$f(z) = \frac{1}{2\pi i} \oint_C \frac{f(z')\, dz'}{z' - z}$$

where C is a closed contour around z and entirely within the ring [see Figure 7-18(a)]. This contour can be deformed [see Figure 7-18(b)] into

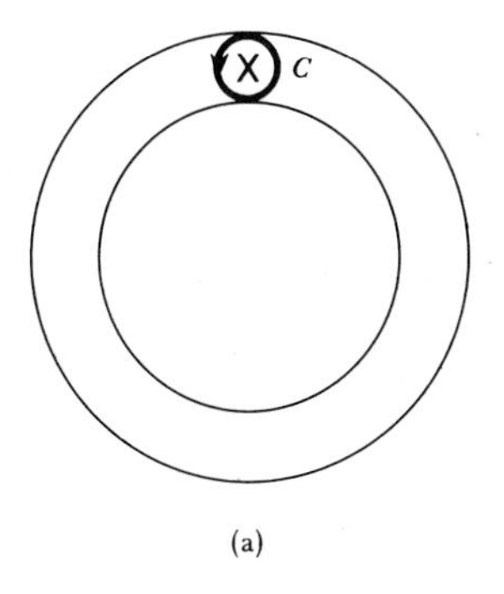

(a)

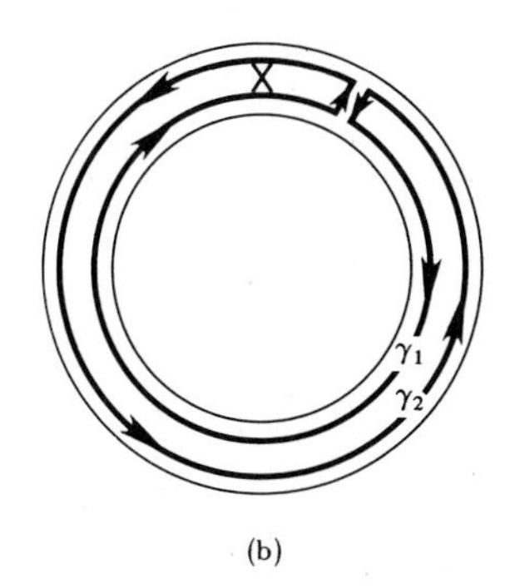

(b)

FIGURE 7-18

two large circles (the integrals along the short connecting pieces cancel each other). Thus we have

$$f(z) = \frac{1}{2\pi i}\left(\int_{\gamma_2} + \int_{\gamma_1}\right)\frac{f(z')\,dz}{z' - z}$$

$$= \frac{1}{2\pi i}\int_{|z'|=r_2-\varepsilon}\frac{f(z')\,dz'}{z' - z} - \frac{1}{2\pi i}\int_{|z'|=r_1+\varepsilon}\frac{f(z')\,dz'}{z' - z}$$

$$= I_2 - I_1$$

Consider the functions defined by I_1 and I_2. Show that I_2 has no singularities for *all* $|z| < r_2$, and that I_1 has no singularities for *all* $|z| > r_1$.

7-4 POLES ON THE CONTOUR

In all of the examples we have studied so far, the poles of the integrand have been explicitly either inside or outside the contour. However, in working physics problems, one occasionally encounters expressions in which the integrand has a pole exactly on the contour. Since a mathematical integral with a pole on the contour does not exist, we must adopt some policy for dealing with these expressions when they arise.

The first thing to do is to look into the physics of the problem to see if this awkward location of the pole results from some approximation. If so, one can decide on which side of the path the pole really lies, and thus see whether its residue should be included. Frequently, as we will see in later chapters applying integral transforms, the boundary conditions set by the physical problem will determine our treatment of such a pole.

CONCEPT OF PRINCIPAL VALUE

For a simple pole on the real axis, it may be helpful to express the answer in terms of a quantity called the *Cauchy principal value*:

$$P\int_a^b \frac{R(x)}{x - x_0}\,dx = \lim_{\delta\to 0}\left[\int_a^{x_0-\delta}\frac{R(x)}{x - x_0}\,dx + \int_{x_0+\delta}^{b}\frac{R(x)}{x - x_0}\,dx\right] \tag{7-26}$$

where δ is positive (we assume $a < x_0 < b$).

The path for the Cauchy principal value integral can form part of a closed contour in which the ends $x_0 \pm \delta$ are joined by a small semicircle centered at the pole (see Figure 7-19). Along this semicircle the integral is easy to evaluate; if we let the radius approach zero, the integrand $f(z)$ behaves like $a_{-1}(z - x_0)^{-1}$. Let

$$z - x_0 = re^{i\theta} \qquad dz = ire^{i\theta}\,d\theta$$

Then

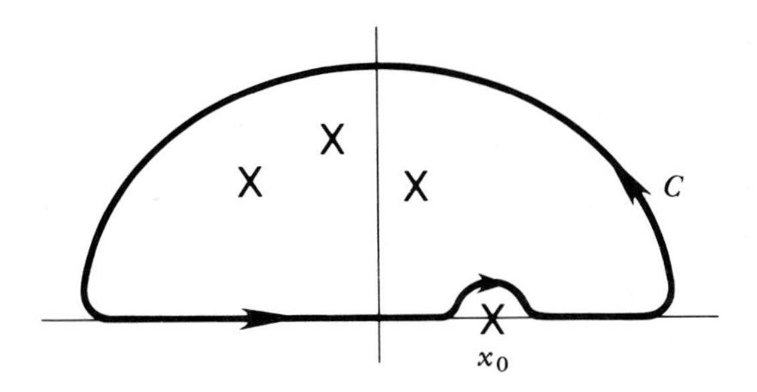

FIGURE 7-19 Illustration of a pole on the real axis

$$\int_{\text{semicircle}} f(z)\,dz \to -\int_0^\pi a_{-1}\,i d\theta = -\pi i a_{-1}$$

where a_{-1} is the residue of $f(z)$ at x_0; and if, as is frequently the case, the large semicircle gives no contribution,

$$\oint_C f(z)\,dz = P\int f(z)\,dz - \pi i\,(\text{residue at } z_0)$$

$$= 2\pi i\left(\sum \text{residues inside } C\right)$$

This gives the result

$$P\int f(z)\,dz = 2\pi i\left(\tfrac{1}{2}\text{ residue at } x_0 + \sum \text{residues inside } C\right) \tag{7-27}$$

Thus the Cauchy principal value is the average of the two results obtained with the pole inside and outside of the contour.

We often have an integral along the real axis with a simple pole just above (or just below) the axis at x_0. We may consider the pole to be on the axis if we make the path of integration miss the pole by going around x_0 on a little semicircle below (or above). Then it follows by reasoning similar to that leading to (7-27) that the integral may be expressed in terms of the Cauchy principal value as follows:

$$\int \frac{R(x)}{x - x_0 \mp i\varepsilon}\,dx = P\int \frac{R(x)}{x - x_0}\,dx \pm i\pi R(x_0)$$

We may express this result in the somewhat symbolic form

$$\frac{1}{x - x_0 \mp i\varepsilon} = P\frac{1}{x - x_0} \pm i\pi\,\delta(x - x_0) \tag{7-28}$$

where $\delta(x - x_0)$ is the Dirac delta function defined in (4-43). Equation (7-28) has meaning only inside an integral sign.

PROBLEMS

7-16 Calculate explicitly

(a) $P\int_{-1}^{1}\frac{a^2\,dx}{x(x+b)}, \qquad |b|>1$

(b) $\int_{-1}^{1}\frac{a^2\,dx}{(x\pm i\varepsilon)(x+b)}, \qquad |b|>1$

7-17 Define a function $f(z)$ by the integral

$$f(z)=\int_1^\infty \frac{g(t)\,dt}{t-z}$$

where $g(t)$ is a *real* function for real values of t, well-behaved along the real axis, which drops off as t goes to infinity. Demonstrate that this $f(z)$ has a branch point at $z=1$, and a cut from $z=1$ to $z=\infty$ along the real axis. What is the discontinuity $f(x+i\varepsilon)-f(x-i\varepsilon)$ across this cut?

7-5 ANALYTIC CONTINUATION

If two functions are each regular in a region R, and have the same values for all points within some subregion, or for all points along an arc of some curve within R (or even for a denumerably infinite number of points having a limit point within R), then the two functions must have the same power series expansion about a point within the region: hence they must be identical functions everywhere in the region. For example, if $f(z)=0$ all along some arc in R, then it is the regular function $f(z)=0$ everywhere in R.

This theorem is useful in extending into the complex plane functions defined on the real axis. For example

$$e^z = 1+z+\frac{z^2}{2!}+\cdots$$

is the *unique* function $f(z)$ that is equal to e^x on the real axis.

The theorem also forms the basis for the procedure of analytic continuation. A power series about z_1 represents a regular function $f_1(z)$ within its circle of convergence, which extends to the nearest singularity. If an expansion of this function is made about a new point z_2, the resulting series will converge in a circle that may extend beyond the circle of convergence of $f_1(z)$. The values of $f_2(z)$ in the extended region are uniquely determined by $f_1(z)$—in fact, by the values of $f_1(z)$ in the common region of convergence of $f_1(z)$ and $f_2(z)$. $f_2(z)$ is said to be the analytic continuation of $f_1(z)$ into the new region. This process may be repeated (with limitations mentioned below) until the entire plane is covered except for singular points by these *elements* of a single function $F(z)$.

Example

In the previous section we found an expansion for the function

$$f(z) = \sqrt{(1 - z^2)(25 - z^2)}$$

that converged for $1 < |z| < 5$. To continue the function outside the ring, given only this Laurent expansion, one could pick a point z_0 as shown in Figure 7-20 and write a Taylor's series expansion about this point. It would

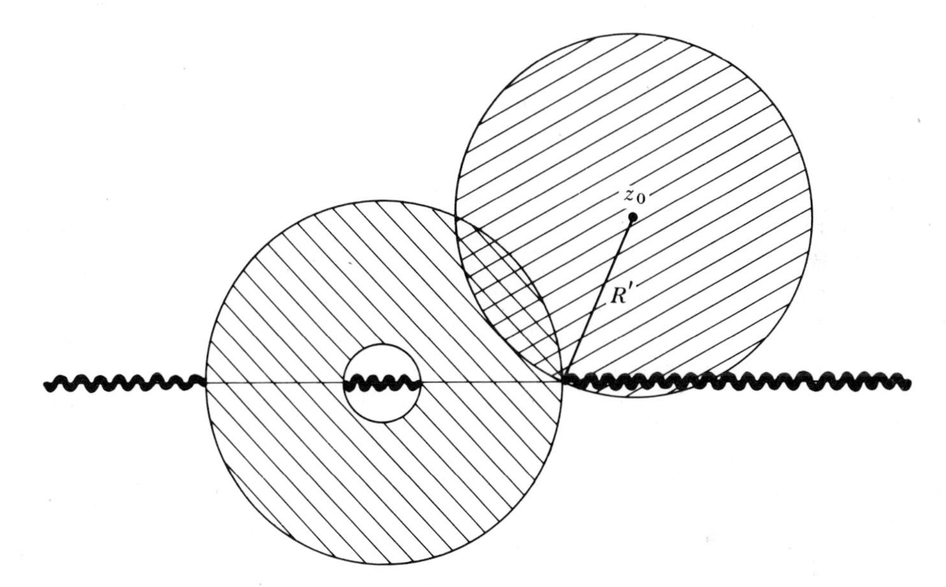

FIGURE 7-20 Analytic continuation of $f(z) = \sqrt{(1 - z^2)(25 - z^2)}$. The annulus marks the region of convergence of a Laurent series expansion about the origin. A Taylor's series expansion about z_0 will converge within the circle. Coefficients in the Taylor's series can be determined from those in the Laurent series by comparison of the *values of the function* in the overlap region. Note that the circle encloses a continuous region of the z-plane that overlaps both Riemann sheets.

converge for all points within a circle of radius R' (the distance from z_0 to the nearest singularity). In particular it would converge for some points within the original ring. From the values of $f(z)$ at these points given by our initial Laurent series, we could evaluate the coefficients in the Taylor's series, and then use this series for all points within the circle about z_0. This allows us to evaluate the function at some points outside the ring.

In theory, the process can be repeated as many times as desired to find the value of $f(z)$ at any point in the complex plane. Notice that the continuations of the original element obtained in this manner may progress from one Riemann sheet to the next, so that the $f(z)$ obtained will not always be single valued.

As discussed here this method is cumbersome and not particularly practical. In actual applications, various tricks are used instead. For example, if it is possible to sum the series into closed form in some region of the plane, then this closed form may be used to continue the function.

Example

$$f_1(z) = 1 + z + z^2 + z^3 + \cdots$$

converges in a circle of radius 1 to

$$F(z) = \frac{1}{1 - z}$$

But $F(z)$ is analytic everywhere except at the simple pole $z = 1$, and no other function analytic outside $|z| = 1$ can coincide with $f_1(z)$ within $|z| < 1$. $F(z)$ is the analytic continuation of $f_1(z)$ into the entire plane.

Similarly the form $f(z) = \sqrt{(1 - z^2)(25 - z^2)}$ can be used to continue the Laurent series discussed above.

Not all functions can be continued indefinitely. The extension may be blocked by a barrier of singularities.

It may also happen that the function $F(z)$ obtained by continuation is multivalued. For example, suppose that after repeating the process described above a number of times, the nth circle of convergence partially overlaps the first one. Then the values of the element $f_n(z)$ in the common region may or may not agree with $f_1(z)$. If they do not agree, then the function $F(z)$ is multivalued, and the "path" along which the continuation was made has encircled one or more branch points.

Frequently in physics problems the function we are interested in is the solution of some (operator) equation for real values of a physical parameter. For example, in both relativistic and nonrelativistic quantum mechanics there are equations relating the scattering amplitude T of two particles to the forces between them; these equations depend on the center-of-mass energy E of the particle pair, which of course is real. If we allow this energy to become complex, the solution of the continued equation is the analytic continuation of $T(E)$ off the real axis. This procedure is useful because the cuts and poles of the scattering amplitude turn out to have direct physical significance; also various complex variable techniques can be used to solve for $T(E)$. Frequently the solution can only be found approximately; in this case it is important to use an approximation technique that does not seriously alter the analyticity properties of the function. The method of Padé approximants (see Problem 7-20) has proved useful.

PROBLEMS

7-18 Suppose we are given the power series expansion

$$f(z) = -\sum_{n=1}^{\infty} \frac{z^n}{n}$$

good for $|z| < 1$.

(a) Derive a power series expansion from which the function can be evaluated at $x = -\frac{3}{2}$. What is the radius of convergence of your series?

(b) Consider an expansion of this function about the point $x = +\frac{1}{2}$. What is the radius of convergence of the series?

(c) Answer the same question as (b) for an expansion about $x = \frac{3}{4}$.

(d) What is the analytic continuation of $f(x)$ to the point $x = +2$?

7-19 One of the many tricks available for analytic continuation is Borel's method. Suppose we have a function $f(z)$ with power series expansion $f(z) = \sum c_n z^n$, which converges for $|z| < 1$. If we form the Borel transform $b(z)$ of f by summing the series $b(z) = \sum c_n z^n/n!$, we have

$$f(z) = \int_0^{\infty} e^{-t} b(zt)\, dt \tag{7-29}$$

within the circle. But equation (7-29) may be well-defined within a larger region than the original power series. In particular, it can be shown that equation (7-29) analytically continues $f(z)$ into a domain R defined as follows: Find each singularity of $f(z)$, and draw a line through it perpendicular to the ray from the origin to the singularity. This gives a polygon R containing the circle $|z| < 1$; if R has a finite number of sides then it is larger than the original circle of convergence.

Apply Borel's method to the series $f(z) = -\sum_{n=1}^{\infty} z^n/n$. What is the region R for this case?

7-20 In the method of Padé approximants, a function is approximated by the ratio of two polynomials

$$f(z) \to \frac{p_0 + p_1 z + \cdots + p_N z^N}{1 + q_1 z + \cdots + q_M z^M} = \frac{P_N(z)}{Q_M(z)}$$

If f can be expanded about the point $z = 0$, the coefficients p_i, q_j are determined by the requirement that

$$f(z) - \frac{P_N(z)}{Q_M(z)} = \mathcal{O}(z^{M+N+1})$$

For functions that approach a constant as $z \to \infty$, it is convenient to take $M = N$. Although the Padé approximant has only poles and no branch points, for high enough M the poles will tend to cluster along the position of the cut of the initial function.

For

$$f(z) = \sqrt{\frac{4 - z}{9 - z}}$$

calculate the approximant $(p_0 + p_1 z)/(1 + q_1 z)$. Compare the poles and zeroes of the approximant with those of the initial function. Compare the values of the function and the approximant at five different values, including $z = 0$ and $z = \infty$. If you have computing facilities available, you might want to study higher-order approximants.

For a thorough discussion of Padé approximants, see Baker (4).

7-21 Consider the function defined by

$$\Gamma(z + 1) = \int_0^\infty t^z e^{-t}\, dt$$

Show that the Γ function is an analytic continuation of the factorial $n! = n(n - 1)(n - 2) \ldots 1$ away from the positive integers. Does $\Gamma(z)$ have poles at any values of z?

SUMMARY

A function $f(z) = U + iV$ is analytic provided

$$\frac{\partial U}{\partial x} = \frac{\partial V}{\partial y} \qquad \frac{\partial U}{\partial y} = -\frac{\partial V}{\partial x}$$

and all first partial derivatives are continuous. If $f(z)$ is not analytic at some point $z = z_0$, it is said to be singular there. Branch points and poles are the most common types of singularities.

The mapping of one complex plane into another given by an analytic function preserves the angles between curves. The individual functions Re f and Im f obey Laplace's equation $\nabla^2\phi = 0$; curves of constant Re f coincide with those curves along which Im f changes most rapidly and vice versa.

Cauchy's theorem tells us that the integral around a closed contour of a function $f(z)$ is zero provided the contour encloses no singularities of $f(z)$. This is the same as saying that the integral of $f(z)$ between two points a and b does not depend on the path taken from a to b, provided all paths under consideration can be continuously deformed into one another without crossing singularities of f.

If the closed contour encloses only poles of f, we can evaluate the integral by

$$\oint_C f(z)\,dz = 2\pi i \sum (\text{residues of poles enclosed})$$

where the residue of an nth order pole is given by

$$\frac{1}{(n-1)!}\left\{\left(\frac{d}{dz}\right)^{n-1}[(z-b)^n f(z)]\right\}\Bigg|_{z=b}$$

Any function regular in a ring about z_0 can be expanded in a Laurent series

$$f(z) = \sum_{n=-\infty}^{\infty} a_n (z-z_0)^n$$

where

$$a_n = \frac{1}{2\pi i}\oint_C \frac{f(z)\,dz}{(z-z_0)^{n+1}}$$

and the contour C is taken within the annulus of regularity. If the region of regularity includes z_0, then there are no negative powers in the expansion; if it includes ∞ there are no positive powers in the expansion.

Within an integral along the real axis, poles near the axis may be treated using

$$\frac{1}{x-x_0 \mp i\varepsilon} = P\frac{1}{x-x_0} \pm i\pi\,\delta(x-x_0)$$

where the symbol P represents the Cauchy principal value

$$P\int_a^b \frac{R(x)\,dx}{x-x_0} = \lim_{\delta\to 0}\left\{\int_a^{x_0-\delta}\frac{R(x)\,dx}{x-x_0} + \int_{x_0+\delta}^{b}\frac{R(x)\,dx}{x-x_0}\right\}$$

CHAPTER 8 Uses of Complex Variable Theory I: Evaluation of Integrals

The theorems developed in Chapter 7 can be applied in both very direct and very sophisticated ways. We feel that the student should acquire a firm grasp of the straightforward applications, as well as some idea of the possibilities inherent in clever variations on these themes. This is the aim of Chapters 8, 9, and 10. Additional applications to partial differential equations are contained in Chapter 12.

Many types of definite real integrals can be evaluated exactly by the theorem of residues, provided one is able to relate the problem to an integral around a closed path in the complex plane. In Section 8-1 we give examples illustrating various techniques for achieving this relation. The student should become proficient with these techniques before moving on to subsequent sections.

Some contour integrals that cannot be evaluated exactly can be approximated very closely by a technique called the method of steepest descent, which we discuss in Section 8-2. Section 8-3 is devoted to the Sommerfeld-Watson transformation, a method for summing series by rewriting them in terms of contour integrals and then evaluating the integrals.

8-1 CONTOUR INTEGRATION

USE OF POLES

One of the most powerful means for evaluating definite integrals is provided by the theorem of residues. Equation (7-13) tells us that if a function $f(z)$ is regular in the region bounded by a closed path C, except for a finite number of poles and isolated essential singularities in the interior of C, then the integral of $f(z)$ along the contour C is

$$\int_C f(z)\,dz = 2\pi i \sum \text{residues}$$

where $\sum$ residues means the sum of the residues at all the poles and essential singularities inside C. This can be applied in a surprising number of cases. The following examples illustrate a number of useful tricks.

In order to apply the theorem of residues, we must first have a closed contour. Create one if necessary!

Example

Consider the integral

$$I = \int_0^\infty \frac{dx}{1 + x^2} = \frac{1}{2}\int_{-\infty}^\infty \frac{dx}{1 + x^2} \tag{8-1}$$

We would like to evaluate this by integrating around the contour shown in Figure 8-1 in the limit as the semicircle approaches infinity. The contribution from the semicircle vanishes, since

$$z = \mathrm{Re}^{i\theta} \qquad dz = i\,\mathrm{Re}^{i\theta}\,d\theta \qquad \frac{1}{1+z^2} \approx \frac{e^{-2i\theta}}{R^2}$$

$$\int \frac{dz}{1+z^2} \approx \frac{i}{R}\int e^{-i\theta}\,d\theta \to 0 \quad \text{as} \quad R \to \infty$$

Hence

$$I = \frac{1}{2}\oint_C \frac{dz}{1+z^2} = \frac{2\pi i}{2}\left(\text{residue of } \frac{1}{1+z^2} \text{ at } z = +i\right)$$

The residue of $1/(1 + z^2) = 1/(z + i)(z - i)$ at $z = i$ is $1/2i$. Thus

$$I = \frac{2\pi i}{2}\,\frac{1}{2i} = \frac{\pi}{2}$$

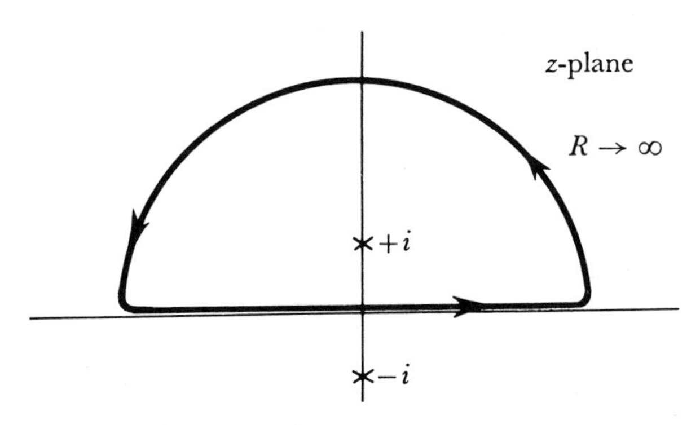

FIGURE 8-1 Contour for the integral (8-1)

Some cases require a little more thought.

Example

$$I = \int_0^\infty \frac{\sin x}{x(a^2 + x^2)}\,dx = \frac{1}{2}\int_{-\infty}^{\infty} \frac{\sin x\,dx}{x(a^2 + x^2)} \tag{8-2}$$

The integrand has poles at $x = \pm ia$, but $x = 0$ is not a singularity. Difficulties arise when we try to close the contour by a semicircle at infinity, since $\sin x = (1/2i)(e^{ix} - e^{-ix})$. The function e^{ix} vanishes as $x \to i\infty$, but e^{-ix} blows up here. Similar considerations apply to the limit $x \to -i\infty$. This suggests breaking the integral into two pieces:

(a)
$$\int_{-\infty}^{\infty} \frac{e^{ix}\,dx}{x(a^2 + x^2)}$$

to be closed in the top half plane, and

(b)
$$\int_{-\infty}^{\infty} \frac{e^{-ix}\,dx}{x(a^2 + x^2)}$$

to be closed in the bottom half plane. However a pole at $x = 0$ has been introduced into each of (a) and (b) by the separation into two pieces.

To escape these problems, adopt the following strategy: First move the contour slightly off the real axis, as shown in Figure 8-2. This will not move the path of integration through any singularities of the integrand,

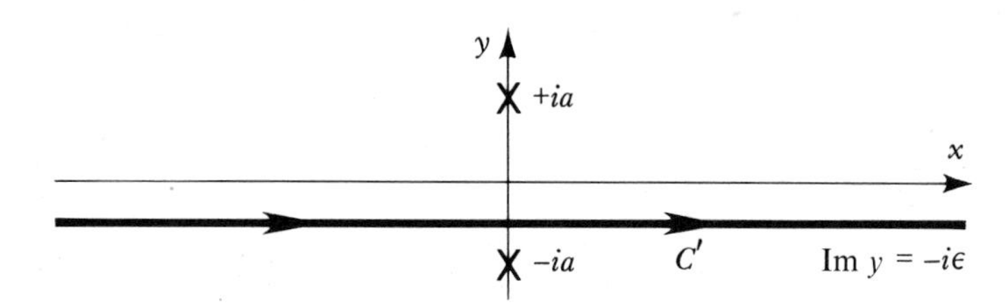

FIGURE 8-2 New contour for the integral (8-2)

$$\frac{\sin x}{x(x^2 + a^2)}$$

and hence the value of the integral will not change, by Cauchy's theorem. This new expression can then be split into two pieces, both of which are well defined.

$$\frac{1}{2}\int_{C'} \frac{\sin x\,dx}{x(a^2 + x^2)} = \frac{1}{4i}\int_{C'} \frac{e^{ix}\,dx}{x(a^2 + x^2)} - \frac{1}{4i}\int_{C'} \frac{e^{-ix}\,dx}{x(a^2 + x^2)}$$

The contour is now closed in the upper half plane for the first piece, enclosing two poles. For the second piece, the contour is closed in the lower half plane; this encloses only one pole. Use of the theorem of residues to evaluate these contributions gives

$$I = \frac{2\pi i}{4i}\left[\frac{1}{a^2} + \frac{e^{-a}}{ia(2ia)}\right] + \frac{2\pi i}{4i}\left[\frac{1}{(-ia)} \frac{e^{-a}}{(-2ia)}\right]$$

$$I = \frac{\pi}{2}\left[\frac{1}{a^2} - \frac{e^{-a}}{a^2}\right] = \frac{\pi}{2a^2}[1 - e^{-a}]$$

Angular integrals are often simplified by the substitution $z = e^{i\theta}$.

Example

$$I = \int_0^{\pi} \frac{d\theta}{a + b\cos\theta} \qquad a > b > 0 \tag{8-3}$$

The integrand is even, so

$$2I = \int_0^{2\pi} \frac{d\theta}{a + b\cos\theta}$$

If we integrate along the unit circle, as in Figure 8-3

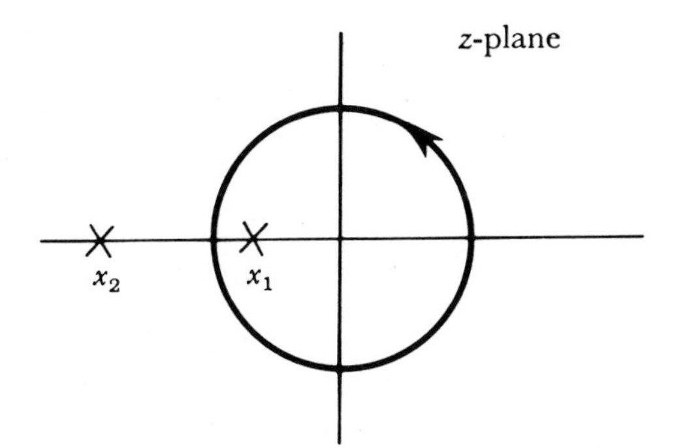

FIGURE 8-3 Contour for the integral (8-3)

$$z = e^{i\theta} \qquad dz = ie^{i\theta}\, d\theta$$

$$\cos\theta = \frac{e^{i\theta} + e^{-i\theta}}{2} = \frac{1}{2}\left(z + \frac{1}{z}\right)$$

Then

$$2I = \int_c \frac{\dfrac{dz}{iz}}{a + \dfrac{b}{2}\left(z + \dfrac{1}{z}\right)}$$

$$= \frac{2}{i} \int_c \frac{dz}{bz^2 + 2az + b}$$

The integrand has two poles, at the roots of the denominator. The product of the roots is $b/b = 1$; one is outside and one inside the unit circle. They are at

$$x_{1,2} = -\frac{a}{b} \pm \sqrt{\frac{a^2}{b^2} - 1}$$

Then

$$2I = \frac{2}{i} 2\pi i \left(\text{residue at } x_1 = -\frac{a}{b} + \sqrt{\frac{a^2}{b^2} - 1} \right)$$

$$I = 2\pi \left(\frac{1}{2bx_1 + 2a} \right) = \frac{\pi}{\sqrt{a^2 - b^2}}$$

BRANCH CUTS

The special properties of branch cuts can often be used to advantage.

Example

$$I = \int_0^\infty \frac{\sqrt{x}\, dx}{1 + x^2} \tag{8-4}$$

Consider the integral

$$I_1 = \oint \frac{\sqrt{z}\, dz}{1 + z^2}$$

along the contour of Figure 8-4. We choose z positive on top of the cut, so that $\sqrt{z} = \sqrt{x}$ here. Thus

$$\underset{\substack{\text{top of}\\ \text{cut}}}{\int_0^\infty} \frac{\sqrt{z}}{1 + z^2}\, dz = I$$

On the bottom of the cut, $\sqrt{z} = \sqrt{xe^{2\pi i}} = -\sqrt{x}$. Hence

$$\underset{\substack{\text{bottom}\\ \text{of}\\ \text{cut}}}{\int_\infty^0} \frac{\sqrt{z}}{1 + z^2}\, dz = \int_0^\infty \frac{\sqrt{x}\, dx}{1 + x^2} = I$$

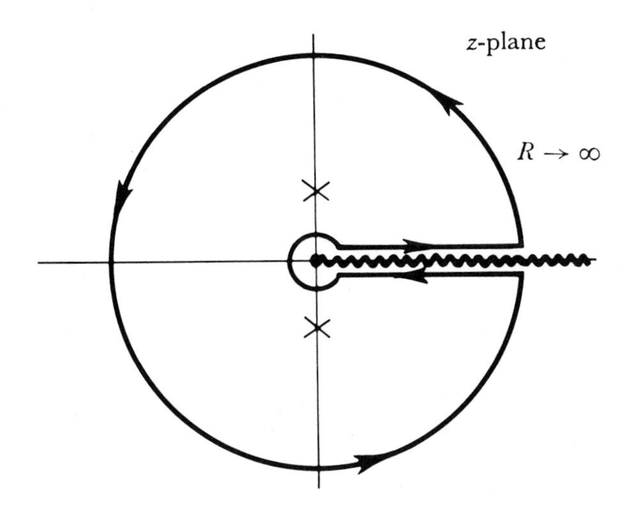

FIGURE 8-4 Contour for the integral (8-4)

The integral around the small circle near the origin vanishes as its radius goes to zero, whereas the integral around the large circle vanishes as its radius goes to infinity. Then

$$\oint \frac{\sqrt{z}\,dz}{1+z^2} = 2I$$

But, using residues,

$$\oint \frac{\sqrt{z}\,dz}{1+z^2} = \pi\sqrt{2}$$

Therefore $I = \pi/\sqrt{2}$.

Sometimes it helps to introduce a branch cut.

Example

$$I = \int_0^\infty \frac{dx}{1+x^3} \tag{8-5}$$

This is not an even function of x, so we cannot extend the contour to $[-\infty, \infty]$. However, consider the replacement $u = x^3$.

$$I = \int_0^\infty \frac{1}{3}\frac{u^{-2/3}\,du}{[1+u]}$$

The integrand now has a branch point at $u = 0$. We take the branch cut from 0 to infinity to lie along the positive real axis, and consider

$$\int_C \frac{u^{-2/3}\,du}{3[1+u]} \tag{8-6}$$

where the contour C is displayed in Figure 8-5.

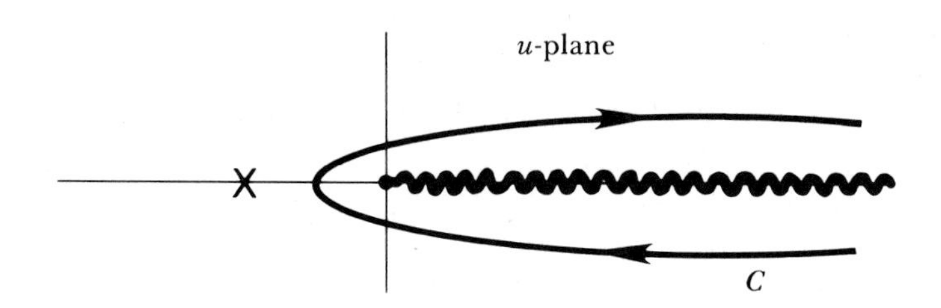

FIGURE 8-5 Contour for the integral (8-6)

Using

$$(u - i\varepsilon)^{-2/3} = e^{2\pi i(-2/3)}(u + i\varepsilon)^{-2/3}$$

we find

$$\int_C \frac{u^{-2/3}\,du}{3[1+u]} = [1 - e^{-4\pi i/3}] \int_{0+i\varepsilon}^{\infty} \frac{u^{-2/3}\,du}{3[1+u]} + \int_{0-i\varepsilon}^{0+i\varepsilon} \frac{(\varepsilon e^{i\theta})^{-2/3}\varepsilon i e^{i\theta}\,d\theta}{3[1+\varepsilon e^{i\theta}]}$$

The integral around the tiny semicircle vanishes like $\varepsilon^{1/3}$ as $\varepsilon \to 0$; hence we obtain

$$I[1 - e^{-4\pi i/3}] = \int_C \frac{u^{-2/3}\,du}{3[1+u]} \tag{8-7}$$

The integral of

$$\frac{u^{-2/3}}{1+u}$$

around a circle at infinity vanishes; we may add this to the right hand side of (8-7) without changing anything. But now I is expressed in terms of an integral around a closed contour; this can be evaluated by the theorem of residues.

$$I[1 - e^{-4\pi i/3}] = \int_{C'' = \{C + \text{contour at infinity}\}} \frac{u^{-2/3}}{3[1+u]}\,du = \frac{2\pi i}{3} e^{-2\pi i/3}$$

$$I = \frac{2\pi}{3\sqrt{3}}$$

There are other ways of introducing a branch cut. Consider

$$\int \frac{\ln z \, dz}{1 + z^3}$$

Again the integrand is many valued; we may cut the plane as shown in Figure 8-6, and define $\ln z$ real ($= \ln x$) just above the cut. Then $\ln z =$

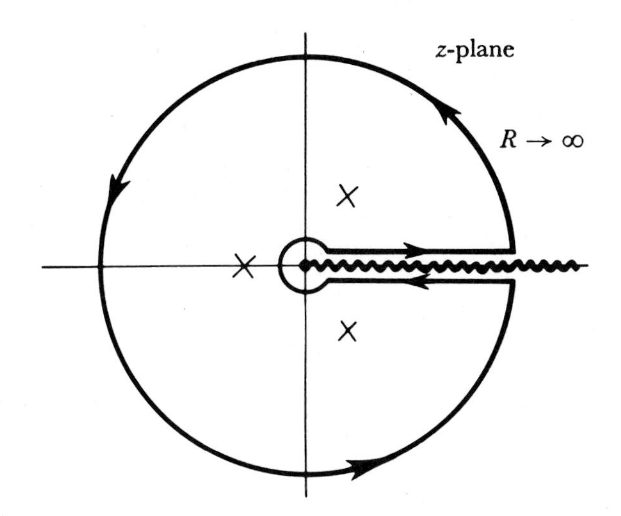

FIGURE 8-6 Contour for the integral (8-8)

$\ln x + 2\pi i$ below the cut, and integrating along the indicated contour,

$$\oint \frac{\ln z \, dz}{1 + z^3} = -2\pi i I \tag{8-8}$$

On the other hand, using the method of residues,

$$\oint \frac{\ln z \, dz}{1 + z^3} = \frac{-4\pi^2 i \sqrt{3}}{9}$$

Thus $I = (2\pi\sqrt{3})/9$, in agreement with the answer obtained above.

Again we must show that the integral on a vanishingly small circle around the branch point is zero. Here this part goes like $r \ln r$. which approaches zero as $r \to 0$.

CLEVER CHOICE OF CONTOUR

A different method equivalent to our original change of variables $u = x^3$ consists of evaluating the integral

$$J = \oint \frac{dz}{1 + z^3}$$

around the contour shown in Figure 8-7. The integration at infinity vanishes

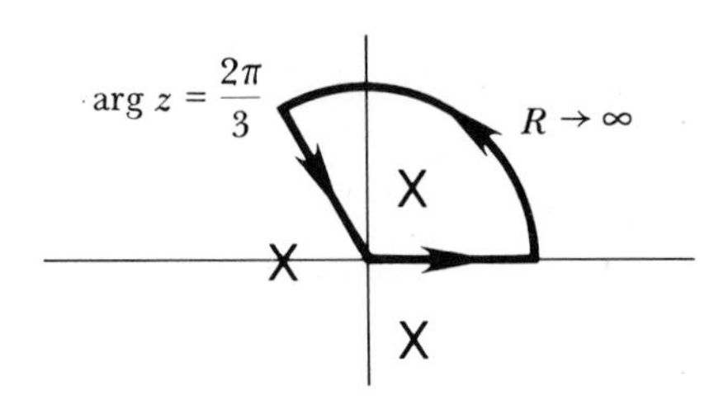

FIGURE 8-7 Alternate contour for evaluation of integral (8-5)

as usual, and the return path along arg $z = 2\pi/3$ has been cleverly chosen such that

$$J = (1 - e^{2\pi i/3})I$$

On the other hand, the integrand of J has a simple pole at $z = e^{\pi i/3}$, and the residue theorem gives

$$J = \frac{2\pi i}{3\, e^{2\pi i/3}}$$

Therefore,

$$I = \frac{2\pi i}{3} \frac{e^{-2\pi i/3}}{1 - e^{2\pi i/3}} = \frac{\pi}{3 \sin \pi/3} = \frac{2\pi}{3\sqrt{3}}$$

as before.

There are other cases in which a clever choice of contour can simplify the situation considerably.

Example

$$I = \int_{-1}^{1} \frac{dx}{\sqrt{1 - x^2}\,(1 + x^2)} \tag{8-9}$$

We consider

$$\oint \frac{dz}{\sqrt{1 - z^2}(1 + z^2)}$$

along the contour of Figure 8-8. On the top side of the cut we get I, and we get another I from the bottom side. Along the circle at infinity, the integral vanishes. Therefore

$$2I = 2\pi i \left[\frac{1}{2i\sqrt{2}} + \frac{1}{2i\sqrt{2}} \right]$$

$$= \pi\sqrt{2}$$

$$I = \frac{\pi}{\sqrt{2}}$$

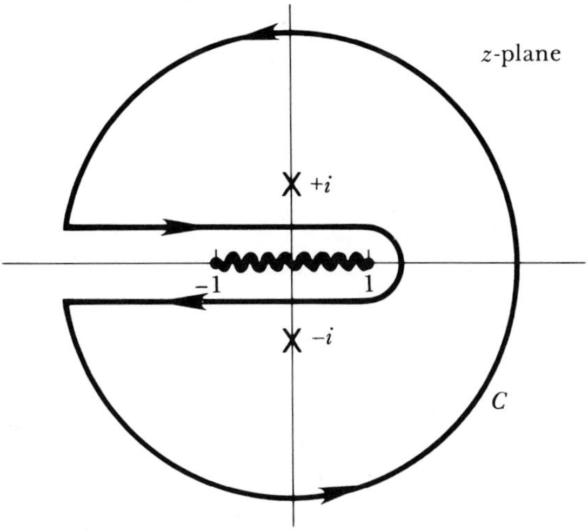

FIGURE 8-8 Contour for the integral (8-9)

BEHAVIOR AT INFINITY

Sometimes the integral around the contour at infinity does not obviously vanish. In these cases, its contribution must be explicitly evaluated.

Example

Consider

$$I = \int_{-1}^{1} \frac{z^2\, dz}{\sqrt{1 - z^2}(1 + z^2)} \tag{8-10}$$

We can evaluate this by using the contour shown in Figure 8-8, since

$$\oint_C \frac{z^2\, dz}{\sqrt{1 - z^2}(1 + z^2)} = 2\pi i[(\text{residue at } z = +i) + (\text{residue at } z = -i)]$$

$$= 2I + \lim_{R\to\infty} \int_{\theta=-\pi}^{\theta=+\pi} \frac{R^2 e^{2i\theta}\, \text{R}e^{i\theta}\, i\, d\theta}{\sqrt{1 - R^2 e^{2i\theta}}(1 + R^2 e^{2i\theta})}$$

$$= 2I - 2\pi$$

where we have chosen that branch of $\sqrt{1-z^2}$ that is positive real above the branch cut. Evaluating the residues, we find

$$2I - 2\pi = -\frac{2\pi}{\sqrt{2}} \qquad I = \pi - \frac{\pi}{\sqrt{2}}$$

Sometimes the desired integral must be obtained by a limiting process from one with the proper behavior at infinity.

Example

$$I = \int_{-\infty}^{\infty} \frac{e^{ax}}{e^x + 1}\,dx \qquad (0 < a < 1) \tag{8-11}$$

We clearly wish to consider

$$\int \frac{e^{az}\,dz}{e^z + 1}$$

along some contour. One choice is shown in Figure 8-9, involving the

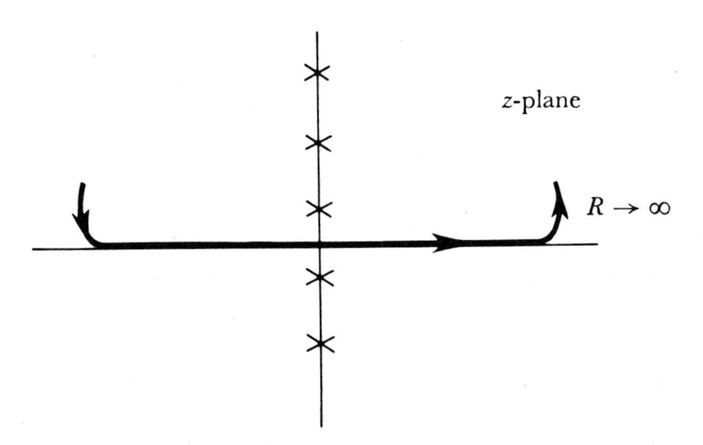

FIGURE 8-9 Possible contour for the integral (8-11)

familiar large semicircle. The integrand is clearly damped exponentially for those portions of the semicircle on which $|x|$ is large; however to ensure damping along the region of the semicircle where $|x|$ is not large but $|y|$ is large, we need to give a a small imaginary part. Then

$$I = 2\pi i \sum \text{residues}$$

There is an infinite number of poles

$$\text{At } z = i\pi, \text{ residue is } -e^{i\pi a}$$

$$\text{At } z = 3i\pi, \text{ residue is } -e^{3i\pi a}, \text{ etc.}$$

The sum over residues forms a geometric series that we can sum because the imaginary part of a puts us within the circle of convergence, $|e^{i\pi a}| < 1$. Hence

$$I = -2\pi i \frac{e^{i\pi a}}{1 - e^{2i\pi a}} = \frac{\pi}{\sin \pi a}$$

Now we can let Im $a \to 0$.

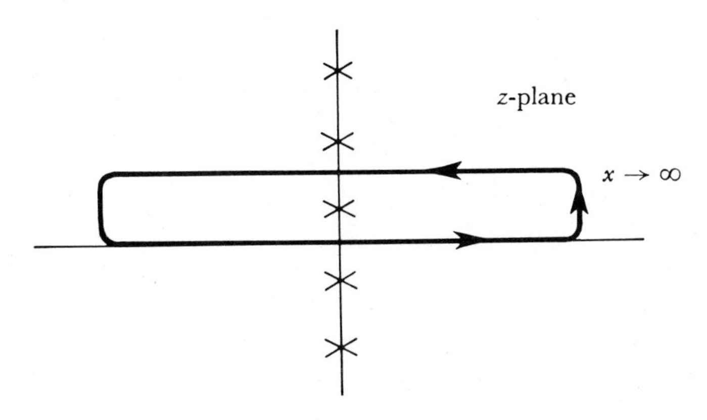

FIGURE 8-10 Another contour for the integral (8-11)

Alternatively, we could use the contour shown in Figure 8-10. Along the real axis, we get I. Along Im $z = 2\pi i$, we get

$$-\int_{-\infty}^{\infty} \frac{e^{a(x+2\pi i)}}{e^{x+2\pi i} + 1} dx = -e^{2\pi i a} I$$

Thus

$$\begin{aligned}
(1 - e^{2\pi i a})I &= 2\pi i \,(\text{residue at } z = \pi i) \\
&= -2\pi i e^{i\pi a} \\
I &= \frac{\pi}{\sin \pi a}
\end{aligned}$$

as before.

If the integral is of the form

$$\int e^{i\omega x} f(x)\, dx$$

the evaluation can be simplified by *Jordan's lemma*:

$$\int_C e^{i\omega x} f(x)\, dx = 0 \tag{8-12}$$

for C a semicircle at infinity along which $\operatorname{Im}(\omega x) > 0$, provided $|f(Re^{i\theta})| < \varepsilon(R) \underset{R\to\infty}{\to} 0$ for values of R, θ on the semicircle. (Note that the bound on $|f|$ must be independent of θ.) A straightforward proof of this can be found in Carrier, Krook, and Pearson (6), p. 81.

Example

Consider a resistance R and inductance L connected in series with a voltage $V(t)$ (Figure 8-11). Suppose $V(t)$ is a voltage impulse; that is, a

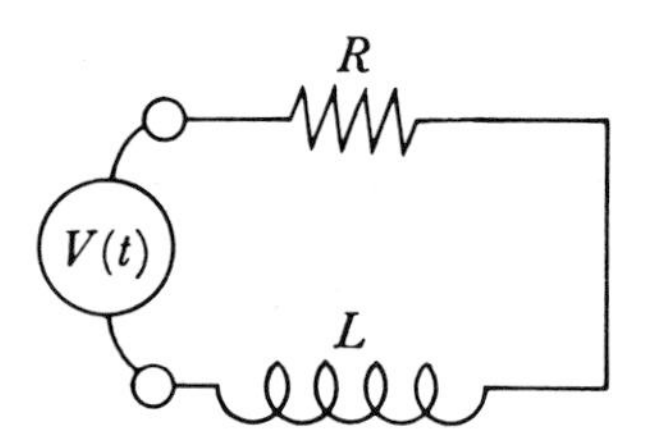

FIGURE 8-11 Series RL circuit

very high pulse lasting for a very short time. As we shall see in Section 9-1, we can write to a good approximation

$$V(t) = \frac{A}{2\pi}\int_{-\infty}^{\infty} e^{i\omega t}\, d\omega = \int_{-\infty}^{\infty} \tilde{V}(\omega) e^{i\omega t}\, d\omega$$

where A is the area under the curve $V(t)$. Here $\tilde{V}(\omega)$ measures the amount of voltage at frequency ω.

The amount of current at frequency ω is then given by

$$\tilde{I}(\omega) = \frac{\tilde{V}(\omega)}{Z(\omega)} = \frac{A}{2\pi(R + i\omega L)}$$

We multiply this by the time variation of each frequency, and sum over frequencies, to find the time variation of the current

$$I(t) = \frac{A}{2\pi}\int_{-\infty}^{\infty} \frac{e^{i\omega t}\, d\omega}{R + i\omega L}$$

Let us evaluate this integral.

If $t < 0$, the integrand is exponentially small for $\operatorname{Im}\omega \to -\infty$. We may complete the contour by a large semicircle in the *lower* half ω-plane, along which the integral vanishes by Jordan's lemma. The contour encloses no singularities, so that $I(t) = 0$.

If $t > 0$, we must complete the contour by a large semicircle in the *upper* half plane. Then

$$I(t) = 2\pi i\left(\frac{A}{2\pi}\right)\frac{e^{-Rt/L}}{iL} = \frac{A}{L}e^{-Rt/L}$$

PROBLEMS

Do the integrals 8-1 to 8-15 by contour integration:

8-1 $\displaystyle\int_0^\infty \frac{dx}{1+x^4}$

8-2 $\displaystyle\int_{-\infty}^\infty \frac{e^{i\omega t}\,d\omega}{\omega^2-\omega_0^2}$ (put poles slightly *above* real axis)

8-3 $\displaystyle\int_{-\infty}^\infty \frac{x^2\,dx}{(a^2+x^2)^2}$

8-4 $\displaystyle\int_{\substack{\text{all}\\\text{space}}} \frac{d^2x}{(a^2+r^2)^3}$ where $d^2x = dx\,dy$ and $r^2 = x^2+y^2$

8-5 $\displaystyle\int_{-1}^{+1} \frac{dx}{\sqrt{1-x^2}(a+bx)}$ $(a > b > 0)$

8-6 $\displaystyle\int_0^\infty \frac{x\,dx}{1+x^5}$

8-7 $\displaystyle\int_0^{2\pi} \frac{\sin^2\theta\,d\theta}{a+b\cos\theta}$ $(a > |b|)$

8-8 $\displaystyle\int_{-\infty}^\infty \frac{\sinh ax}{\sinh \pi x}\,dx$ $|a| < |\pi|$

8-9 $\displaystyle\int_0^\infty \frac{(\ln x)^2}{1+x^2}\,dx$

8-10 $\displaystyle\int_0^\infty \frac{dx}{1+x^2+x^4}$

8-11 $\displaystyle\int_0^\infty \frac{dx}{(a+bx^2)^3}$ $(a > 0, b > 0)$

8-12 $\displaystyle\int_0^\infty \frac{x^2\,dx}{(a^2+x^2)^3}$

8-13 $\displaystyle\int_0^{2\pi} \frac{d\theta}{(a + b\cos\theta)^2} \qquad (a > b > 0)$

8-14 $\displaystyle\int_0^{\infty} \frac{\ln x\, dx}{(x+1)^2}$

8-15 $\displaystyle\int_0^{\infty} \frac{x^3 \sin x\, dx}{(b^2 + x^2)}$

8-2 METHOD OF STEEPEST DESCENT

The discussion in Section 8-1 tends to give the impression that almost any integral can be evaluated by the theorem of residues. This is not true. Many contour integrals cannot be evaluated in closed form. The special properties of analytic functions of a complex variable do, however, provide us with additional tools to use in approximating otherwise intractable integrals. In this section we discuss one such method, the method of steepest descent.

Any contour integral

$$\int_C f(z)\, dz = \int_{s_1}^{s_2} f(z) \frac{dz}{ds}\, ds$$

can be interpreted as a summing up of values of $f(z)$ at points z along some path C. In general the functions Re $f(z)$ and Im $f(z)$ will fluctuate a great deal along the path. [See Figure 8-12, which shows the values of Re(e^{-z^2}) along the

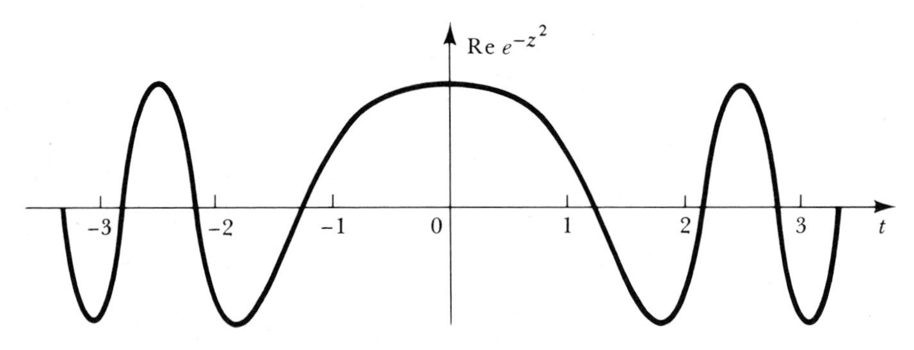

FIGURE 8-12 Values of Re (e^{-z^2}) along the path $x = y$ in the complex z-plane

path $x = y$. The values are plotted in the parameter t, arc length along the curve.] Sums of values like these are hard to approximate. If, however, the values were very small except in one region [Figure 8-13 shows the values of Re(e^{-z^2}) along the path $y = 0$], we could get a good approximation to the value of the entire contour integral by just integrating over the hump.

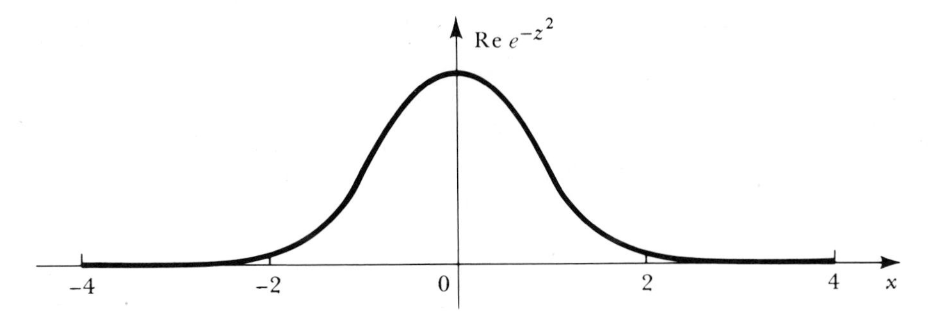

FIGURE 8-13 Values of Re (e^{-z^2}) along the path $y = 0$

Cauchy's theorem tells us that in many cases we can deform a contour without changing the value of the integral; hence it might be possible to deform the given contour until we reach a situation like that shown in Figure 8-13. What conditions must the integrand $f(z)$ satisfy in order that such a hump exist along any path? Clearly the point $z_0 = z(s_0)$ should be a relative maximum of both the real and imaginary parts of $f(z)\,dz/ds$ for points along the path; that is

$$\frac{d}{ds}\left[f(z)\frac{dz}{ds}\right]\bigg|_{s_0} = \frac{df}{dz}\left(\frac{dz}{ds}\right)^2 + f(z)\frac{d^2z}{ds^2}\bigg|_{s_0} = 0$$

For simplicity, let us consider only paths such that

$$\frac{dz}{ds}\bigg|_{s_0} \neq 0 \qquad \frac{d^2z}{ds^2}\bigg|_{s_0} = 0$$

These paths can be approximated by straight lines near s_0. In this case, the condition that z_0 be a point of relative maximum along the curve is that

$$\frac{df}{dz}\bigg|_{z_0} = 0$$

Provided that the integrand $f(z)$ has a first derivative that vanishes at some point $z = z_0$, we must next determine which path to take through this point. The path should be chosen so values of the integrand along the path are principally contributed by the region about z_0 (as in Figure 8-13). We must also develop some simple technique for approximating the contribution of this hump. Clearly this technique will depend on the general form of the integrand.

We consider integrals of the form

$$I = \int_C e^{\alpha h(z)}\,dz \tag{8-13}$$

The condition that the first derivative of the integrand vanish at $z = z_0$ becomes

$$h'(z_0) = 0$$

For nearby points (which are supposed to give most of the contribution), we can approximate $h(z)$ by a Taylor's series:

$$\begin{aligned} h(z) &\approx h(z_0) + h'(z_0)(z - z_0) + \tfrac{1}{2}h''(z_0)(z - z_0)^2 \\ &= h(z_0) + \tfrac{1}{2}h''(z_0)(z - z_0)^2 \end{aligned}$$

Thus

$$I \approx \int_{C_{\text{new}}} e^{\alpha h(z_0)}\, e^{\alpha h''(z_0)(z - z_0)^2/2}\, dz \tag{8-14}$$

The contour C_{new} is to be chosen such that the integrand decreases in absolute value as rapidly as possible on either side of z_0, to approximate the situation in Figure 8-13. For α a real number, this means that the real part of $h(z)$ must fall off as rapidly as possible; i.e., C_{new} must run along the line where $\operatorname{Re} h(z)$ changes most rapidly. Hence the path is called a *path of steepest descent.* We proved in Chapter 7 that lines where $\operatorname{Re} h(z)$ changes most rapidly are lines where $\operatorname{Im} h(z)$ is constant. For these lines the integrand

$$e^{\alpha h(z)} = e^{\alpha \operatorname{Re} h(z)}\, e^{i\alpha \operatorname{Im} h(z_0)}$$

has a constant phase equal to the phase at z_0.

The orientation of the path of steepest descent is chosen such that

$$h''(z_0)(z - z_0)^2$$

is purely real (if it had an imaginary part the integrand's phase would change and the path would not be the one desired). This expression must also be negative, in order that the integrand fall off on either side of z_0. Hence if we set

$$z - z_0 = \rho\, e^{i\theta}$$

and

$$h''(z_0) = |h''(z_0)|\, e^{i\varphi}$$

then

$$e^{2i\theta}\, e^{i\varphi} = e^{i\pi(1 + 2n)}$$

and

$$\theta = \frac{\pi}{2}(1 + 2n) - \frac{\varphi}{2} \tag{8-15}$$

determines the angle at which our path should go "over the hump." Notice that the direction is defined only up to an angle π; the proper angle must be chosen from the two possibilities (8-15) by a detailed study of the topography of the surface given by $\operatorname{Re} h(z)$ in the region near z_0. Two possible cases are illustrated for an example in Figure 8-14; they correspond to the two different topographies shown in Figure 8-15.

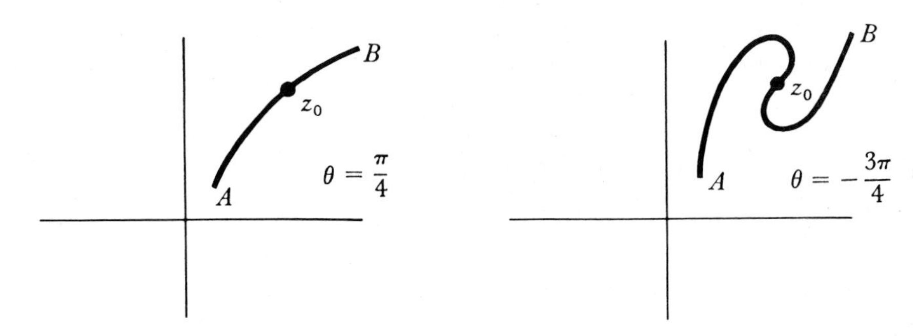

FIGURE 8-14 Alternative possibilities for steepest descent contours

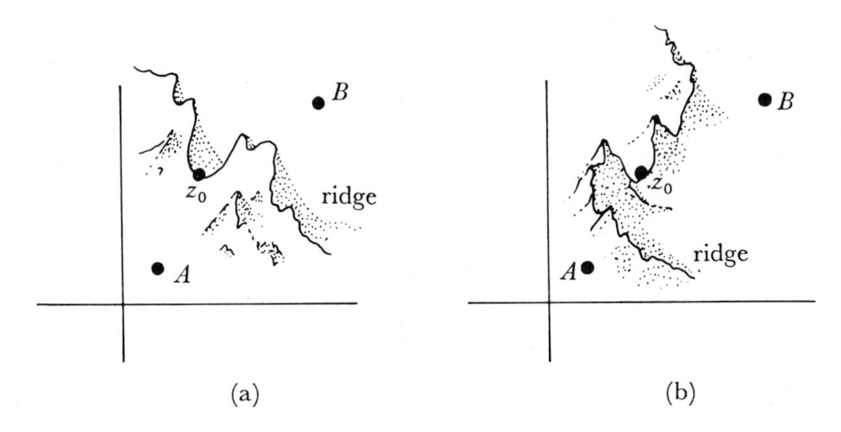

FIGURE 8-15 Alternative "mountain ranges" for the two cases of Figure 8-14

To get a better geometrical understanding of this uniqueness of the path of steepest descent, examine Figure 8-16. Here the surface Re $h(z)$ is plotted for a typical function $h(z)$. As we discovered in Chapter 7, the only points where Re $h(z)$ has a relative maximum are poles and essential singularities. At regular points, the equation ∇^2 Re $h(z) = 0$ dictates that the curvatures along perpendicular directions have opposite signs. Thus if Re $h(z)$ is to have a relative maximum along some contour, it will have a relative minimum along the perpendicular contour. This forces the surface Re $h(z)$ into the shape of a saddle centered on z_0; z_0 is called a *saddle point*. Hence if the contour of integration is moved even slightly from the path of steepest descent, the hump in Re $h(z)$ becomes considerably fatter, fluctuations in Im $h(z)$ become important, and it is much more difficult to formulate an approximation for the integral.

Let us now apply this to evaluate (8-14)

$$I \approx e^{\alpha h(z_0)} \int_{\text{hump}} e^{-\alpha/2\,|h''(z_0)|\,\rho^2}\, e^{i\theta}\, d\rho \approx e^{\alpha h(z_0)}\, e^{i\theta} \int_{-\infty}^{\infty} e^{-\alpha/2\,|h''(z_0)|^2\,\rho^2}\, d\rho$$

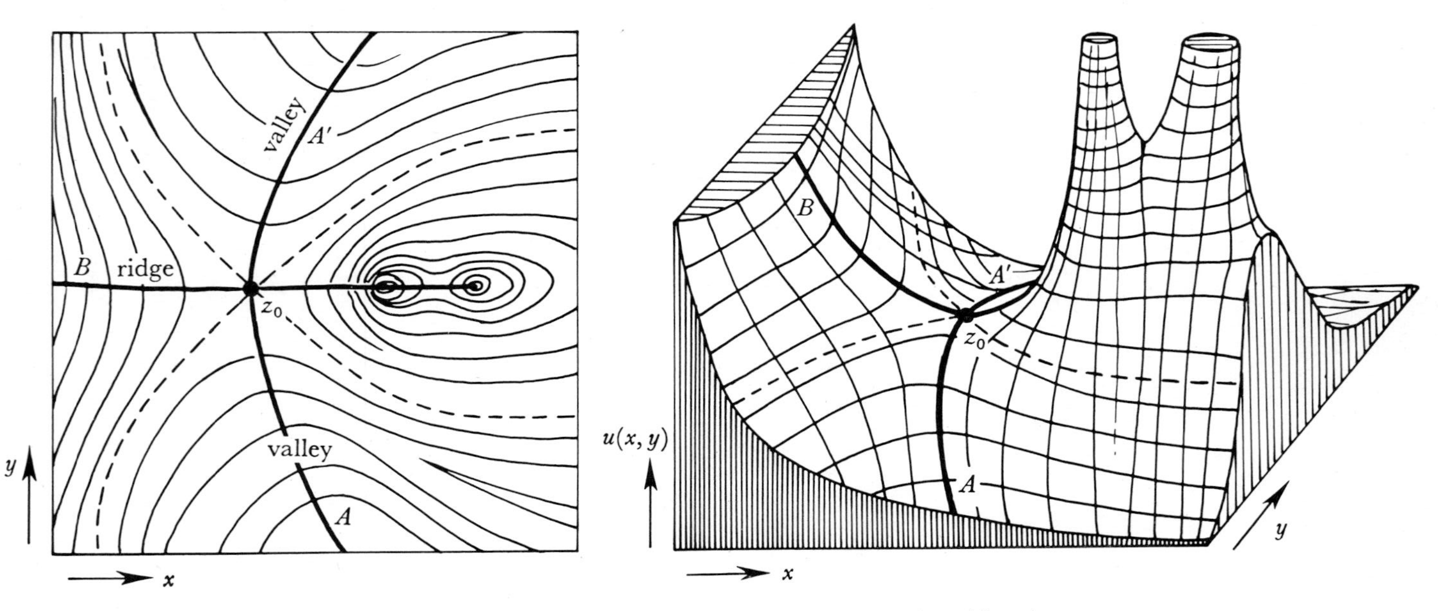

FIGURE 8-16 Topography of the surface $u = \text{Re } h(z)$ near the saddle point z_0 for a typical function $h(z)$. The heavy solid curves follow the centers of the ridges and valleys from the saddle point, and the dashed curves follow level contours, $u = u(x_0, y_0) = \text{constant}$. The curve AA' is the path of steepest descent.

We can extend the integral to infinity since the contribution from the tails of a Gaussian distribution is quite small. (Functions of the form $A\,e^{-Bx^2}$ are called *Gaussian* functions, because of their occurrence in the least-squares method of data analysis, which originated with Gauss.) Applying (6-2) we find

$$I \approx e^{i\theta}\, e^{\alpha h(z_0)}\, 2\left(\frac{1}{2}\right)\sqrt{\frac{2\pi}{\alpha\,|\,h''(z_0)\,|}} = \sqrt{\frac{2\pi}{\alpha\,|\,h''(z_0)\,|}}\; e^{\alpha h(z_0)}\, e^{i\,(\pm\pi/2\,-\,\varphi/2)} \tag{8-16}$$

Example

The *gamma function* is defined by an integral

$$\Gamma(z+1) = \int_0^\infty t^z e^{-t}\, dt = \int_0^\infty e^{-t+z\ln t}\, dt \tag{8-17}$$

Let us find an approximation for large $z = \gamma\, e^{i\beta}$. Casting the integral into the form (8-13), we find (for $\alpha = 1$),

$$h(t) = -t + z\ln t$$

$$h'(t) = -1 + \frac{z}{t}$$

$$h'(t_0) = 0 \quad \text{for} \quad t_0 = z$$

$$h''(t_0) = -\frac{z}{t_0^2} = -\frac{1}{z} = \frac{e^{i(\pi-\beta)}}{\gamma}$$

Substitution into (8-16) yields

$$\begin{aligned}\Gamma(z+1) &\approx \sqrt{2\pi\gamma}\, e^{-z+z\ln z}\, e^{i(\pm\pi/2\,-\,\pi/2\,+\,\beta/2)} \\ &= \sqrt{2\pi}\, z^{z+1/2}\, e^{-z}(\pm 1)\end{aligned} \tag{8-18}$$

In this case we can determine the phase just by examination of the answer: if z is large and positive ($\beta = 0$), then we see from (8-17) that $\Gamma(z+1)$ is positive. Hence we must choose the positive sign in (8-18), and

$$\Gamma(z+1) \approx \sqrt{2\pi}\, z^{z+1/2}\, e^{-z} \tag{8-19}$$

PROBLEMS

8-16 Find an asymptotic approximation for large x (real and positive) to the function

$$f_n(x) = \int_C e^{-ix\sin t + int}\, dt$$

where C is the contour shown in Figure 8-17.

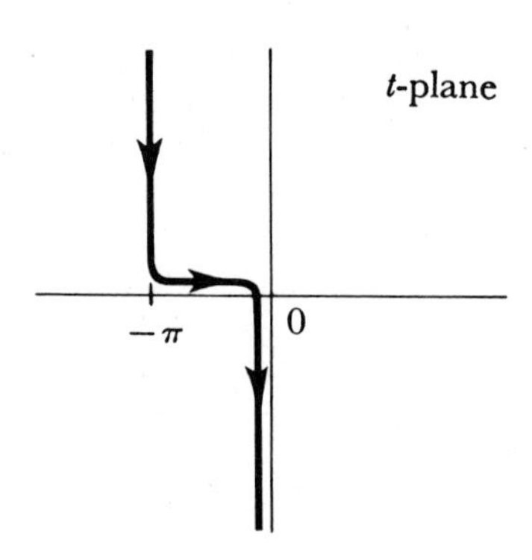

FIGURE 8-17

8-17 Evaluate $I(x) = \int_0^\infty dt\, e^{xt - e^t}$ approximately for large positive x.

8-18 Energy in a star is produced by nuclear reactions. The number of collisions with CM kinetic energy in the interval from E to $E + dE$ is

$$Ne^{-E/kT}\, E\, dE$$

per unit time, where k is Boltzmann's constant, T is the temperature, and N is a constant. The probability that a collision with CM kinetic energy E will result in a nuclear reaction is

$$Me^{-\alpha/\sqrt{E}}$$

where M and α are constants. Find an approximate expression for the total number of nuclear reactions per unit time, assuming

$$\left(\frac{kT}{\alpha^2}\right)^{1/3} \ll 1$$

8-3 SOMMERFELD-WATSON TRANSFORMATION

In our evaluation of integral (8-11) we calculated an integral by summing a series. It is also possible to sum certain series by evaluating contour integrals.

Consider the integral

$$I = \oint \frac{f(z)\, dz}{\sin \pi z} \tag{8-20}$$

around the contour of Figure 8-18, where $f(z)$ has several isolated singularities (indicated by crosses in the figure) and goes to zero at least as fast as $|z|^{-1}$ as $|z| \to \infty$.

On the one hand, the integral I may be evaluated by summing the residues at the zeros of sin πz, indicated by dots in Figure 8-18. The result is

$$I = 2\pi i \sum_{n=-\infty}^{\infty} \frac{1}{\pi} (-)^n f(n) \tag{8-21}$$

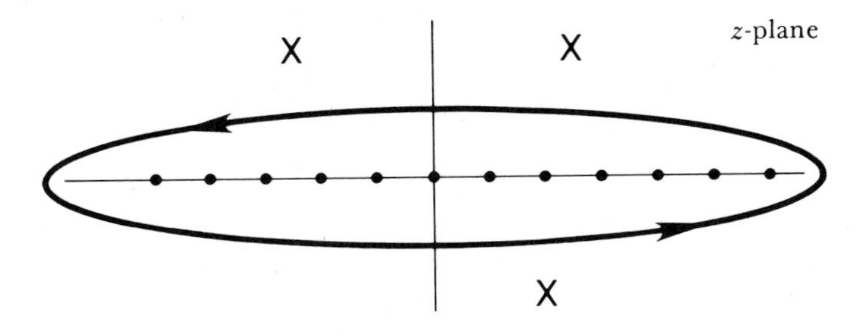

FIGURE 8-18 Original contour for the integral (8-20)

On the other hand, the integral along the contour of Figure 8-19 is clearly

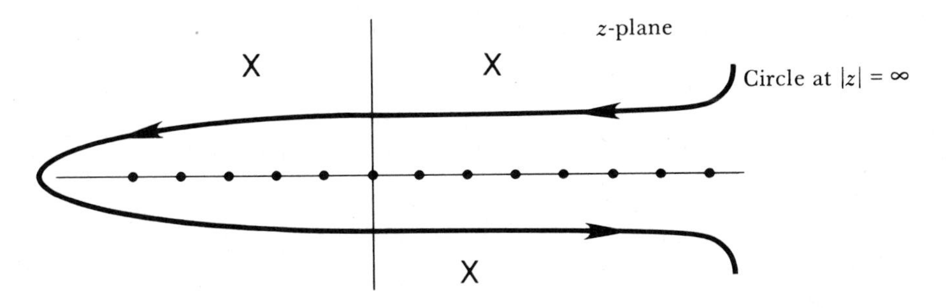

FIGURE 8-19 Deformed contour for the integral (8-20)

the same as that along the contour of Figure 8-18 since the integral around the circle at $|z| = \infty$ vanishes. The singularities enclosed by the contour of Figure 8-19 are now the singularities of $f(z)$; if the locations and residues are denoted by z_k and R_k, respectively, then

$$I = -2\pi i \sum_k \frac{R_k}{\sin \pi z_k}$$

Comparison with (8-21) gives the summation formula

$$\sum_{n=-\infty}^{\infty} (-)^n f(n) = -\pi \sum_k \frac{R_k}{\sin \pi z_k} \tag{8-22}$$

This device, which converts an infinite sum into a contour integral that is subsequently deformed, is known as a *Sommerfeld-Watson transformation.* [See Sommerfeld (31), Appendix to Chapter VI, or Watson (33).]

Example

Consider the series

$$S(x) = \frac{\sin x}{a^2+1} - \frac{2\sin 2x}{a^2+4} + \frac{3\sin 3x}{a^2+9} - + \cdots$$

$$= \sum_{n=1}^{\infty} (-)^{n+1} \frac{n \sin nx}{a^2+n^2}$$

$$= -\frac{1}{2} \sum_{n=-\infty}^{\infty} (-)^n \frac{n \sin nx}{a^2+n^2}$$

This is just the left side of equation (8-22), if we set

$$f(z) = -\frac{1}{2}\frac{z \sin xz}{a^2+z^2}$$

(Note that we must require $|x| < \pi$ in order that the large circle in Figure 8-19 make no contribution to the integral.) Now the locations and residues of the poles of $f(z)$ are

$$z = ia \qquad R = -\frac{\sin(iax)}{4}$$

$$z = -ia \qquad R = \frac{\sin(iax)}{4}$$

Hence

$$-\frac{1}{2} \sum_{n=-\infty}^{\infty} (-1)^n \frac{n \sin nx}{a^2+n^2} = \frac{\pi \sinh ax}{2 \sinh a\pi}$$

Example

As another example, consider the "partial wave series"

$$g(x) = \sum_{l=0}^{\infty} (2l+1) a_l P_l(x)$$

which converges within a small ellipse in the complex x-plane (Problem 10-18). Suppose we desire to study $g(x)$ for large values of x; some new means must be found to represent the analytic continuation of the series into this region.

$$g(x) = -\frac{1}{2i} \oint_C \frac{a_l (2l+1) P_l(-x)\, dl}{\sin \pi l} \tag{8-23}$$

where C is shown in Figure 8-20. P_l is an analytic function of l for fixed

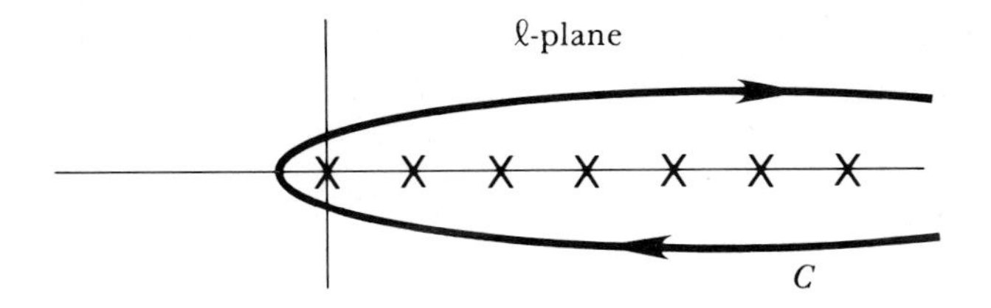

FIGURE 8-20 Contour for the integral (8-23)

x [see equation (10-27)] and has no poles as a function of l; so the only poles of the integrand are those of $a_l/\sin \pi l$. Suppose a_l is such that the integral at infinity vanishes; we can then deform the contour to that shown in Figure 8-21.

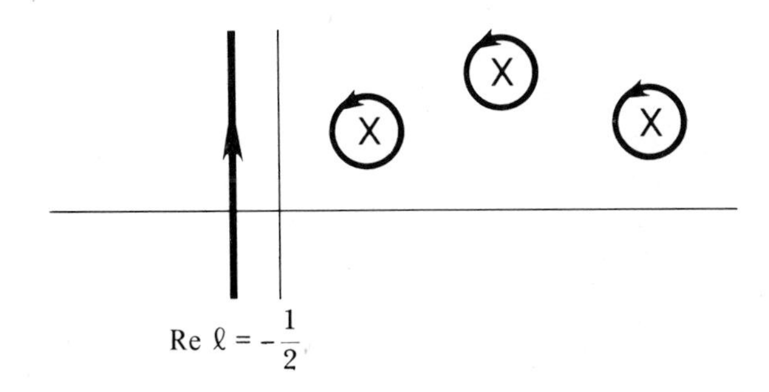

FIGURE 8-21 A different contour for (8-23)

If a_l behaves like

$$\frac{R_i}{l - l_i}$$

near the poles, the theorem of residues yields

$$g(x) = \pi \sum_i \frac{(2l_i + 1)R_i(l_i)P_{l_i}(-x)}{\sin \pi l_i} + \frac{1}{2i} \int_{-(1/2)-i\infty}^{-(1/2)+i\infty} \frac{a_l(2l + 1)P_l(-x)\, dl}{\sin \pi l}$$

Now the behavior as $x \to \infty$ can be easily read off from the behavior of the P_l's:

$$g(x) \sim (-x)^{l_i}$$

for the pole l_i with the largest real part.

PROBLEMS

8-19 (a) Consider the contour integral $\oint f(z) \operatorname{ctn} \pi z \, dz$ around a suitable large contour, and obtain thereby a formula for the sum

$$\sum_{n=-\infty}^{\infty} f(n)$$

(b) Evaluate

$$g(a) = \sum_{n=-\infty}^{\infty} \frac{1}{n^2 + a^2}$$

8-20 Evaluate

$$S = \sum_{n=0}^{\infty} \frac{(-1)^n}{n!} \frac{1}{\Gamma(\alpha_1 - n + 1)} \frac{1}{(n - \alpha_1)}$$

SUMMARY

In evaluating integrals by use of the theorem of residues, one should keep the following basic points in mind:

Choose the contour so that evaluation is as easy as possible.

Carefully define angles from branch cuts.

Check to see whether integrals at infinity or around branch points vanish. If they do not vanish, they must be evaluated explicitly.

Angular integrals can be made into path integrals by the substitution $z = e^{i\theta}$.

Jordan's lemma: $\int_C e^{i\omega x} f(x)\, dx = 0$ for C a semicircle at infinity along which $\operatorname{Im}(\omega x) > 0$, provided $\lim_{|x| \to \infty} f(x) = 0$ uniformly with $\arg x$ for those x on the semicircle.

Contour integrals can often be approximated by the method of steepest descent. If the integral is of the form

$$I = \int_C e^{\alpha h(z)}\, dz$$

with α real, then the method gives

$$I \approx \sqrt{\frac{2\pi}{\alpha\, |h''(z_0)|}}\; e^{\alpha h(z_0)}\, e^{i(\pm \pi/2 - \varphi/2)}$$

where z_0 is a stationary point of $h(z)$ and a maximum of $\operatorname{Re} h$, and φ is defined by $h''(z_0) = |h''(z_0)|\, e^{i\varphi}$. If there is more than one stationary point of $h(z)$, we may have to sum over their contributions. The technique is most successful if α is large, so that the values of the integrand fall off quickly about z_0.

Sommerfeld-Watson transforms may be used to sum series. If the series represents a function, the transform may help define its analytic continuation beyond the region of convergence of the series.

CHAPTER 9 Uses of Complex Variable Theory II: Transform Techniques and Conformal Mapping

The two applications of complex variables discussed in this chapter are basic tools every physicist should be able to use.

Our treatment in Section 9-1 of Fourier and Laplace transforms concentrates on their usefulness for solving differential equations. The basic properties of transforms are developed and used to solve some simple problems. However, the procedure of transforming an entire problem, solving it in the transform space, and then inverting the solution, is also helpful in solving integral and partial differential equations. Thus the methods developed here will be applied and expanded in Chapters 11 and 12.

Conformal mapping techniques, treated in Section 9-2, often considerably simplify the solutions of boundary value problems. We discuss the basic concepts involved for conditions of the type $\phi =$ constant and $\partial\phi/\partial n =$ constant on the boundary surface. Real skill with this method comes only after a great deal of practice; students desiring additional practice problems will find a very good set in Churchill (7), Chapter 9.

9-1 INTEGRAL TRANSFORMS

Fourier and Laplace transforms are very useful tools for solving differential and integral equations. In this section we develop the elements of transform theory and discuss some applications.

FOURIER TRANSFORMS

In Section 4-3 we demonstrated that a function of one variable can be expressed in terms of its frequency components as

$$f(x) = \frac{1}{2\pi}\int_{-\infty}^{\infty} \tilde{f}(\omega)e^{i\omega x}\, dx$$

Since $e^{i\omega x}$ is an eigenfunction of the operator d/dx, this expansion of $f(x)$ is particularly convenient for use in differential equations. For example, the equation

$$\alpha \frac{d^2 f}{dx^2} + \beta \frac{df}{dx} + \gamma f = S(x)$$

can be reexpressed as

$$\frac{1}{2\pi} \int_{-\infty}^{\infty} e^{i\omega x}(-\alpha\omega^2 + i\omega\beta + \gamma)\tilde{f}(\omega)\, d\omega = \frac{1}{2\pi} \int_{-\infty}^{\infty} \tilde{S}(\omega) e^{i\omega x}\, d\omega$$

Using the fact that the functions $e^{i\omega x}$ form an orthogonal basis, we can equate their coefficients to find

$$\tilde{f}(\omega) = \frac{\tilde{S}(\omega)}{-\alpha\omega^2 + i\beta\omega + \gamma}$$

and application of formula (4-46) then yields

$$f(x) = \frac{1}{2\pi} \int_{-\infty}^{\infty} \frac{\tilde{S}(\omega) e^{i\omega x}\, d\omega}{-\alpha\omega^2 + i\beta\omega + \gamma}$$

The function $\tilde{f}(\omega)$ is called the Fourier transform of $f(x)$; this is a reciprocal relationship so that $f(x)$ can equally well be called the Fourier transform of $\tilde{f}(\omega)$. The two relations (4-46) and (4-47) are then termed the Fourier inversion formulae:

$$f(x) = \frac{1}{2\pi} \int_{-\infty}^{\infty} \tilde{f}(\omega) e^{i\omega x}\, d\omega \tag{9-1a}$$

$$\tilde{f}(\omega) = \int_{-\infty}^{\infty} f(x) e^{-i\omega x}\, dx \tag{9-1b}$$

Because of our increased sophistication with integrals, we are now able to derive (9-1b) directly from (9-1a). Beginning with (9-1a), we define

$$I(\omega) = \int_{-\infty}^{\infty} f(x) e^{-i\omega x}\, dx = \frac{1}{2\pi} \int_{-\infty}^{\infty} \tilde{f}(\omega') \left[\int_{-\infty}^{\infty} e^{i\omega' x - i\omega x}\, dx\, d\omega' \right]$$

The integral in brackets may be defined by use of a convergence factor:

$$\begin{aligned} J(\omega - \omega') &= \lim_{a \to 0} \int_{-\infty}^{\infty} e^{-a|x|} e^{ix(\omega' - \omega)}\, dx \\ &= 2 \lim_{a \to 0} \int_{0}^{\infty} e^{-a|x|} \cos x(\omega' - \omega)\, dx \\ &= \lim_{a \to 0} \frac{2a}{a^2 + (\omega - \omega')^2} = \begin{cases} 0 & \omega \neq \omega' \\ \infty & \omega = \omega' \end{cases} \end{aligned}$$

The infinity at $\omega = \omega'$ can be further defined by calculating

$$\int_{-\infty}^{\infty} J(\omega - \omega')\, d\omega = 2 \lim_{a \to 0} \int_{-\infty}^{\infty} \frac{a\, d\omega}{a^2 + (\omega - \omega')^2} = \frac{2a(2\pi i)}{2ia} = 2\pi$$

Hence we conclude that

$$\int_{-\infty}^{\infty} e^{i\omega' x}\, e^{-i\omega x}\, dx = 2\pi\delta(\omega - \omega')$$

and thus

$$\tilde{f}(\omega) = \int_{-\infty}^{\infty} f(x) e^{-i\omega x}\, dx$$

as we derived from the limiting process with the Fourier series. Clearly, only those $f(x)$ are transformable which vanish sufficiently rapidly at infinity to make this integral well defined.

We should note at this point that the position of the 2π is quite arbitrary; one often defines things more symmetrically in terms of a modified transform function $\bar{\tilde{f}}(\omega)$:

$$f(x) = \sqrt{\frac{1}{2\pi}} \int_{-\infty}^{\infty} \bar{\tilde{f}}(\omega) e^{ix\omega}\, d\omega \qquad \bar{\tilde{f}}(\omega) = \frac{1}{\sqrt{2\pi}} \int_{-\infty}^{\infty} f(x) e^{-ix\omega}\, dx \tag{9-2}$$

Example

Suppose we want to find a particular solution to the inhomogeneous equation

$$\frac{d\psi}{dx} + \alpha\psi = \frac{1}{x^2 + \beta^2} \tag{9-3}$$

Fourier transform both sides of the equation.

$$iy\,\phi(y) + \alpha\phi(y) = \int_{-\infty}^{\infty} \frac{e^{-iyx}}{x^2 + \beta^2}\, dx$$

$$\varphi(y) = \frac{1}{(iy + \alpha)} \int_{-\infty}^{\infty} \frac{e^{-iyx'}\, dx'}{x'^2 + \beta^2}$$

$$\psi(x) = \frac{1}{2\pi} \int_{-\infty}^{\infty} \frac{e^{iyx}}{(iy + \alpha)}\, dy \int_{-\infty}^{\infty} \frac{e^{-iyx'}}{(x'^2 + \beta^2)}\, dx'$$

$$= \frac{1}{2\pi} \int_{-\infty}^{\infty} \frac{dx'}{x'^2 + \beta^2} \int_{-\infty}^{\infty} \frac{e^{iy(x - x')}\, dy}{iy + \alpha}$$

Perform the y integration first. For $x - x' < 0$, the contour should be closed by a semicircle in the lower half plane (which contributes zero because of Jordan's lemma). Let us assume $\alpha > 0$ for definiteness. Then there are no poles in the lower half plane and this integral vanishes. For $x - x' > 0$ the contour should be completed in the upper half plane. We thus find

$$\psi(x) = \int_{-\infty}^{x} \frac{dx'}{x'^2 + \beta^2} e^{-\alpha(x - x')} \tag{9-4}$$

Recall that the solution of an inhomogeneous equation may be expressed in terms of a Green's function. In general, if $G(x, x')$ is the solution to $\theta_x G = \delta(x - x')$ where θ_x is a differential operator in x, then

$$\psi(x) = \int_{-\infty}^{\infty} S(x')G(x, x')\, dx'$$

is the solution to

$$\theta_x \psi = S(x)$$

Hence (9-4) allows us to read off the Green's function for (9-3):

$$G(x, x') = \begin{cases} e^{-\alpha(x - x')} & x > x' \\ 0 & x < x' \end{cases}$$

The definition (2-17) for the Green's function shows that it may take different forms depending on the boundary conditions forced on the eigenfunctions. Let us determine the counterpart of this for our Fourier transform integrals.

Consider the one-dimensional wave equation

$$\frac{\partial^2 \psi(x, t)}{\partial x^2} - \frac{1}{C^2} \frac{\partial^2 \psi(x, t)}{\partial t^2} = S(x) e^{-i\omega t}$$

Because the inhomogeneous term has periodic time dependence, we assume that ψ is also of this form and write $\psi(x, t) = \psi(x) e^{-i\omega t}$. With $k^2 = \omega^2/C^2$ our equation becomes

$$\frac{d^2 \psi(x)}{dx^2} + k^2 \psi(x) = S(x)$$

Fourier transforming both sides leads to

$$\varphi(y) = \frac{1}{(k^2 - y^2)} \int_{-\infty}^{\infty} dx'\, e^{-ix'y} S(x')$$

and hence

$$\psi(x) = \int_{-\infty}^{\infty} dx'\, S(x') G(x, x')$$

where

$$G(x, x') = \frac{1}{2\pi} \int_{-\infty}^{\infty} \frac{dy\, e^{iy(x-x')}}{(k^2 - y^2)} \tag{9-5}$$

is the Green's function desired.

A little thought shows that (9-5) is not really well defined—there are poles on the contour of integration. To define what we mean by the Green's function, we must specify a method of handling the poles. There are four possible ways of displacing the poles from the real axis, shown in Figure 9-1. Each of these corresponds to a different Green's function:

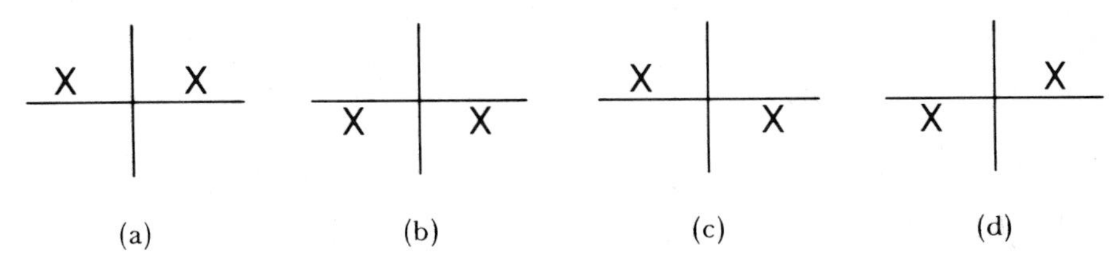

FIGURE 9-1 Possible ways of handling the poles in integral (9-5)

method (a) of displacing the poles gives

$$G(x, x') = \begin{cases} +\dfrac{1}{k} \sin k(x - x') & x > x' \\ 0 & x < x' \end{cases}$$

(b)

$$G(x, x') = \begin{cases} 0 & x > x' \\ -\dfrac{1}{k} \sin k(x - x') & x < x' \end{cases}$$

(c)

$$G(x, x') = \begin{cases} \dfrac{i}{2k} e^{-ik(x-x')} & x > x' \\ \dfrac{i}{2k} e^{ik(x-x')} & x < x' \end{cases}$$

(d)

$$G(x, x') = \begin{cases} -\dfrac{i}{2k} e^{ik(x-x')} & x > x' \\ -\dfrac{i}{2k} e^{-ik(x-x')} & x < x' \end{cases}$$

These can be interpreted physically by multiplying by $e^{-i\omega t}$ to restore the time dependence of the waves: (a) and (b) correspond to standing waves, (c) corresponds to a wave traveling into the point $x = x'$ (i.e., we have a sink at that point that sucks in waves), and (d) corresponds to a wave traveling out from $x = x'$. Each of these is the solution for a different set of physical boundary conditions.

PHYSICAL INTERPRETATION OF THE TRANSFORM

In addition to its value in solving differential equations, the Fourier transform has an interesting and useful physical interpretation. As we showed in Section 4-3, the function $\tilde{f}(\omega)$ can be thought of as a measure of the amount of frequency ω present in $f(x)$. From equations (4-55) and (4-56),

$$\int_{-\infty}^{\infty} dx\, f_1^*(x) f_2(x)\, dx = \frac{1}{2\pi}\int_{-\infty}^{\infty} \tilde{f}_1^*(\omega)\tilde{f}_2(\omega)\, d\omega$$

we derive

$$\int_{-\infty}^{\infty} dx\, |f(x)|^2 = \frac{1}{2\pi}\int_{-\infty}^{\infty} |\tilde{f}(\omega)|^2\, d\omega \tag{9-6}$$

which is usually referred to as Parseval's theorem.

Example

$$f(t) = \begin{cases} 0 & (t < 0) \\ e^{-t/T}\sin\omega_0 t & (t > 0) \end{cases}$$

This function is shown in Figure 9-2. $f(t)$ might represent the displacement

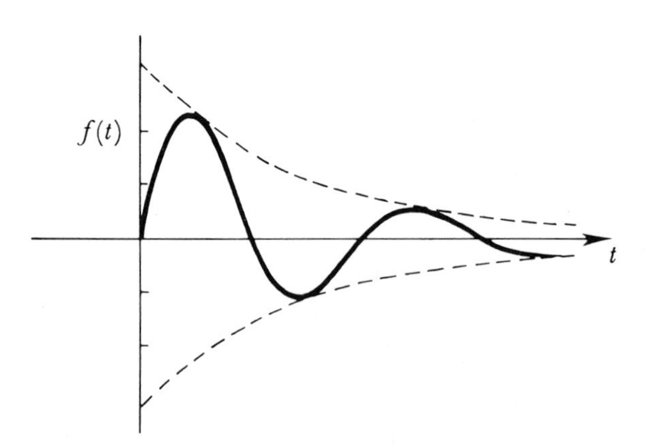

FIGURE 9-2 A damped sine wave

of a damped harmonic oscillator, or the electric field in a radiated wave, or the current in an antenna, for example.

The Fourier transform of $f(t)$ is

$$g(\omega) = \int_{-\infty}^{\infty} f(t)e^{-i\omega t}\,dt$$

$$= \int_{0}^{\infty} e^{-t/T}e^{-i\omega t}\sin\omega_0 t\,dt$$

$$= \frac{1}{2}\left(\frac{1}{\omega + \omega_0 - \dfrac{i}{T}} - \frac{1}{\omega - \omega_0 - \dfrac{i}{T}}\right)$$

We may interpret the physical meaning of $g(\omega)$ with the help of Parseval's theorem (9-6). For example, if $f(t)$ is a radiated electric field, the radiated power is proportional to $|f(t)|^2$ and the total energy radiated is proportional to $\int_0^\infty |f(t)|^2\,dt$. This is equal to

$$\frac{1}{2\pi}\int_{-\infty}^{\infty} |g(\omega)^2|\,d\omega$$

by Parseval's theorem. Then $|g(\omega)|^2$ may be interpreted as the energy radiated per unit frequency interval (times some constant).

Let us assume that T is rather large ($\omega_0 T \gg 1$). Then our "frequency spectrum" $g(\omega)$ is sharply peaked near $\omega = \pm\omega_0$. For example, near $\omega = \omega_0$,

$$g(\omega) \approx -\frac{1}{2}\frac{1}{\omega - \omega_0 - \dfrac{i}{T}}$$

$$|g(\omega)| \approx \frac{1}{2}\frac{1}{\sqrt{(\omega - \omega_0)^2 + \dfrac{1}{T^2}}}$$

When $\omega = \omega_0 \pm 1/T$, the "amplitude" $g(\omega)$ is down by a factor $\sqrt{\frac{1}{2}}$, and the radiated energy $|g(\omega)|^2$ is down by a factor $\frac{1}{2}$. In other words, the width Γ at half (power) maximum is given by $\Gamma \approx 2/T$. The energy spectrum $|g(\omega)|^2$ is sketched in Figure 9-3.

This result is a typical *uncertainty principle* and is very closely related to the Heisenberg uncertainty principle in quantum mechanics. The length of time (T) during which something oscillates is inversely proportional to the width Γ, which is a measure of the "uncertainty" in frequency.

Fourier transforms may easily be generalized to more than one dimension. For example, in three-dimensional space we have the transform pair

$$\varphi(\mathbf{k}) = \int d^3x f(x)e^{-i\mathbf{k}\cdot\mathbf{x}}$$

$$f(\mathbf{x}) = \int \frac{d^3k}{(2\pi)^3}\, \varphi(\mathbf{k}) e^{i\mathbf{k}\cdot\mathbf{x}} \tag{9-7}$$

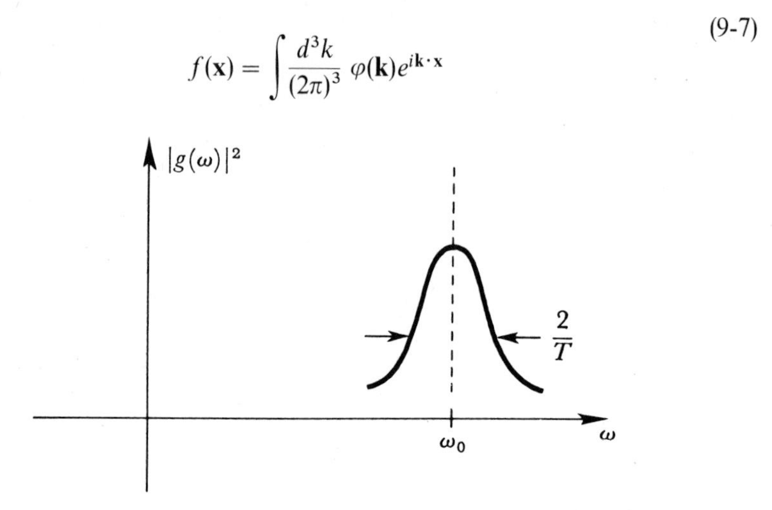

FIGURE 9-3 Energy spectrum for the damped oscillation of Figure 9-2

From these we may deduce the integral representation

$$\delta(\mathbf{x}) = \int \frac{d^3k}{(2\pi)^3}\, e^{i\mathbf{k}\cdot\mathbf{x}} \tag{9-8}$$

where the three-dimensional delta function is defined by

$$\delta(\mathbf{x}) = 0 \qquad \mathbf{x} \neq 0$$

$$\int d^3x\, \delta(\mathbf{x}) = 1 \qquad \text{provided origin is inside region of integration} \tag{9-9}$$

$$\int d^3x\, f(\mathbf{x})\delta(\mathbf{x}-\mathbf{y}) = f(\mathbf{y})$$

Such transform pairs are of interest in quantum mechanics. If $f(\mathbf{x})$ is the wave function of a particle, the Fourier transform $\varphi(\mathbf{k})$ is the so-called "wave function in momentum space." $|f(\mathbf{x})|^2$ and $|\varphi(\mathbf{k})|^2$ are the probability distributions for the position and momentum, respectively.

Example

$$f(\mathbf{x}) = \left(\frac{2}{\pi a^2}\right)^{3/4} e^{-r^2/a^2} = Ne^{-r^2/a^2} \qquad (r = |\mathbf{x}|)$$

This is a wave function that gives a Gaussian probability distribution, $|f(\mathbf{x})|^2$, centered at $r = 0$, and normalized so that $\int d^3x\, |f(\mathbf{x})|^2 = 1$. The Fourier transform is

$$\varphi(\mathbf{k}) = N \int d^3x\, e^{-r^2/a^2}\, e^{-i\mathbf{k}\cdot\mathbf{x}}$$

Introduce polar coordinates, with the z-axis along $\mathbf{k}$. Let $\cos\theta = \alpha$. Then

$$\begin{aligned}
\varphi(\mathbf{k}) &= 2\pi N \int_0^\infty r^2\,dr \int_{-1}^{+1} d\alpha\, e^{-r^2/a^2}\, e^{-ikr\alpha} \\
&= \frac{4\pi}{k} N \int_0^\infty r\,dr\, e^{-r^2/a^2} \sin kr \\
&= \frac{4\pi}{k}\frac{N}{2i} \int_{-\infty}^\infty r\,dr\, e^{-r^2/a^2}\, e^{ikr} \\
&= \frac{2\pi}{ik} N e^{-k^2a^2/4} \int_{-\infty}^\infty r\,dr \exp\left[-\frac{1}{a^2}\left(r - \frac{ika^2}{2}\right)^2\right] \\
&= \frac{2\pi}{ik} N e^{-k^2a^2/4} \int_{-\infty}^\infty \left(y + \frac{ika^2}{2}\right) dy\, e^{-y^2/a^2} \\
&= \frac{2\pi}{ik} N e^{-k^2a^2/4} \frac{ika^2}{2} a\sqrt{\pi}
\end{aligned}$$

Finally, remembering $N = (2/\pi a^2)^{3/4}$, we obtain

$$\varphi(\mathbf{k}) = (2a^2\pi)^{3/4}\, e^{-k^2a^2/4}$$

Note that the Fourier transform of a Gaussian is another Gaussian. The narrower the distribution in $\mathbf{x}$ (that is, the smaller the value of a), the broader is the distribution in $\mathbf{k}$. The widths Δx and Δk are roughly inverse to each other,

$$\Delta x\, \Delta k \simeq 1$$

The prescription $\mathbf{p} = \hbar\mathbf{k}$ from quantum mechanics ($\mathbf{p}$ = momentum) converts this into another quantum mechanical uncertainty relation

$$\Delta x\, \Delta p \simeq \hbar$$

The probability distribution $|\phi(\mathbf{k})|^2$ found above is a Gaussian centered at $k = 0$. We might expect that a similar distribution centered at an arbitrary $\mathbf{k}_0$ could be obtained, starting with the same probability distribution in position $|f(\mathbf{r})|^2$. This may be done by simply multiplying $f(x)$ by $ae^{i\mathbf{k}_0\cdot\mathbf{x}}$ as the reader can easily verify.

LAPLACE TRANSFORMS

We are often interested in functions whose Fourier transforms do not exist. For example, the simple function $f(x) = x^2$ has a Fourier Transform integral that does not converge. For many functions, the trouble at $x \to +\infty$ may be remedied by multiplication with a factor e^{-cx} if c is real and larger than some minimum value α. This factor may make the behavior at $x \to -\infty$ worse, but we are often interested in a function only for positive x. Thus we can take care of the behavior for negative x by a second factor, the (Heaviside) step function:

$$H(x) = \begin{cases} 0 & x < 0 \\ 1 & x > 0 \end{cases}$$

[Note that $(d/dt)H(t) = \delta(t)$, as can be shown by integrating the δ-function.]

The function $f(x)e^{-cx}H(x)$ now has the Fourier transform:

$$g(y) = \int_{-\infty}^{\infty} f(x)e^{-cx}H(x)e^{-ixy}\,dx = \int_0^{\infty} f(x)e^{-cx}\,e^{-ixy}\,dx$$

The inverse transform is

$$f(x)e^{-cx}H(x) = \frac{1}{2\pi}\int_{-\infty}^{\infty} g(y)e^{ixy}\,dy$$

It is conventional to introduce a new variable

$$s = c + iy$$

and to define $F(s) \equiv g(y)$. The above two integrals become

$$F(s) = \int_0^{\infty} f(x)e^{-sx}\,dx \tag{9-10}$$

$$f(x)H(x) = \frac{1}{2\pi i}\int_C F(s)e^{sx}\,ds \tag{9-11}$$

where the path of integration C is upward along the straight line, $\text{Re}\, s = c =$ constant (Figure 9-4).

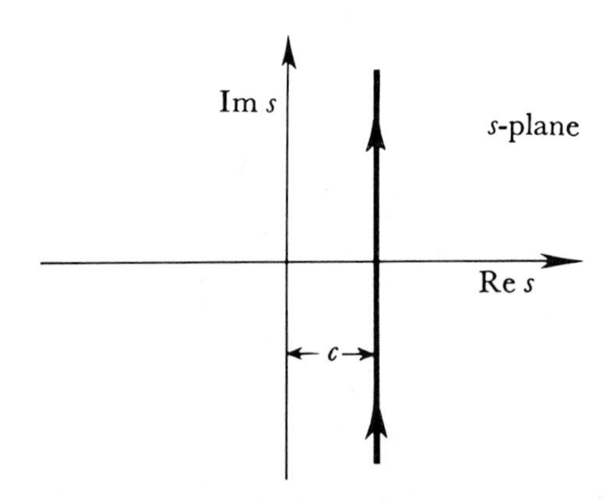

FIGURE 9-4 Contour for the Laplace inversion integral (9-11)

$F(s)$, as given by (9-10), is called the *Laplace transform* of $f(x)$. The integral exists only in the right half s-plane, $\text{Re}\, s > \alpha$, where α is the minimum limit for c mentioned above. In this region, $F(s)$ is analytic; $F(s)$ may usually be defined in the left half plane by analytic continuation.

The second integral (9-11) is called the Laplace *inversion integral.* Note that for $x > 0$ it gives $f(x)$ (or more precisely $\frac{1}{2}[f(x+) + f(x-)]$) but for $x < 0$ it automatically gives zero; this is because the contour C may be closed by the addition of a large semicircle on the *right*, where $F(s)$ is analytic.

We shall hereafter omit the $H(x)$ in (9-11), and write simply

$$f(x) = \frac{1}{2\pi i} \int_{c-i\infty}^{c+i\infty} F(s) e^{sx} \, ds \tag{9-12}$$

with the understanding that all functions $f(x)$ that are to be Laplace transformed vanish for negative arguments.

Example

$$f(x) = 1$$

$$F(s) = \int_0^\infty f(x) e^{-sx} \, dx = \int_0^\infty e^{-sx} \, dx = \frac{1}{s}$$

where the integral exists for $\text{Re}\, s > 0$ (that is, $\alpha = 0$). Note that $F(s)$ has a singularity (a simple pole in this example) on the limiting line, $\text{Re}\, s = 0$.

We may verify the inversion formula:

$$f(x) = \frac{1}{2\pi i} \int_{c-i\infty}^{c+i\infty} F(s) e^{sx} \, ds = \frac{1}{2\pi i} \int_{c-i\infty}^{c+i\infty} \frac{e^{sx}}{s} \, dx$$

where $c > 0$.

If $x > 0$, we complete the contour by a large semicircle to the *left*, and

$$f(x) = 1$$

If $x < 0$, we complete the contour to the *right*, and

$$f(x) = 0$$

There are many physical situations in which a system is set into motion at $t = 0$ with particular initial conditions, and its subsequent behavior is described by some differential equation of motion. These situations are natural for the application of Laplace transforms, both because of the half-infinite time domain, and because the Laplace transform of a derivative involves the initial conditions automatically:

$$\int_0^\infty e^{-sx} \frac{df}{dx} \, dx = e^{-sx} f \Big|_0^\infty + s \int_0^\infty e^{-sx} f(x) \, dx$$

If we write $\mathscr{L}(f)$ for the Laplace transform of $f(x)$, then this can be rewritten as

$$\mathscr{L}\left(\frac{df}{dx}\right) = s\mathscr{L}(f) - f(0) \tag{9-13}$$

Similarly

$$\mathscr{L}\left(\frac{d^2f}{dx^2}\right) = s^2\mathscr{L}(f) - sf(0) - f'(0) \tag{9-14}$$

Example

Find the current in the circuit of Figure 9-5 if the switch is closed at time $t = 0$, and the initial charge on the condenser is Q_0.

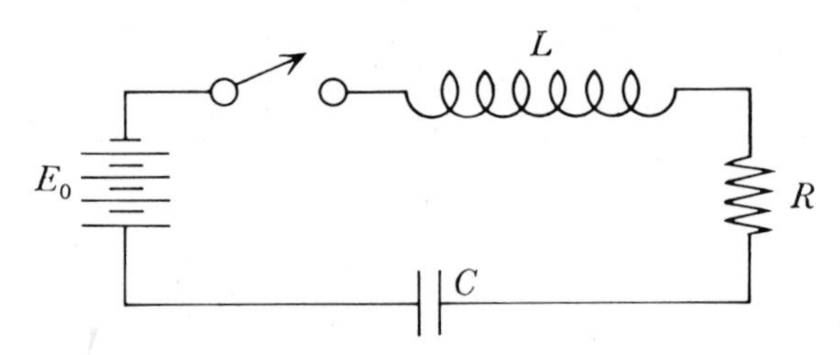

FIGURE 9-5 Series *RLC* circuit

The equation for the circuit is

$$RI + L\frac{dI}{dt} + \frac{Q}{C} = E_0$$

which we can rewrite in terms of Q, the charge on the capacitor:

$$L\frac{d^2Q}{dt^2} + R\frac{dQ}{dt} + \frac{Q}{C} = E_0$$

Strictly speaking, we should write $E_0H(t)$, where $H(t)$ is the step function, but in dealing with Laplace transforms we assume everything vanishes for $t < 0$ anyway.

Now take the Laplace transform of both sides:

$$L[s\bar{Q}(s) - sQ_0 - Q'(0)] + R[s\bar{Q}(s) - Q_0] + \frac{1}{C}\bar{Q}(s) = \frac{E_0}{s}$$

Solving for $\bar{Q}(s)$ gives (by simple algebra!)

$$Q(s) = \frac{\dfrac{E_0}{s} + sLQ_0 + LQ'(0) + RQ_0}{Ls^2 + Rs + \dfrac{1}{C}}$$

What we actually want is the current in the circuit, $I(t)$. This has transform $\bar{I}(s) = s\bar{Q}(s) - Q_0$, so we obtain

$$\bar{I}(s) = \frac{E_0 + s^2 L Q_0 + RsQ_0 + sLQ'(0)}{Ls^2 + Rs + \dfrac{1}{C}} - Q_0$$

$$= \frac{E_0 + sLQ'(0) - \dfrac{1}{C} Q_0}{Ls^2 + Rs + \dfrac{1}{C}}$$

At $t = 0$ there is no current in the circuit (why?), so $Q'(0) = 0$. We then invert this transform by calculating

$$I(t) = \frac{1}{2\pi i} \int_c \frac{\left(E_0 - \dfrac{1}{C} Q_0\right)}{L(s - \lambda_1)(s - \lambda_2)} e^{st}\, ds$$

where the poles $\lambda_1 = -R/2L + \frac{1}{2}\sqrt{(R/L)^2 - 4/LC}$ and $\lambda_2 = -R/2L - \frac{1}{2}\sqrt{(R/L)^2 - 4/LC}$ are both in the left half plane. On performing the integral, we find

$$I(t) = \frac{\left(E_0 - \dfrac{1}{C} Q_0\right)}{L(\lambda_1 - \lambda_2)} (e^{\lambda_1 t} - e^{\lambda_2 t})$$

$$= \frac{\left(E_0 - \dfrac{1}{C} Q_0\right)}{L\sqrt{\left(\dfrac{R}{L}\right)^2 - \dfrac{4}{LC}}} e^{-(R/2L)t}\, 2i \sin \sqrt{\frac{1}{LC} - \frac{1}{4}\left(\frac{R}{L}\right)^2}\, t$$

$$= \frac{\left(E_0 - \dfrac{1}{C} Q_0\right) e^{-(R/2L)t}}{L\sqrt{\dfrac{1}{LC} - \dfrac{1}{4}\left(\dfrac{R}{L}\right)^2}} \sin \sqrt{\frac{1}{LC} - \frac{1}{4}\left(\frac{R}{L}\right)^2}\, t$$

Example

Consider the coupled pendulums shown in Figure 9-6. Assume the initial conditions $x_1 = x_2 = 0$, $\dot{x}_1 = v$, $\dot{x}_2 = 0$ at $t = 0$. Newton's equations are

$$\begin{aligned} m\ddot{x}_1 &= -\frac{mg}{l} x_1 + k(x_2 - x_1) \\ m\ddot{x}_2 &= -\frac{mg}{l} x_2 + k(x_1 - x_2) \end{aligned} \tag{9-15}$$

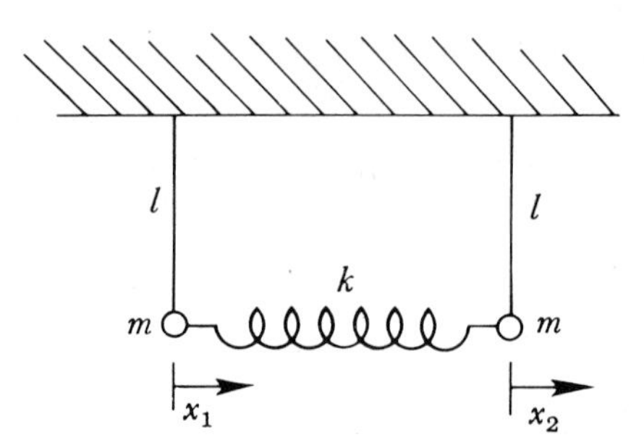

FIGURE 9-6 Coupled pendulums

Let $\mathscr{L}[x_i(t)] = F_i(s)$. Then the Laplace transforms of our two differential equations (9-15) are

$$m(s^2 F_1 - v) = -\frac{mg}{l} F_1 + k(F_2 - F_1)$$

$$ms^2 F_2 = -\frac{mg}{l} F_2 + k(F_1 - F_2)$$

We must now solve these simultaneous algebraic equations for F_1 and F_2. We find

$$F_1(s) = \frac{v\left(s^2 + \dfrac{g}{l} + \dfrac{k}{m}\right)}{\left(s^2 + \dfrac{g}{l} + 2\dfrac{k}{m}\right)\left(s^2 + \dfrac{g}{l}\right)}$$

$$= \frac{v}{2}\left(\frac{1}{s^2 + \dfrac{g}{l} + 2\dfrac{k}{m}} + \frac{1}{s^2 + \dfrac{g}{l}}\right)$$

Therefore,

$$x_1(t) = \frac{v}{2}\left(\frac{\sin\sqrt{\dfrac{g}{l} + 2\dfrac{k}{m}}\,t}{\sqrt{\dfrac{g}{l} + 2\dfrac{k}{m}}} + \frac{\sin\sqrt{\dfrac{g}{l}}\,t}{\sqrt{\dfrac{g}{l}}}\right)$$

Similarly, we can work out $x_2(t)$. Note that $\sqrt{g/l + 2(k/m)}$ and $\sqrt{g/l}$ are the (angular) frequencies of the two normal modes.

BASIC PROPERTIES

We shall summarize some basic properties of Fourier and Laplace transforms:

1. Both transforms are *linear*; i.e., the transform of $\alpha f(x) + \beta g(x)$ equals α times the transform of $f(x)$ plus β times the transform of $g(x)$.

2. *Derivatives*: We shall write $\mathscr{L}[f(x), s]$ or $\mathscr{L}[f(x)]$ for the Laplace transform of $f(x)$, and similarly $\mathscr{F}[f(x)]$ for the Fourier transform. Then our relation (9-13) may be written

$$\mathscr{L}[f'(x)] = s\mathscr{L}[f(x)] - f(0) \tag{9-16}$$

Similarly,

$$\mathscr{L}[f''(x)] = s^2\mathscr{L}[f(x)] - sf(0) - f'(0) \tag{9-17}$$

etc. (Note that 0 really means $0+$, the limit as zero is approached from the *positive* side.) For Fourier transforms the integrated parts vanish, and

$$\mathscr{F}[f'(x)] = iy\mathscr{F}[f(x)] \text{ etc.} \tag{9-18}$$

3. *Integrals*:

$$\begin{aligned}\mathscr{L}\left[\int_0^x f(t)\,dt\right] &= \int_0^\infty dx\, e^{-sx} \int_0^x f(t)\,dt \\ &= \int_0^\infty dt \int_t^\infty dx\, f(t)e^{-sx} \\ &= \frac{1}{s}\int_0^\infty dt\, f(t)e^{-st}\end{aligned}$$

Thus

$$\mathscr{L}\left[\int_0^x f(t)\,dt\right] = \frac{1}{s}\mathscr{L}[f(t)] \tag{9-19}$$

For Fourier transforms, things are not quite so simple. Suppose $g(x) = \int f(x)\,dx$ is an indefinite integral of $f(x)$. Then, from (9-18),

$$\mathscr{F}[f(x)] = iy\mathscr{F}[g(x)]$$

However, we cannot immediately conclude that

$$\mathscr{F}[g(x)] = \frac{\mathscr{F}[f(x)]}{iy}$$

Why not? Consider the equation $xf(x) = g(x)$. Can we conclude that

$$f(x) = \frac{g(x)}{x}$$

No, not at $x = 0$. The general result is

$$f(x) = \frac{g(x)}{x} + C\,\delta(x)$$

where C is an arbitrary constant. Thus

$$\mathscr{F}\left[\int f(x)\,dx\right] = \frac{\mathscr{F}[f(x)]}{iy} + C\,\delta(y) \tag{9-20}$$

This arbitrariness is also obvious from the fact that $\int f(x)\,dx$ is uncertain to within an arbitrary additive constant C, and

$$\mathscr{F}[C] = 2\pi C\,\delta(y)$$

4. *Translation*:

$$\mathscr{F}[f(x+a)] = \int_{-\infty}^{\infty} f(x+a)e^{-ixy}\,dx$$

$$= \int_{-\infty}^{\infty} f(x)e^{-iy(x-a)}\,dx$$

Therefore,

$$\mathscr{F}[f(x+a)] = e^{iay}\mathscr{F}[f(x)] \tag{9-21}$$

For Laplace transforms, we must be a little more careful. Consider the cases $a > 0$, $a < 0$ separately.

For $a > 0$, $f(x + a)$ is shown in Figure 9-7(b).

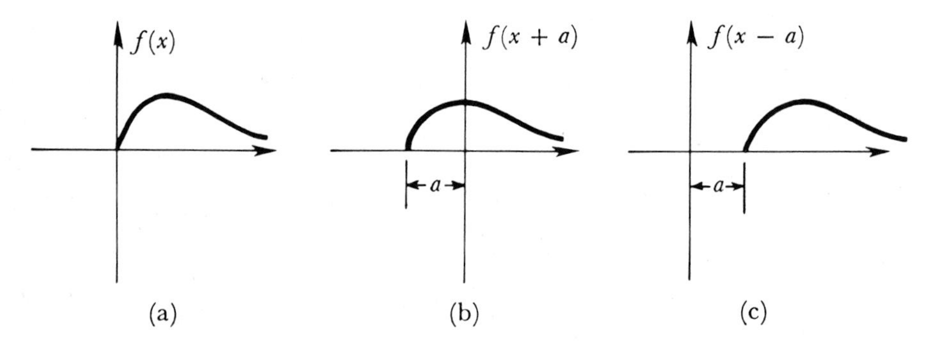

FIGURE 9-7 $f(x)$ translated to the left, $f(x + a)$, and to the right, $f(x - a)$

Since the Laplace transform ignores $f(x)$ for $x < 0$ (in fact, assumes it is zero), we must chop off some of our function, and

$$\mathscr{L}[f(x+a)] = \int_0^{\infty} f(x+a)e^{-sx}\,dx = \int_a^{\infty} f(x)e^{-s(x-a)}\,dx$$

so that

$$\mathscr{L}[f(x+a)] = e^{as}\left\{\mathscr{L}[f(x)] - \int_0^a f(x)e^{-sx}\,dx\right\} \qquad a > 0 \tag{9-22}$$

On the other hand, $f(x - a)$ is shown in Figure 9-7(c) and

$$\mathscr{L}[f(x-a)] = \int_0^{\infty} f(x-a)e^{-sx}\,dx$$

$$= \int_{-a}^{\infty} f(x) e^{-s(x+a)}\, dx$$

Therefore,

$$\mathscr{L}[f(x-a)] = e^{-as}\mathscr{L}[f(x)] \qquad a > 0 \tag{9-23}$$

5. *Multiplication by an exponential*: The following two formulas are easily verified:

$$\mathscr{F}[e^{\alpha x}f(x); y] = \mathscr{F}[f(x); y + i\alpha] \tag{9-24}$$

$$\mathscr{L}[e^{\alpha x}f(x); s] = \mathscr{L}[f(x); s - \alpha] \tag{9-25}$$

6. *Multiplication by a power of* x: If

$$g(y) = \int_{-\infty}^{\infty} f(x) e^{-ixy}\, dx$$

then

$$g'(y) = -i\int_{-\infty}^{\infty} x f(x) e^{-ixy}\, dx$$

Thus

$$\mathscr{F}[xf(x)] = i\frac{d}{dy}\mathscr{F}[f(x)] \tag{9-26}$$

An analogous result is easily shown to hold for Laplace transforms

$$\mathscr{L}[xf(x)] = -\frac{d}{ds}\mathscr{L}[f(x)] \tag{9-27}$$

7. *Convolution theorems*: Let $f_1(x)$, $f_2(x)$ be two arbitrary functions. We define their *convolution* (*Faltung* in German) to be

$$g(x) = \int_{-\infty}^{\infty} f_1(y) f_2(x-y)\, dy \tag{9-28}$$

What is the Fourier transform of such a convolution? A straightforward change of variable shows that

$$\mathscr{F}[g(x)] = \mathscr{F}[f_1(x)]\mathscr{F}[f_2(x)] \times \text{constant} \tag{9-29}$$

The value of the constant in (9-29) depends on our convention for Fourier transforms; that is, whether we use (9-1) or (9-2) and, if (9-1), which of $f(x)$ and $\tilde{f}(\omega)$ is considered the original function and which the Fourier transform. The student is urged to evaluate the constant in (9-29) for at least one convention. In any case, the important result is that, to within some constant, the Fourier transform of a convolution is the product of the Fourier transforms of the "factors" of the convolution.

An analogous result holds for Laplace transforms; if

$$g(x) = \int_0^x dt\, f_1(t) f_2(x-t)$$

then

$$\mathscr{L}[g(x)] = \mathscr{L}[f_1(x)]\mathscr{L}[f_2(x)]$$

An interesting converse relation holds for Laplace transforms. Suppose

$$\mathscr{L}[f_1] = g_1(s) \qquad \text{and} \qquad \mathscr{L}[f_2] = g_2(s)$$

where the Laplace integrals for $g_1(s)$ and $g_2(s)$ exist for Re $s > \alpha_1$ and Re $s > \alpha_2$, respectively. Then the Laplace transform of the product $f_1 f_2$ is given by

$$\mathscr{L}[f_1 f_2] = \frac{1}{2\pi i}\int_{c-i\infty}^{c+i\infty} g_1(z) g_2(s-z)\, dz$$

where the path of integration is along the line Re $z = c$, with (Re $s - \alpha_2) > c > \alpha_1$. This result may be obtained by substituting the Laplace inversion integral for $f_1(z)$ in the integral $\mathscr{L}[f_1 f_2]$.

The corresponding relation for Fourier transforms is

$$\mathscr{F}[f_1 f_2] = \frac{1}{2\pi}\int_{-\infty}^{\infty} g_1(z) g_2(y-z)\, dz$$

Again, the factor $1/2\pi$ depends on one's convention for Fourier transforms; see remarks after equation (9-29).

Example

These methods can be used to calculate the current in the circuit of Figure 9-5 in a somewhat more direct way. We begin with the equation

$$RI + L\frac{dI}{dt} + \frac{Q}{C} = E_0$$

and write all quantities in terms of the current $I(t)$, the particular unknown desired.

$$RI + L\frac{dI}{dt} + \frac{1}{C}\left[Q_0 + \int_0^t I(t')\, dt'\right] = E_0$$

Now take the Laplace transform of both sides. Let $L[I(t)] = i(s)$:

$$Ri(s) + [si(s) - I(0)] + \frac{1}{C}\left[\frac{Q_0}{s} + \frac{i(s)}{s}\right] = \frac{E_0}{s}$$

Solving for $i(s)$ gives

$$i(s) = \frac{E_0 - \dfrac{Q_0}{C}}{L} \frac{1}{(s + a)^2 + b^2}$$

where

$$a = \frac{R}{2L} \qquad b = \sqrt{\frac{1}{LC} - \frac{R^2}{4L^2}} \qquad [\text{and } I(0) = 0]$$

To invert the transform, we note (see Problem 9-7) that

$$\mathscr{L}(\sin \lambda t) = \frac{\lambda}{s^2 + \lambda^2}$$

and hence [from equation (9-25)]

$$\mathscr{L}[e^{-at} \sin bt] = \frac{b}{(s + a)^2 + b^2}$$

Therefore

$$I(t) = \frac{E_0 - \dfrac{Q_0}{C}}{L} \cdot \frac{e^{-at} \sin bt}{b}$$

Application of transform techniques to partial differential equations is discussed in Chapter 12.

OTHER TRANSFORM PAIRS

Suppose $f(x)$ is an even function. Then

$$\begin{aligned} g(y) &= \int_0^\infty f(x) e^{-ixy}\, dx + \int_{-\infty}^0 f(x) e^{-ixy}\, dx \\ &= \int_0^\infty f(x)(e^{ixy} + e^{-ixy})\, dx \qquad (9\text{-}30) \\ &= 2 \int_0^\infty f(x) \cos xy\, dx \end{aligned}$$

Now note that $g(y)$ is even. Therefore

$$f(x) = \frac{1}{\pi} \int_0^\infty g(y) \cos xy\, dy \qquad (9\text{-}31)$$

$f(x)$ and $g(y)$, which now need only be defined for positive x and y, are called *Fourier cosine transforms* of each other.

By considering the Fourier transform of an odd function, we similarly obtain the relations between *Fourier sine transforms*

$$f(x) = \frac{1}{\pi}\int_0^\infty g(y)\sin xy\,dy \qquad g(y) = 2\int_0^\infty f(x)\sin xy\,dx \tag{9-32}$$

We may symmetrize by putting $\sqrt{2/\pi}$ before each integral in (9-30), (9-31), and (9-32) if we wish.

Other useful transform pairs are:

Fourier-Bessel transforms: (or *Hankel transform*)

$$g(k) = \int_0^\infty f(x)J_m(kx)x\,dx$$

$$f(x) = \int_0^\infty g(k)J_m(kx)k\,dk$$

Mellin transform:

$$\varphi(z) = \int_0^\infty t^{z-1}f(t)\,dt$$

$$f(t) = \frac{1}{2\pi i}\int_{-i\infty}^{i\infty} t^{-z}\varphi(z)\,dz$$

Hilbert transform:

$$g(y) = \frac{1}{\pi}P\int_{-\infty}^{\infty}\frac{f(x)\,dx}{x-y}$$

$$f(x) = \frac{1}{\pi}P\int_{-\infty}^{\infty}\frac{g(y)\,dy}{y-x}$$

P denotes that the (Cauchy) *principal value* of the integral is to be taken. We do not use these other transforms (except for Hilbert transforms) in the remainder of the book; the interested reader should consult the Bateman manuscript (13) and references listed therein.

PROBLEMS

9-1 A linear system is driven by a periodic input $f(t)$, such that $f(t+T) = f(t)$. The response of the system is such that a sinusoidal input of angular frequency ω is multiplied by $(\omega_0/\omega)^2$, unless $\omega = 0$, in which case no output occurs. The output can be written in the form

$$g(t) = \frac{1}{T}\int_0^T G(t-t')f(t')\,dt'$$

Find the function $G(t)$. (*Note*: $T = 2\pi/\omega_0$.)

Problems 9-2 through 9-6 involve Fourier transforms.

9-2 A linear system has a response $G(\omega)e^{-i\omega t}$ to the input signal $e^{-i\omega t}$ (ω arbitrary). If the input $f(t)$ has the particular form

$$f(t) = \begin{cases} 0 & (t < 0) \\ e^{-\lambda t} & (t > 0) \end{cases}$$

where λ is some fixed constant, the output is observed to be

$$F(t) = \begin{cases} 0 & (t < 0) \\ (1 - e^{-\alpha t})e^{-\lambda t} & (t > 0) \end{cases}$$

where α is another fixed constant.

(a) Find $G(\omega)$.

(b) Find the response of the system to the input $f(t) = A\,\delta(t)$.

9-3 Solve the equation

$$\frac{d^2\psi}{dx^2} + \alpha\frac{d\psi}{dx} = \delta(x - x') \qquad (-\infty < x < +\infty)$$

and discuss the role of different boundary conditions in determining the answer.

9-4 Find the Fourier transform of the wave function for a $2p$ electron in hydrogen:

$$\psi(\mathbf{x}) = \frac{1}{\sqrt{32\pi a_0^5}}\, z e^{-r/2a_0}$$

where a_0 = radius of first Bohr orbit and z is a rectangular coordinate. What momentum value is the electron most likely to have?

9-5 Consider a one-dimensional quantum mechanical scattering problem, in which a plane wave incident from the left scatters on a delta function potential. The Schrödinger equation for this situation is

$$-\frac{\hbar^2}{2m}\frac{d^2\psi}{dx^2} + V\,\delta(x - x_0)\psi = E\psi$$

Use Fourier transforms to find the reflected and transmitted waves if the incident wave is normalized to e^{ikx}.

Hint: recall that if $xf(x) = 1$, we have $f(x) = 1/x + A\,\delta(x)$ where A is an arbitrary constant.

9-6 Consider a simple one-loop circuit with a resistance R and an inductance L in series with a source $E(t) = E_0 \sin(\omega_0 t)e^{-\alpha|t|}$ (see Figure 9-8). The source $E(t)$ is assumed to have been operating according to this formula since $t = -\infty$, and to continue until $t = +\infty$. Find the current in the circuit for $t < 0$ and $t > 0$.

Problems 9-7 through 9-15 involve Laplace transforms.

9-7 In practice, if one does a lot of manipulation involving transforms, it is convenient to have a table handy to invert certain commonly occurring

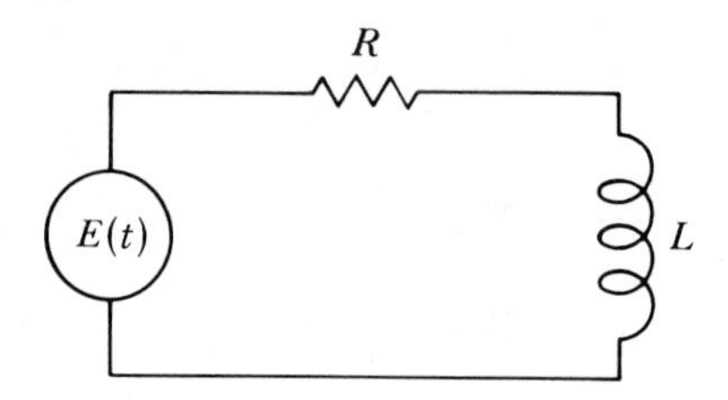

FIGURE 9-8 Series *RL* circuit

Laplace transforms. These tables are available in the standard handbooks for engineers and scientists. For your own use, check the table below:

$f(x)$	$F(s) = \int_0^\infty e^{-sx} f(x)\, dx$
1	$\frac{1}{s}$
$\delta(x - x_0)(x_0 > 0)$	e^{-sx_0}
$\sin \lambda x$	$\frac{\lambda}{s^2 + \lambda^2}$
$\cos \lambda x$	$\frac{s}{s^2 + \lambda^2}$
x^n	$\frac{n!}{s^{n+1}}$
$e^{-\lambda x}$	$\frac{1}{s + \lambda}$

Any $F(s)$ that is a rational function in s may then be decomposed into partial fractions, and the fractions inverted by use of the table. This allows one to utilize the transforms without actually performing any contour integrals at all.

9-8 Find the Laplace transform $\mathscr{L}[f(x)]$ of the function sketched in Figure 9-9.

9-9 Of what function is

$$\frac{1}{(s^2 + 1)(s - 1)}$$

the Laplace transform?

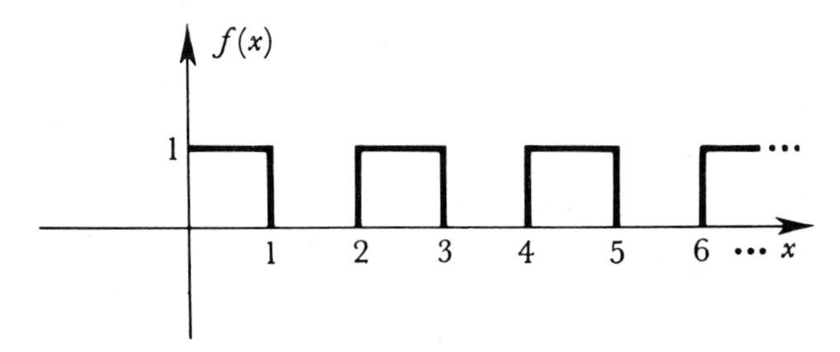

FIGURE 9-9

9-10 Three radioactive nuclei decay successively in series, so that the numbers $N_i(t)$ of the three types obey the equations

$$\frac{dN_1}{dt} = -\lambda_1 N_1$$

$$\frac{dN_2}{dt} = \lambda_1 N_1 - \lambda_2 N_2$$

$$\frac{dN_3}{dt} = \lambda_2 N_2 - \lambda_3 N_3$$

If initially $N_1 = N, N_2 = 0, N_3 = n$, find $N_3(t)$ by using Laplace transforms.

9-11 A function $f(x)$ has the series expansion

$$f(x) = \sum_{n=0}^{\infty} \frac{c_n x^n}{n!}$$

Write the function $g(y) = \sum_{n=0}^{\infty} c_n y^n$ in closed form in terms of $f(x)$.

9-12 By using the integral representation

$$J_0(x) = \frac{1}{2\pi} \int_0^{2\pi} \cos (x \cos \theta) \, d\theta$$

find the Laplace transform of $J_0(x)$.

Check your result by transforming the differential equation obeyed by $J_0(x)$ and solving it in the transform space.

9-13 Consider the system shown in Figure 9-10. A piston of mass m is attached

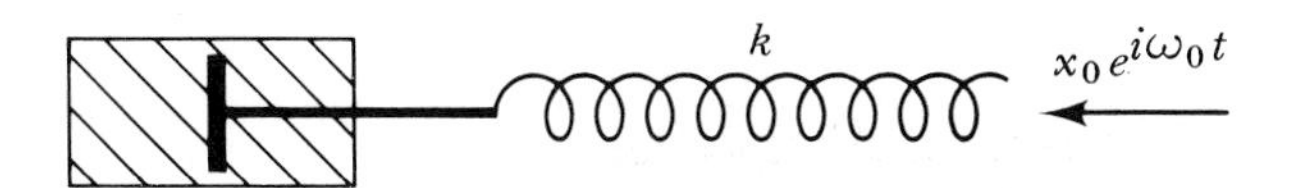

FIGURE 9-10 Oscillator with driving force and damping

to a massless spring with constant k. The piston moves in a viscous fluid, which provides a retarding force proportional to its velocity. The other end of the spring is moved back and forth sinusoidally with frequency ω_0, and amplitude x_0. Assume that at $t = 0$ the piston is at rest at its equilibrium position. Using Laplace transforms, find the subsequent time dependence of the motion. Identify the steady-state and transient pieces of the solution.

9-14 Suppose the battery in Figure 9-5 is replaced with a sinusoidal generator of frequency ω_0. Find the current in the circuit after the switch is closed.

9-15 Consider the circuit shown in Figure 9-11. At $t = 0$ the condenser is

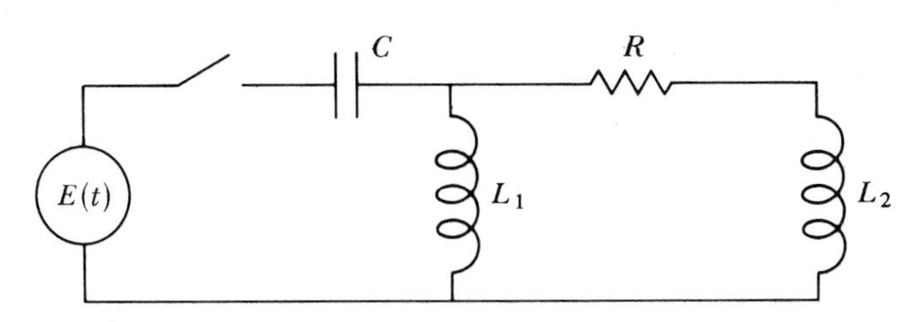

FIGURE 9-11 Circuit for Problem 9-15

charged to Q_0, and no currents are flowing. The switch is then closed. For a source $E(t) = E_0 \sin \omega_0 t$, find the currents for $t > 0$.

9-2 CONFORMAL MAPPING

In Chapter 7 we saw that if $f = u + iv$ is an analytic function of z, then

$$\nabla^2 u = \nabla^2 v = 0$$

and the angle between two curves in the f-plane is the same as the corresponding angle in the z-plane. These facts are a great help in finding solutions to the equation

$$\frac{\partial^2 \phi}{\partial x^2} + \frac{\partial^2 \phi}{\partial y^2} = 0$$

with particular boundary conditions.

This equation arises naturally in electrostatics and magnetostatics. Maxwell's equations in the most general case take the form (in *cgs* units)

$$\vec{\nabla} \cdot \vec{E} = 4\pi\rho$$

$$\vec{\nabla} \times \vec{B} = \frac{4\pi}{c} \vec{J} + \frac{1}{c} \frac{\partial \vec{E}}{\partial t}$$

$$\vec{\nabla} \times \vec{E} = -\frac{1}{c} \frac{\partial \vec{B}}{\partial t}$$

$$\vec{\nabla} \cdot \vec{B} = 0$$

However if there is no time dependence, $\partial\vec{E}/\partial t$ and $\partial\vec{B}/\partial t$ are zero and we have $\vec{\nabla}\times\vec{E}=0$. This allows us to write $\vec{E}=-\vec{\nabla}\phi$ for some potential $\emptyset$. The equation $\vec{\nabla}\cdot\vec{E}=4\pi\rho$ then becomes Poisson's equation $\nabla^2\phi=-4\pi\rho$, and in a region where there are no free charges, we have $\nabla^2\phi=0$. If we have no currents, as well as a static situation, then $\vec{\nabla}\times\vec{B}=0$ and the magnetic field can also be expressed in terms of a magnetic potential $\vec{B}=-\vec{\nabla}\phi_m$, with $\nabla^2\phi_m=0$. In general the equation $\nabla^2\phi=0$ involves variation in all three coordinates x, y, and z. We then need the methods discussed in Chapter 12. For some simple cases, however, the potential has no variation in the third coordinate and the equation reduces to (9-33).

Suppose we are to solve equation (9-33) subject to the condition $\varphi=0$ on some curve in the xy-plane. If we can find a mapping f such that the curve goes into the line $\text{Re}\,f=0$ (or $\text{Im}\,f=0$) then $\varphi=\text{Re}\,f$ (or $\varphi=\text{Im}\,f$) is the desired function. Hence a knowledge of mappings may considerably decrease the effort in solving a problem.

Example

Find a solution of $\nabla^2\varphi(x, y)=0$ in the first quadrant that obeys the boundary condition $\varphi=0$ along the axes $x=0$, $y=0$. Let $z=x+iy$ and $\zeta=z^2$. Then the first quadrant of the z-plane becomes the upper half ζ-plane, with the boundaries of our problem transformed into the real ζ-axis, as shown in Figure 9-12. It is trivial to find an analytic function

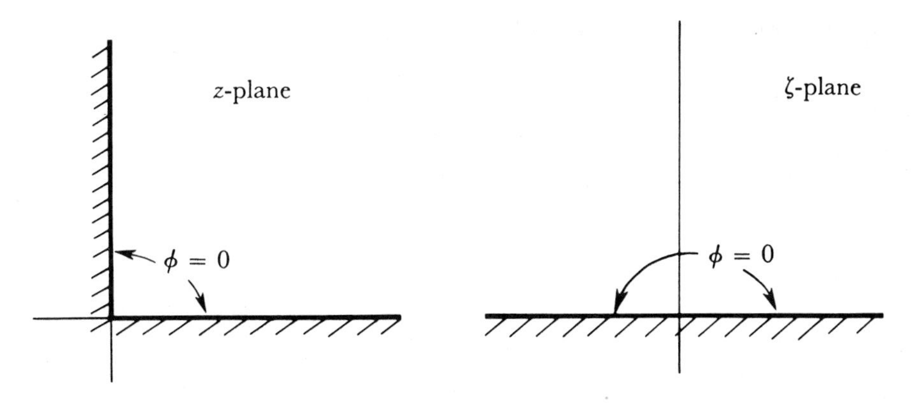

FIGURE 9-12 Application of the conformal transformation $\zeta=z^2$

of ζ whose imaginary part vanishes on the real axis, namely, $f(\zeta)=c\zeta$, where c is an arbitrary real constant. This is also an analytic function of z, whose imaginary part is thus a solution to our original problem

$$\varphi=\text{Im}\,cz^2=2cxy \tag{9-34}$$

As it stands, this example is rather academic, although the solution φ might represent the electrostatic potential near the inside corner of a bent conducting sheet. However, we can apply the result to a problem of considerable practical importance in the design of "optical" systems for focusing charged particles, including the alternating gradient synchrotron. The alternating gradient focusing systems, [E. Courant et al. (9)] make use of magnetic (or electric) fields having the property

$$B_x = Cy \qquad B_y = Cx \qquad (C = \text{constant}) \tag{9-35}$$

If our solution (9-34) for φ is interpreted as a magnetostatic potential, the resulting field $\mathbf{B} = -\nabla\varphi$ is of just the form (9-35). To realize the field in practice, one places a pole piece along one of the hyperbolic equipotential surfaces and puts it at the proper potential. By symmetry, we can place similar pole pieces in the other three quadrants with potentials of alternating sign. The result is a *quadrupole magnet* such as is shown in Figure 9-13. Such a magnet is *focusing* for particles having orbits displaced in the

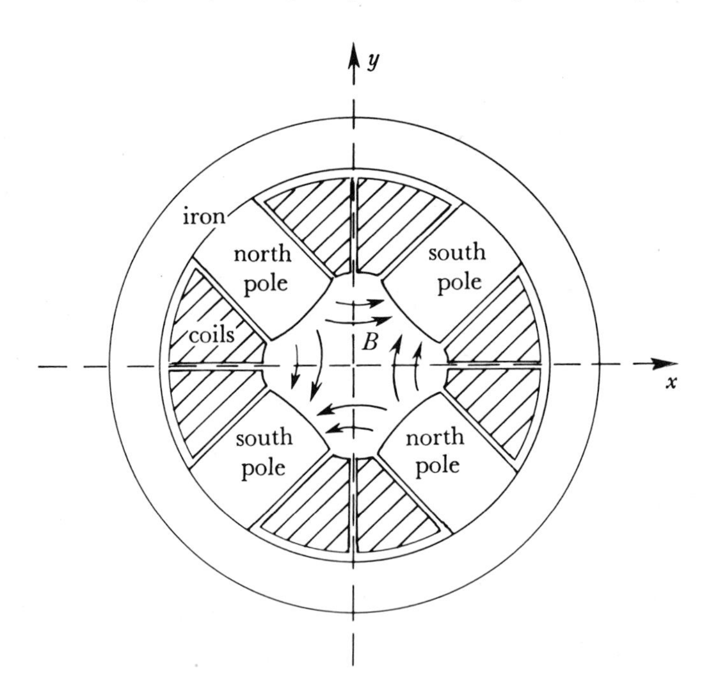

FIGURE 9-13 The cross-section of a quadrupole focusing magnet with hyperbolic pole pieces for producing a field $B_x = C_y$, $By = Cx$. The particles move normal to the plane of the paper.

$\pm y$ directions and *defocusing* for orbits displaced in the $\pm x$ directions. Two such magnets in succession, rotated 90° with respect to each other about their common axis, can give net focusing in both directions. (We are assuming the particles to be *positively* charged and moving *into* the plane of the figure. The student should be able to verify the focusing and defocusing properties, just by using the right-hand rule for forces on moving charges in magnetic fields.)

Returning to conformal transformations, if $V = \operatorname{Im} f(z)$ represents the function of physical interest obeying Laplace's equation, what is the significance of U? The qualitative significance of U should be obvious from the fact that the curves $U = \text{constant}$ and $V = \text{constant}$ are mutually orthogonal. For example, if V is the electrostatic potential, then the field lines, or lines of force, will lie along curves $U = \text{constant}$. Furthermore, the Cauchy-Riemann equations imply

$$U(Q) - U(P) = -\int_P^Q (\nabla V)_n \, ds$$

where P and Q are any two points, and the integral is along any path connecting P and Q. ds is the element of path length and $(\nabla V)_n$ is the component of the gradient normal to the path, the positive direction being to the right as one proceeds from P to Q. Similarly,

$$V(Q) - V(P) = \int_P^Q (\nabla U)_n \, ds$$

If one of the functions U, V represents the potential, the other function is conventionally called the *stream function.*

Now if V is the electrostatic potential, $-(\nabla V)_n = E_n =$ the component of electric field normal to the path. If the path lies along the surface of a conductor (that is, along the intersection of the conductor surface with the x, y plane), then E_n is the electric field normal to the conducting surface, which is proportional to the surface charge density by Gauss's law in electrostatics. Specifically, in mks units, the surface charge density is $\sigma = \varepsilon_0 E_n$, while in cgs units $\sigma = E_n/4\pi$.

By integrating σ along the conducting surface, we may find the total charge in this region (per unit distance in the z-direction). If we define $C(P, Q)$ to be the surface charge between P and Q per unit length normal to the x, y plane, then we have shown (in mks units)

$$C(P, Q) = -\varepsilon_0 \int_P^Q (\nabla V)_n \, ds = \varepsilon_0 [U(Q) - U(P)] \tag{9-36}$$

Example

Use the conformal transformation $\zeta = a \sin z$ to find the potential produced by an infinitely long conducting strip of width $2a$ and charge λ per unit length. The mapping $\zeta = a \sin z$ is shown in Figure 9-14.

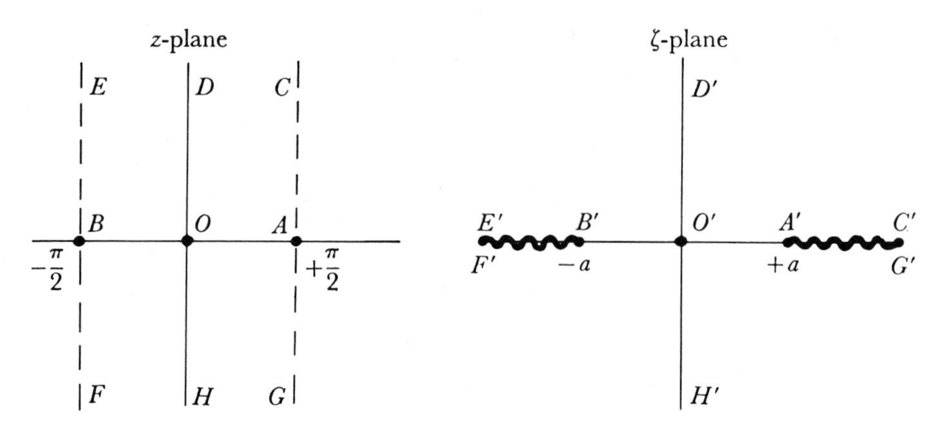

FIGURE 9-14 The mapping $\zeta = a \sin z$. Note that the ζ-plane is "cut" along the real axis for $|\text{Re}\,\zeta| > a$. The vertical strip $-\pi/2 \leq \text{Re}\,z \leq \pi/2$ maps onto the entire ζ-plane, as do infinitely many other vertical strips of width π in the z-plane.

Consider the charged strip to be infinitely long in the direction normal to the ζ-plane, and to occupy the real axis between $-a$ and a in that plane, as in Figure 9-14. The electrostatic potential is constant along the strip, of course, and from symmetry we know that the lines $O'D'$, $B'E'$, and $A'C'$ all lie along lines of force, or field lines. Then we may restate our electrostatics problem as a mathematical exercise: find an analytic function $f(\zeta)$ whose imaginary part (the electrostatic potential) is constant for ζ on the real axis between $-a$ and a, and whose real part (the stream function) is constant for ζ on the real axis outside this strip, and also on the positive and negative imaginary axes (i.e., these are lines of force).

This problem is not easy to solve by inspection, but the conformal transformation $\zeta = a \sin z$ makes it trivial, when we look at the z-plane of Figure 9-14. We now want an analytic function $F(z)[=f(\zeta)]$ whose imaginary part is constant on BA, and whose real part is constant on BE, OD, and AC. The solution is obvious, namely $F(z) = Kz$, where K is a real, but otherwise arbitrary, constant. In terms of the variable ζ, we have

$$f(\zeta) = F(z) = Kz = K \sin^{-1} \frac{\zeta}{a} \tag{9-37}$$

Now we evaluate the constant K, using the fact that the total charge per unit length on the strip is λ. We have already seen [equation (9-36)] that the total charge per unit length on the *top* of the strip, between A' and B' in Figure 9-12, is just ε_0 times the increase in the real part of $f(\zeta)$ as we go from A' to B'. From (9-37), this increase is

$$K \sin^{-1}(-1) - K \sin^{-1}(1) = -K\pi$$

Therefore the total charge per unit length on the top of the strip is

$-K\pi\varepsilon_0$. But, by symmetry, half of the total charge is on the top side of the strip; the other half, of course, is on the bottom. Therefore

$$-K\pi\varepsilon_0 = \frac{1}{2}\lambda \Rightarrow K = -\frac{\lambda}{2\pi\varepsilon_0}$$

With this value of $K, f(\zeta)$ equals $-(\lambda/2\pi\varepsilon_0) \sin^{-1} \zeta/a$, and the electrostatic potential is

$$V = \operatorname{Im} f(\zeta) = -\frac{\lambda}{2\pi\varepsilon_0} \operatorname{Im} \sin^{-1}\frac{\zeta}{a}$$

What do the equipotentials look like? These are lines of constant $\operatorname{Im} f(\zeta)$, or, from (9-37) lines of constant $\operatorname{Im} z$. If we write $z = x + iy$ and $\zeta = \xi + i\eta$, our conformal transformation becomes

$$\xi + i\eta = a \sin(x + iy)$$

or

$$\xi = a \sin x \cosh y$$

$$\eta = a \cos x \sinh y$$

The equipotentials are lines of constant y; eliminating x from these two equations gives

$$\frac{\xi^2}{a^2 \cosh^2 y} + \frac{\eta^2}{a^2 \sinh^2 y} = 1$$

or, since $V = \operatorname{Im} f(\zeta) = \operatorname{Im} F(z) = Ky = -(\lambda/2\pi\varepsilon_0)y$, the equation for the equipotentials may be written

$$\frac{\xi^2}{a^2 \cosh^2\left(\frac{2\pi\varepsilon_0 V}{\lambda}\right)} + \frac{\eta^2}{a^2 \sinh^2\left(\frac{2\pi\varepsilon_0 V}{\lambda}\right)} = 1$$

The equipotentials are ellipses.

Electrostatics is not the only branch of physics where Laplace's equation is of importance. In steady state heat flow, the temperature obeys

$$\nabla^2 T = 0$$

In irrotational fluid flow with no sources or sinks, we can define a function φ (the velocity potential) such that

$$\nabla^2 \varphi = 0$$

and $\mathbf{v} = \nabla\varphi$. For fluid flow the stream function has contours of equal magnitude along the direction given by the velocity $\mathbf{v} = \nabla\varphi$; this is the origin of the name stream function.

These problems provide us with a different kind of boundary condition:

$$\frac{\partial T}{\partial n} = 0$$

at a boundary if there is no heat flow through the boundary (i.e., if the boundary is a perfect insulator); and

$$\frac{\partial \varphi}{\partial n} = 0$$

for the velocity potential at a wall, since the velocity must always be parallel to the wall. This type of boundary condition also fits naturally into the conformal transformation method. Because angles between curves are not changed under conformal transformations, the temperature (or velocity potential) will be required to have zero normal derivative at the boundary in the transformed plane. This is easy to accomplish if, for example, $T = \text{Im}\, f(z)$, and the boundary transforms into a contour $\text{Re}\, F(z) = \text{constant}$.

Example

Find the temperature in the wedge-shaped region illustrated in Figure 9-15

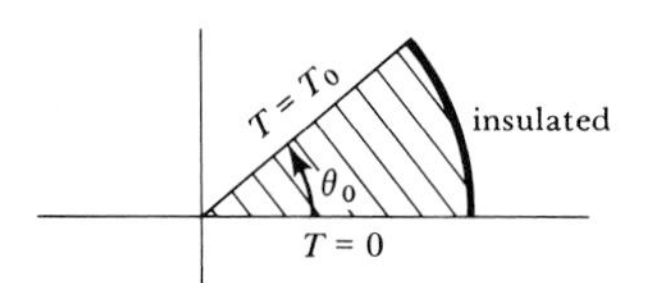

FIGURE 9-15 Region and boundary conditions for a heat flow problem

subject to the conditions specified: $T = 0$ on one edge, $T = T_0$ on the other edge, and insulation on the curved side.

If we first apply the transformation

$$\omega = z^{\pi/\theta_0}$$

the wedge is spread out into a semicircle. This is a convenient first step, as it allows us to use the entire upper half plane as a domain for further transformations. Transformations may easily be classified and remembered according to their action on the upper half plane, so one almost always begins by putting the problem into "standard" form, i.e., mapping the region under consideration into the upper half plane.

The transformation

$$t = \ln w = \frac{\pi}{\theta_0} \ln z = \frac{\pi}{\theta_0} \ln r + i\pi \frac{\theta}{\theta_0}$$

maps the wedge into the region shown in Figure 9-16. Clearly the identification

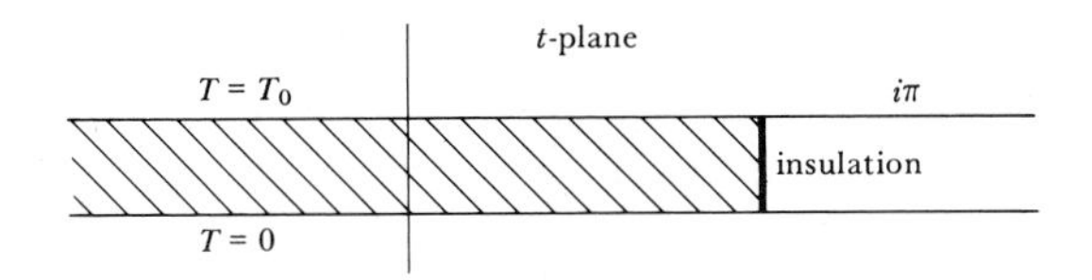

FIGURE 9-16 Image of Figure 9-13 under $t = (\pi/\theta_0) \ln z$

$$T = \frac{T_0}{\pi} \operatorname{Im} t = \frac{T_0 \theta}{\theta_0}$$

leads to a temperature function that satisfies all the boundary conditions as well as Laplace's equation. Thus

$$T = \frac{T_0}{\theta_0} \tan^{-1}\left(\frac{y}{x}\right)$$

PROBLEMS

9-16 For fluid flow at a right angle corner, find the equation for the stream lines shown in Figure 9-17 and plot a few.

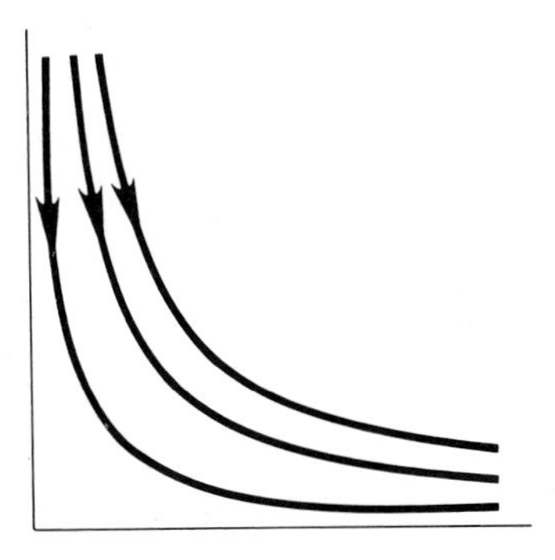

FIGURE 9-17

9-17 Show that the velocity potential for fluid flow past a right circular cylinder of radius 1 is given by

$$\varphi = K\left(r + \frac{1}{r}\right) \cos \varphi$$

($r = 0$ at the center of the circle).

9-18 Consider the action of the map

$$w = z + \frac{1}{z}$$

on the following domain: a circle with its center at a point x_0 on the x-axis, $0 < x_0 < 1$, and passing through the point $x = -1$ (see Figure 9-18). Plot some points and show that the image of the circle is shaped

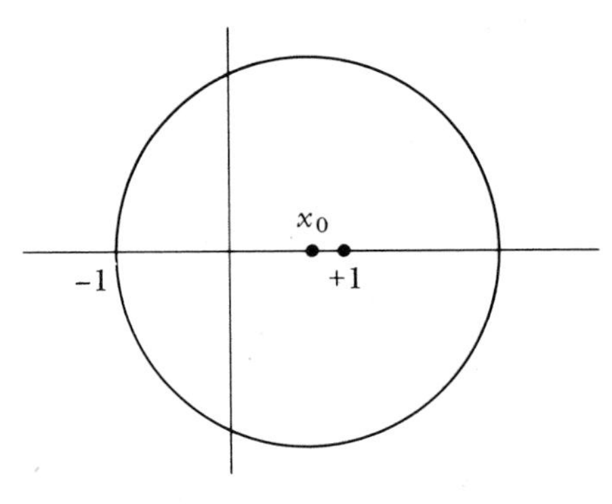

FIGURE 9-18

something like the cross section of an airplane wing (the Joukowski airfoil).

9-19 Using the results of Problems 9-17 and 9-18, find the air velocity around an airplane wing if the velocity at a very large distance from the wing is some constant $\mathbf{v}$.

Hint: If $\phi = \text{Re } F(z)$ is the velocity potential, then

$$\frac{dF}{dz} = \frac{\partial \text{ Re } F}{\partial x} + i\frac{\partial \text{ Im } F}{\partial x} = \frac{\partial \text{ Re } F}{\partial x} - i\frac{\partial \text{ Re } F}{\partial y}$$

may be interpreted as a two-component vector $(\partial\phi/\partial x, -\partial\phi/\partial y)$ related to the velocity vector $\vec{\mathbf{v}} = \vec{\nabla}\phi = (\partial\phi/\partial x, \partial\phi/\partial y)$. This simplifies calculation of the velocity once the appropriate conformal transformation has been found.

9-20 Verify that the solution of the example of Figure 9-14 approaches the potential of a line charge λ at large distances; i.e., for large values of $|\zeta|$.

9-21 Use the method of conformal transformations to find the electrostatic potential arising from a series of coplanar charged strips, each of width $2a$ separated by spaces of width $2b$, and each with charge λ per unit length. (See Figure 9-19.)

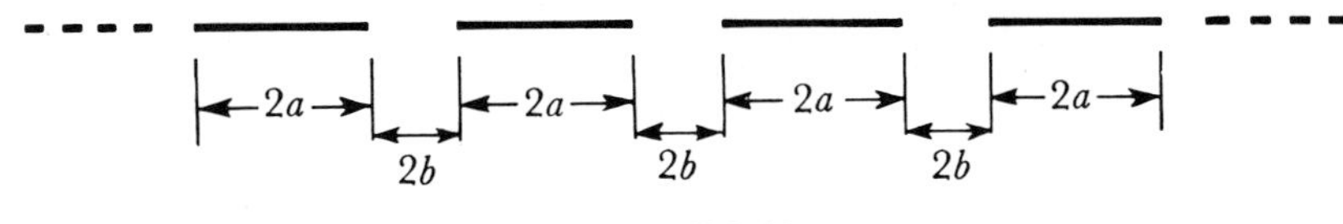

FIGURE 9-19

9-22 Line charges $\pm\lambda$ are placed at the points $(0, \pm d)$, respectively. If $\lambda \to \infty$, $d \to 0$ in such a way that $p = 2\lambda d =$ constant, find the function $W(z)$ whose real part gives the electrostatic potential in the xy-plane.

9-23 Consider a line charge λ at $(0, a)$ between the two grounded planes $y = 0$ and $y = b\ (0 < a < b)$. Find the potential everywhere between the planes. (*Hint:* The transformations $\zeta = e^{z\pi/b}$ is useful.)

9-24 (a) Consider the problem of finding the potential distribution in the upper half plane, if the potential along the x-axis is fixed thus:

$$\phi(y=0) = \begin{cases} 0 & x < -a \\ V_0 & -a < x < a \\ 0 & x > a \end{cases}$$

Show that this is the same problem as finding the potential due to two line charges, if we exchange the roles of potential and stream function. The solution is

$$\phi = \frac{V_0}{\pi}\left(\tan^{-1}\frac{y}{x-a} - \tan^{-1}\frac{y}{x+a}\right)$$

(b) With the aid of the solution to part (a), and the conformal transformation

$$\zeta = \sinh\frac{\pi z}{d}$$

find the temperature distribution in the semi-infinite metal plate shown in Figure 9-20 if the three faces are maintained at the indicated temperatures. Note that $\nabla^2 T = 0$ inside the metal.

9-25 Find the temperature distribution inside a long wall with sides maintained at fixed temperatures as shown in Figure 9-21 by first mapping the wall into the upper half plane as in Figure 9-22(a) and then into the strip shown in Figure 9-22(b). Compare with your result from Problem 9-24.

9-26 Consider an array of many parallel semi-infinite conducting plates, alternately charged to potentials $+V_0$ and $-V_0$ as shown in Figure 9-23.

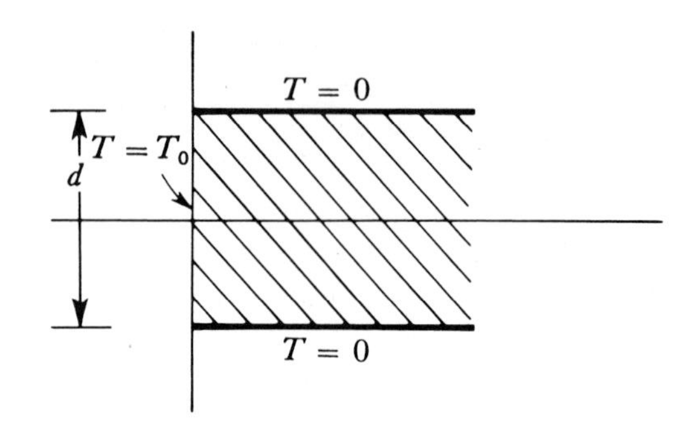

FIGURE 9-20

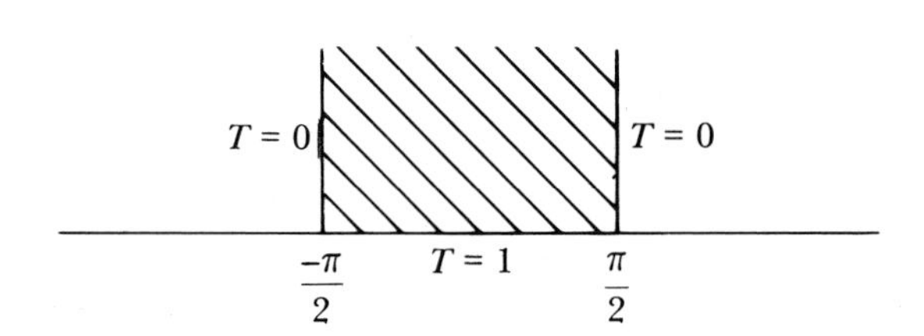

FIGURE 9-21

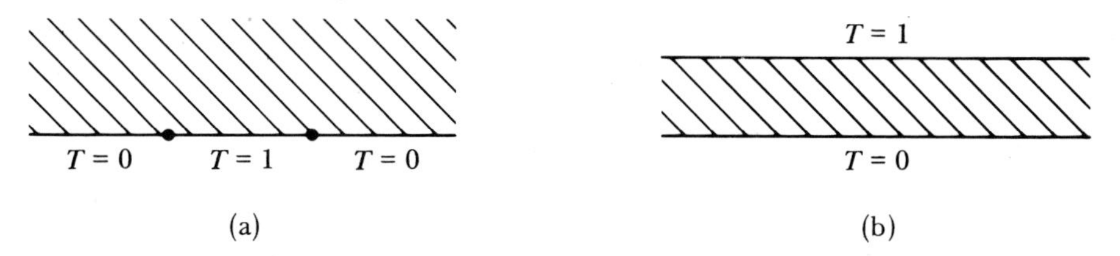

FIGURE 9-22

(a) Show that the conformal transformation

$$\zeta = \frac{2V_0}{\pi} \sin^{-1} e^{\pi z/d}$$

reduces the problem to a trivial one in the ζ-plane.

(b) Find the potential distribution from this array.

9-27 *The Schwartz Transformation*

(a) Consider a mapping $\zeta = \zeta(z)$ such that

$$\frac{d\zeta}{dz} = A(z - a_1)^{s_1}(z - a_2)^{s_2}$$

where a_1, a_2, s_1, s_2 are arbitrary real constants, while A is an arbitrary complex constant. Show that, as z moves along the real axis, ζ also moves in a straight line, except for sudden changes of direction when

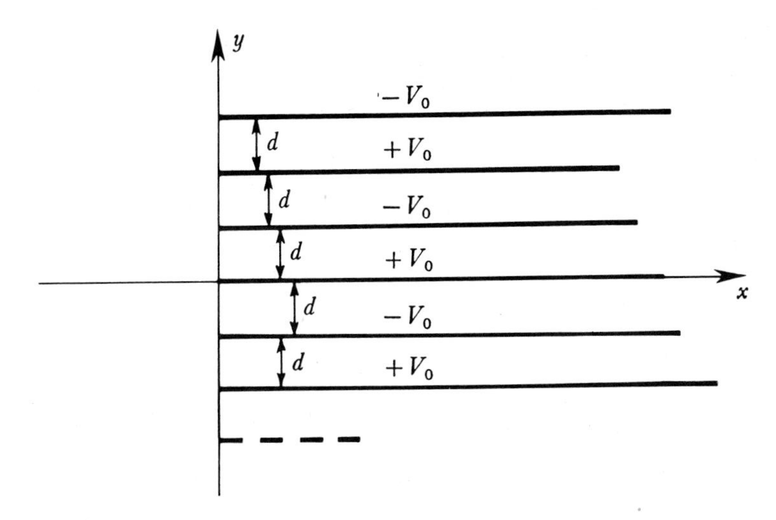

FIGURE 9-23

$z = a_1$ or $z = a_2$. As z passes (infinitesimally above) a_1 from left to right, show that ζ turns to the right through an angle πs_1, and similarly by πs_2 when z passes above a_2 (see Figure 9-24). Thus the

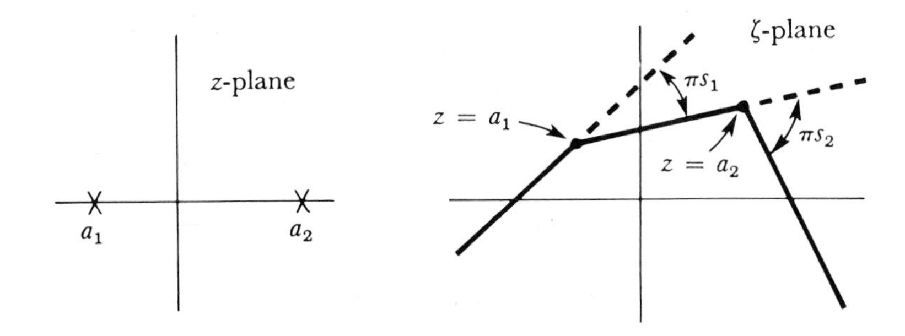

FIGURE 9-24

upper half z-plane maps into a region bounded by an arbitrary polygon in the ζ-plane. This method can clearly be generalized to an arbitrary number of corners.

(b) Apply the method of (a) to find the mapping $\zeta(z)$ that carries the upper half z-plane into the interior of the rectangular trough that is shaded in Figure 9-21. Compare your result with Figure 9-14 and the associated example.

(c) Apply the method of (a) to find the mapping that carries Figure 9-22(a) into Figure 9-22(b), and compare with the result you obtained in Problem 9-25 by more intuitive methods.

SUMMARY

The Fourier transform $\mathscr{F}[f(x)]$ is related to $f(x)$ by

$$g(y) = \mathscr{F}[f(x)] = \int_{-\infty}^{\infty} f(x)e^{-iyx}\,dx \qquad f(x) = \frac{1}{2\pi}\int_{-\infty}^{\infty} g(y)e^{iyx}\,dx$$

Its properties under differentiation and integration of f are simple:

$$\mathscr{F}[f'(x)] = iy\mathscr{F}[f(x)] \qquad \mathscr{F}\left[\int f(x)\,dx\right] = \frac{\mathscr{F}[f(x)]}{iy} + C\delta(y)$$

Similarly the Laplace transform $\mathscr{L}[f(x)]$ is related to f by

$$F(s) = \mathscr{L}[f(x)] = \int_{0}^{\infty} f(x)e^{-sx}\,dx \qquad f(x)H(x) = \frac{1}{2\pi i}\int_{C} F(s)e^{sx}\,dx$$

where $H(x)$ is the Heaviside step function and C is a contour parallel to the imaginary s-axis and to the right of all singularities of $F(s)$ The Laplace transform also has simple properties:

$$\mathscr{L}[f'(x)] = s\mathscr{L}[f(x)] - f(0) \qquad \mathscr{L}\left[\int_0^x f(t)\,dt\right] = \frac{1}{s}\mathscr{L}[f(t)]$$

This simple behavior under differentiation and integration often allows us to solve a complicated differential equation by algebraic manipulation in the transform space. (Remember to transform the boundary conditions also!) Complications sometimes arise in inverting the transform, however.

The real and imaginary parts of an analytic function $f = U + iV$ satisfy Laplace's equation and therefore may provide solutions to the problem $\nabla^2\varphi = 0$ in two dimensions with particular boundary conditions. The boundary conditions determine which function f is applicable. If $\phi = 0$ along the boundary, then we look for a mapping such that the boundary curve goes into the line $\text{Im } f = 0$. ($\text{Re } f = 0$). Then $\phi = \text{Im } f = V$ ($\phi = U$) is the desired solution. If $\partial\phi/\partial n = 0$ along the boundary, the boundary curve must map into the line $\text{Im } f = \text{constant}$ ($\text{Re } f = \text{constant}$). Then the solution is $\phi = \text{Re } f = U$ ($\phi = \text{Im } f = V$).

CHAPTER 10 Uses of Complex Variable Theory III: Special Topics

In this chapter we discuss in some detail three different places where use of complex variable techniques leads to results of use in advanced physics problems. These topics are meant to give the student some idea of the broad range of applicability of the tools learned in previous chapters. The treatment is not exhaustive; rather we try to give a solid grasp of the basic ideas with perhaps one or two excursions into more exotic areas.

Dispersion relations are integral equations that must be satisfied by analytic functions. In practice one can often measure only the imaginary part of an analytic function such as a scattering amplitude; dispersion relations may then be used to calculate the real part. Section 10-1 is devoted to the derivation of these equations, and to related applications of Cauchy's theorem.

Many functions defined by series or other means can be re-expressed in terms of contour integrals. Often the integral form is more convenient for analytic continuation in one of the parameters of the function, or for manipulations involving the generating function of a set of polynomials. In Section 10-2 we discuss properties of the gamma function, Legendre functions, Bessel functions, and the function $\mathrm{Ei}(x)$ and $\mathrm{erf}(x)$, which can be obtained by manipulations with various integral representations.

Finally, we apply the method of steepest descent to derive (in Section 10-3) the WKB connection formulae. These relate approximate solutions of differential equations in regions where the approximations take different forms. The connection formulae are used in quantum mechanics to relate the wave functions inside a region where the potential is important to the waves outside.

10-1 DISPERSION RELATIONS

Cauchy's integral formula (7-14) expresses the remarkable result that the values of an analytic function $f(z)$ along a closed curve within its region of regularity determine the values of $f(z)$ everywhere inside this curve. Namely,

$$f(z) = \frac{1}{2\pi i} \int_C \frac{f(\zeta)}{(\zeta - z)} d\zeta$$

where C is any closed contour within and on which $f(z)$ is regular and z is any point inside C. This fact can be used to derive integral relations between the real and imaginary parts of $f(x)$, for x real, provided $f(z)$ and C satisfy certain criteria. These integral relations, called dispersion relations, often provide important constraints on functions chosen to represent physical quantities.

Consider first the elementary situation where $f(z)$ has no singularities in the upper half plane and $f(z) \to 0$ when $|z| \to \infty$. We then apply Cauchy's

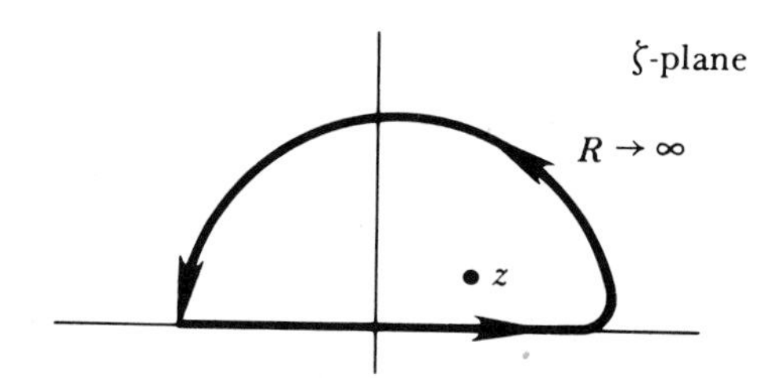

FIGURE 10-1 Contour used to derive the dispersion relations (10-2)

formula with a contour consisting of the real axis and a large upper semicircle, as in Figure 10-1. The integral on the semicircle vanishes because, as $R \to \infty$,

$$\left| \int_{\text{semicircle}} \frac{f(\zeta)}{(\zeta - z)} d\zeta \right| \le |f(\zeta)|_{\max} \left| \int \frac{Re^{i\theta} i \, d\theta}{Re^{i\theta}} \right| = |\pi i| \, |f(\zeta)|_{\max} \to 0$$

where $|f(\zeta)|_{\max}$ is the maximum magnitude of $f(\zeta)$ on the semicircle, and this tends to zero as $R \to \infty$. Thus

$$f(z) = \frac{1}{2\pi i} \int_{\text{real axis}} \frac{f(\zeta)}{\zeta - z} d\zeta \qquad \text{Im } z > 0 \tag{10-1}$$

Suppose further that the function of physical interest is $f(z)$ on the real axis, or, in case there is a branch line, the limit as z approaches the real axis from above. That is, calling this function $F(x)$,

$$F(x) = \lim_{\varepsilon \to 0} f(x + i\varepsilon)$$

where ε is positive. Then it follows from (10-1) and (7-28) that

$$2\pi i F(x) = \lim_{\varepsilon \to 0} \int_{-\infty}^{\infty} \frac{f(x')}{x' - x - i\varepsilon}\, dx' = P \int_{-\infty}^{\infty} \frac{F(x')}{x' - x}\, dx' + \pi i F(x)$$

where P designates the Cauchy principal value.

Thus

$$F(x) = \frac{1}{\pi i} P \int_{-\infty}^{\infty} \frac{F(x')}{x' - x}\, dx'$$

and, equating real and imaginary parts,

$$\begin{aligned} \text{Re}\, F(x) &= \frac{1}{\pi} P \int_{-\infty}^{\infty} \frac{\text{Im}\, F(x')}{x' - x}\, dx' \\ \text{Im}\, F(x) &= -\frac{1}{\pi} P \int_{-\infty}^{\infty} \frac{\text{Re}\, F(x')}{x' - x}\, dx' \end{aligned} \tag{10-2}$$

These are *dispersion relations* for the function of physical interest $F(x)$.

The name (dispersion relations) comes from the fact that relations of this sort were first derived by Kronig and by Kramers in the theory of optical and X-ray dispersion. In optics, a dispersion relation is an integral relationship between the refractive part and the absorptive part of the refractive index at different frequencies. Since these are the real and imaginary parts, respectively, of the (complex) refractive index, we apply the term dispersion relation to any such integral relationship between the real and imaginary parts of a function of a complex variable.

Dispersion relations now have extensive applications in the theory of elementary particle interactions, where they are applied to scattering amplitudes. This application is closely related to the original one since the refractive index of light is closely related to its scattering amplitude. Other applications are found in electrical circuit analysis. For example, see Seshu and Balabanian (30), Chapter 7.

In this context, we can understand the physical basis of some of the assumptions made in deriving (10-2). The "response" $\varphi(t)$ of a linear system to an "input" $g(t)$ may be written in the form

$$\varphi(t) = \int dt'\, K(t - t') g(t')$$

where $K(t)$ is the Green's function of the system. The requirement of *causality*, namely, that no response shall precede its input, means that $K(t) = 0$ for $t < 0$. The functions $f(z)$ being discussed are Fourier transforms of such Green's functions, and the condition $K(t) = 0$ for $t < 0$ implies that the Fourier transform $f(z) = \int dt\, e^{itz} K(t)$ of $K(t)$ has no singularities for $\text{Im}\, z > 0$, and that $f(z) \to 0$ for $|z| \to \infty$ in the upper half z-plane.

Example

If the function $f(z)$ has no singularities within the upper half plane and drops off as $|z| \to \infty$, then our evaluation of its Fourier transform is straightforward:

$$K(t) = \frac{1}{2\pi} \int_{-\infty}^{\infty} e^{-izt} f(z)\, dz$$

For $t < 0$, we close the contour of integration in the upper half z-plane. By Jordan's lemma we add nothing by including the semicircle at infinity. Because there are no singularities in the upper half z-plane, the contour can be shrunk to zero using Cauchy's theorem, and

$$K(t) = 0 \qquad t < 0$$

Conversely, any function $f(z)$ with no singularities in the upper half plane, such that $|f(z)| \to 0$ as $|z| \to \infty$, can be written as

$$f(z) = \int_0^{\infty} K(t) e^{izt}\, dt$$

Example

Consider a wave of definite frequency propagating in a homogeneous dielectric medium. It takes the form

$$e^{-i\omega[t-(n/c)z]}$$

where $n = \text{Re}(n) + i\,\text{Im}(n)$ is the complex index of refraction. The imaginary part of n allows for absorption of the wave by the medium. For a small dielectric slab of thickness d, the electric field coming out is related to the electric field going in by

$$E_{\text{out}}^{\omega}(d) = E_{\text{in}}^{\omega}(o)\, e^{i\omega(n/c)d}$$

Now if we have, for a general field which is a superposition of many frequencies, this response

$$E_{\text{out}}(t) = \int K(t-t') E_{\text{in}}(t')\, dt'$$

then for each frequency the convolution theorem tells us that

$$E_{\text{out}}(\omega, d) = \tilde{K}(\omega, d)\, E_{\text{in}}(\omega, o)$$

Hence $\tilde{K}(\omega, d) = e^{i\omega(n/c)d}$, and our Green's function for the system is

$$K(t-t') = \frac{1}{2\pi} \int_{-\infty}^{\infty} e^{i\omega(n/c)d}\, e^{-i(t-t')\omega}\, d\omega \tag{10-3}$$

Causality for this case requires that the signal not propagate more rapidly than the speed of light. Hence $K(t - t') = 0$ if $t < t' + d/c$. By rewriting the integrand of equation (10-3) in the form

$$e^{i\omega(d/c)(n-1)}e^{-i\omega[t-t'-(d/c)]}$$

we see that this causality condition requires that $e^{i\omega(d/c)(n-1)}$ have no singularities in the upper half ω-plane. Further arguments [see, for example Nussenzweig (28) p. 19] can then be used to show that $n(\omega)$ itself has no singularities in the upper half plane. Hence the Kramers-Kronig relation between Re n and Im n can be written.

The refractive index of a medium can be related to the scattering amplitude in the medium by

$$n = 1 + \frac{2\pi N}{k^2} f(0)$$

where N is the number of scattering centers per unit volume. Hence the forward scattering amplitude of light by the medium will also obey a dispersion relation.

Example

Consider the quantum mechanical scattering, in three dimensions, of a wave from a small source. As $r \to \infty$, the net wave for each frequency (incident plus scattered) takes the form

$$e^{ikz-i\omega t} + f(\omega, \theta)\frac{e^{ikr}}{r} e^{-i\omega t}$$

where $k = k(\omega)$ in general [see, for example, Merzbacher's *Quantum Mechanics* (26) or any other quantum mechanics book at that level]. For simplicity let us consider a beam of light so that $\omega = ck$. Suppose that we have scattering of a well-defined wave packet made up of many frequencies in order that the wave be localized in space. If $Q(\omega)$ is a measure of the amount of each frequency present, we require that

$$\int_{-\infty}^{\infty} Q(\omega)\, e^{i(\omega/c)z-i\omega t}\, d\omega = 0 \qquad \text{if } z > tc$$

Clearly, $Q(\omega)$ has no singularities in the upper half ω-plane.

Suppose that the scattering potential acts over a very small distance, so we may neglect the difference between times of scattering from different parts of the potential. Clearly, in this case, the outgoing wave

$$\int Q(\omega) f(\omega, \theta) \frac{e^{i\omega(r/c)}}{r} e^{-i\omega t} \, d\omega$$

must not appear for $t < r/c$. Hence $Q(\omega) f(\omega, \theta)$ must not have any singularities in the upper half plane. Since $Q(\omega)$ has no singularities there, we see that $f(\omega, \theta)$ cannot have any either. More generally, one can show [see Hilgevoord (17) p. 62] that regardless of the range of the scatterer the forward amplitude $f(\omega, 0)$ has no singularities in the upper half plane and can be shown to obey a dispersion relation. Further developments of these ideas lead to the dispersion relations used in particle physics.

In most applications, relations (10-2) are not in a form suitable for comparison with experiment. The reason is that the variable x usually has physical significance only for $x > 0$ or even $x > x_1 \geq 0$, so that the integrals above extend over a large "nonphysical" region. For example, x may be the energy of a scattering particle, which cannot be less than the rest mass. In this situation, some symmetry of $f(z)$ may be used to relate the integral over a nonphysical region to that over a physical region.

A simple case occurs if the physical range of x is $x \geq 0$ and if $f(z)$ has "even" or "odd" symmetry, defined by $f(-z) = \pm f^*(z^*)$ respectively. For the "even" symmetry,

$$F(x) = \frac{1}{\pi i} P \int_\infty^0 \frac{F^*(x')}{x' + x} \, dx' + \frac{1}{\pi i} P \int_0^\infty \frac{F(x')}{x' - x} \, dx'$$

$$\text{Re } F(x) = \frac{2}{\pi} P \int_0^\infty \frac{x' \text{ Im } F(x')}{(x')^2 - x^2} \, dx'$$

$$\text{Im } F(x) = -\frac{2}{\pi} P \int_0^\infty \frac{x \text{ Re } F(x')}{(x')^2 - x^2} \, dx'$$

The optical dispersion relations have this form, where x is the optical frequency and $F(x)$ is the refractive index. Similar results hold for the "odd" case.

Thus far, we have focused attention on dispersion relations in their historical form. A related point is the way in which certain analytic functions may be specified uniquely by describing their singularities and behavior at infinity. Along a branch line, the information required is simply the discontinuity across the branch line, that is, the difference in the values of the function on opposite sides of the cut.

Some of the conditions under which such a representation is possible will become clear in the examples below.

Example

Suppose we have the following information about a function $f(z)$:

1. $f(z)$ is analytic everywhere except for a pole of residue 1 at $z = 0$, and branch lines from $+1$ to $+\infty$ and -1 to $-\infty$ along the real axis.

2. $f(z) \to 0$ as $|z| \to \infty$.

3. $f(z)$ is real on the real axis from -1 to $+1$.

It follows from the *Schwartz reflection principle* (p. 159) that $f(z^*) = f^*(z)$.

Now apply the Cauchy formula along the contour shown in Figure 10-2. As in our previous example, the integral around the large circle

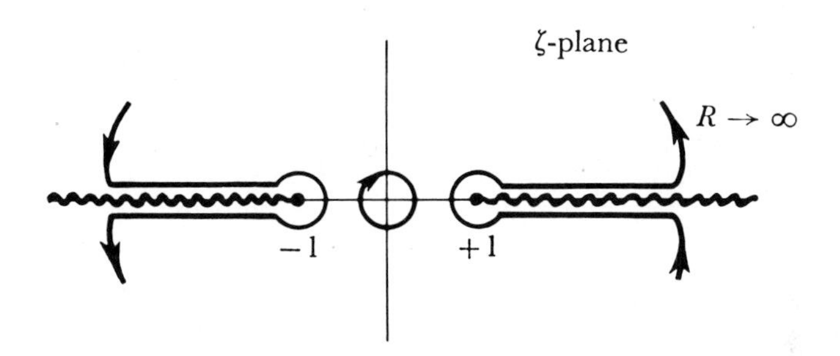

FIGURE 10-2 Contour used to obtain equation (10-4)

vanishes as $R \to \infty$. We shall assume that near the branch points ± 1, $f(z)$ behaves in such a way that the integrals on the two little circles around these points also vanish. Then

$$2\pi i f(z) = 2\pi i\left(\frac{1}{z}\right) + \lim_{\varepsilon \to 0}\left[\int_1^\infty \frac{f(x' + i\varepsilon)\,dx'}{x' + i\varepsilon - z} - \int_1^\infty \frac{f(x' - i\varepsilon)\,dx'}{x' - i\varepsilon - z} + \int_{-\infty}^{-1} \frac{f(x' + i\varepsilon)\,dx'}{x' + i\varepsilon - z} - \int_{-\infty}^{-1} \frac{f(x' - i\varepsilon)\,dx'}{x' - i\varepsilon - z}\right]$$

Combining the first two integrals and the last two, we see that only the discontinuity of $f(z)$ across the cuts is involved, as mentioned before. The discontinuity is supplied in this example by the reflection symmetry $f(x' - i\varepsilon) = f^*(x' + i\varepsilon)$, from which

$$f(x' + i\varepsilon) - f(x' - i\varepsilon) = 2i \operatorname{Im} f(x' + i\varepsilon)$$

Thus

$$f(z) = \frac{1}{z} + \frac{1}{\pi}\int_1^\infty \frac{\operatorname{Im} F(x')}{x' - z}\,dx' + \frac{1}{\pi}\int_{-\infty}^{-1} \frac{\operatorname{Im} F(x')}{x' - z}\,dx' \qquad (10\text{-}4)$$

where $F(x) = \lim_{\varepsilon \to 0} f(x + i\varepsilon)$ as before. This completes the representation of $f(z)$ in terms of its prescribed singularities.

If $F(x)$ is the function of physical interest, we can write a dispersion relation for it by going to the limit $z = x + i\varepsilon \to x$.

$$F(x) = \frac{1}{x} + \frac{1}{\pi} P \int_1^\infty \frac{\operatorname{Im} F(x')}{x' - x} dx' + \frac{1}{\pi} P \int_{-\infty}^{-1} \frac{\operatorname{Im} F(x')}{x' - x} dx' + \frac{1}{\pi} \pi i \operatorname{Im} F(x)$$

If $|x| > 1$, the last term comes from the singularity in one of the integrals. If $|x| < 1$, $\operatorname{Im} F(x) = 0$ by assumption, and we make no error by including the last term. In any case, we obtain

$$\operatorname{Re} F(x) = \frac{1}{x} + \frac{1}{\pi} P \int_{-\infty}^{-1} \frac{\operatorname{Im} F(x')}{x' - x} dx' + \frac{1}{\pi} P \int_1^\infty \frac{\operatorname{Im} F(x')}{x' - x} dx \qquad (10\text{-}5)$$

We do not obtain by this procedure the "other" dispersion relation giving $\operatorname{Im} F(x)$ in terms of an integral over $\operatorname{Re} F(x)$.

If the physical region of the variable x is $x > 1$, we still need a symmetry principle to relate $\operatorname{Im} F(x)$ for $x < -1$ to $\operatorname{Im} F(x)$ for $x > 1$. This symmetry relation must be found in the physics of the problem.

In obtaining the representation (10-4) we made use of the special contour chosen. However, the dispersion relation (10-5) could also have been obtained

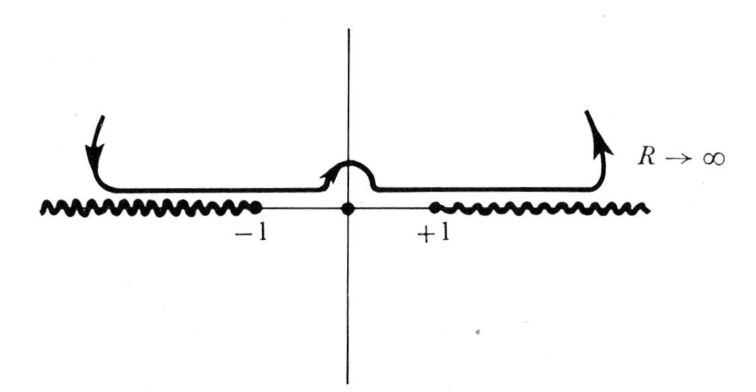

FIGURE 10-3 Alternative integration contour

by using Cauchy's formula along the contour shown in Figure 10-3, which gives

$$2\pi i f(z) = P \int_{-\infty}^{\infty} \frac{f(x')}{x' - z} dx' - \pi i \left(\frac{1}{-z} \right)$$

Let $z = x + i\varepsilon$ and $\varepsilon \to 0$. Then

$$2F(x) = \frac{1}{x} + \frac{1}{\pi i} \left[P \int_{-\infty}^{\infty} \frac{F(x')}{x' - x} dx' + \pi i F(x) \right]$$

$$F(x) = \frac{1}{x} + \frac{1}{\pi i} P \int_{-\infty}^{\infty} \frac{F(x')}{x' - x} dx'$$

This result contains (10-5) as its real part.

Example

Find a function $f(z)$ that has the following properties:

1. $f(z)$ is analytic except for a branch line from $z = 0$ to $+\infty$ along the real axis, and a simple pole of residue 1 at $z = -1$.
2. $f(z) \to 0$ as $|z| \to \infty$.
3. $f(z)$ is real on the negative real axis.
4. For $x > 0$, $\text{Im } f(x + i\varepsilon) = 1/(1 + x^2)$.

We apply Cauchy's formula along the contour shown in Figure 10-4.

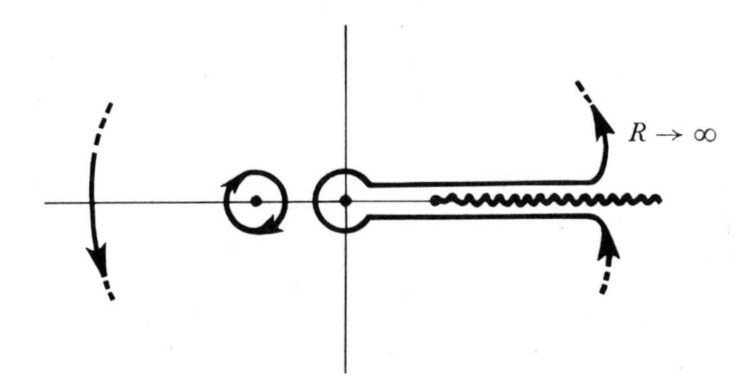

FIGURE 10-4 Integration contour for example

The result may be written down from our work in the previous example.

$$f(z) = \frac{1}{z+1} + \frac{1}{\pi}\int_0^{\infty} \frac{dx'}{(1 + x'^2)(x' - z)}$$

The integral is straightforward and yields the result

$$f(z) = \frac{1}{z+1} - \frac{1}{\pi}\frac{\ln z}{(1 + z^2)} - \frac{1}{(1 + z^2)}\left(\frac{z}{2} - i\right) \tag{10-6}$$

A little care is required with the multivalued function $\ln z$. The answer given in (10-6) assumes that $\ln z$ is real on the top side of the cut.

SUBTRACTIONS

We conclude this section with a description of a modification that may be made in order to obtain dispersion relations for a function that does not satisfy the condition $f(z) \to 0$ as $|z| \to \infty$.

If $(1/z)f(z) \to 0$ as $|z| \to \infty$ and $(1/z)f(z)$ also satisfies other conditions such as those in the above examples, then a dispersion relation for $f(z)$ may be obtained from one for $(1/z)f(z)$.

A convergence factor $1/(z - x_0)$ will serve just as well as $1/z$, of course.

The pole introduced by this factor at $z = x_0$ [or a pole already present in $f(z)$] may be removed by subtracting a pole with the same residue at the same point. For example, a dispersion relation for the function $[f(z) - f(x_0)]/(z - x_0)$ corresponding to (10-2) is

$$\text{Re}\left\{\frac{1}{x - x_0}[F(x) - F(x_0)]\right\} = \frac{1}{\pi} P \int_{-\infty}^{\infty} \frac{\text{Im } F(x') - \text{Im } F(x_0)}{(x' - x_0)(x' - x)} dx'$$

Generally, $\text{Im } F(x_0) = 0$, in which case

$$\text{Re } F(x) = \text{Re } F(x_0) + \frac{(x - x_0)}{\pi} P \int_{-\infty}^{\infty} \frac{\text{Im } F(x')}{(x' - x_0)(x' - x)} dx' \tag{10-7}$$

This is called a subtracted dispersion relation. The subtraction procedure improves the convergence of the dispersion integral, as is easily seen in the above example.

PROBLEMS

10-1 By writing a dispersion relation for $\ln [f(z)]$, derive relations similar to equation (10-2) for the magnitude and phase of the function.

10-2 In Chapter 8, we discussed the application of transform techniques to circuit problems. The Laplace transformed loop equations take the form

$$Z_{ij}(s)I_j(s) = E_i(s)$$

so

$$I_j(s) = (Z^{-1})_{jk} E_k(s)$$

Where are the poles of $I_j(s)$ in the s-plane? In the frequency plane? What does this tell us about the relation between the magnitude and phase of the current (see Problem 10-1)?

10-3 A function $f(z)$ has the following properties:

(a) $f(z)$ is analytic except for: (1) a branch line from 0 to $+\infty$ along the real axis; (2) a simple pole of residue 2 at $z = -2$.

(b) $f(z) \to 0$ as $|z| \to \infty$.

(c) $f(z)$ is real on negative real axis.

(d) For $x > 0$, $\text{Im } f(x + i\varepsilon) = x/(1 + x^2)$. Find the function $F(x) = f(x + i\varepsilon)$.

10-4 Consider the subtracted dispersion relation (10-7). Suppose that, in fact, $|F(x)| \to 0$ as $|x| \to \infty$.

(a) Derive from (10-7) the "sum rule"

$$\text{Re } F(0) = \frac{1}{\pi} P \int_{-\infty}^{\infty} \frac{\text{Im } F(x')\, dx'}{x'}$$

(b) How can this sum rule be obtained from the unsubtracted dispersion relation (10-2)?

(c) Suppose now that $F(x)$ approaches zero at infinity so rapidly that $|xF(x)| \to 0$ as $|x| \to \infty$. Derive a further sum rule by analogy with part (a).

Such relations are called "superconvergence relations" because of the rapid convergence at infinity that is imposed on $F(x)$. See, for example, V. de Alfaro, S. Fubini, G. Rossetti, and G. Furlan, *Physics Letters* **21**: 576 (1966). (The "superconvergence relation" was coined after this reference appeared.)

10-2 CONTOUR INTEGRAL REPRESENTATIONS OF SPECIAL FUNCTIONS

Some of the special functions of mathematical physics are defined by integrals. For example the gamma function, used to illustrate the method of steepest descent in Chapter 8, is defined for Re $z > 0$ by

$$\Gamma(z) = \int_0^\infty x^{z-1} e^{-x}\, dx \tag{10-8}$$

Integration by parts gives

$$\Gamma(z) = (z-1)\,\Gamma(z-1) \tag{10-9}$$

Thus, from $\Gamma(1) = 1$, we obtain

$$\Gamma(2) = 1,\ \Gamma(3) = 2,\ \Gamma(4) = 6, \ldots, \Gamma(n) = (n-1)!$$

The gamma function may be analytically continued into the entire complex plane by means of the recursion relation (10-9); notice that this definition introduces simple poles at $z = 0, -1, -2 \ldots.$

$$\Gamma(1+\varepsilon) = \varepsilon\Gamma(\varepsilon) \underset{\varepsilon\to 0}{\to} 1$$

$$\Gamma(1+\varepsilon) = \varepsilon(\varepsilon-1)\Gamma(\varepsilon-1) \underset{\varepsilon\to 0}{\to} 1$$

etc.

A related integral of some interest is encountered if we consider

$$\Gamma(r)\Gamma(s) = \int_0^\infty x^{r-1}e^{-x}\, dx \int_0^\infty y^{s-1}e^{-y}\, dy$$

Let $x + y = u$:

$$\Gamma(r)\Gamma(s) = \int_0^\infty du \int_0^u dx\, x^{r-1}(u-x)^{s-1}e^{-u}$$

Let $x = ut$. Then

$$\Gamma(r)\Gamma(s) = \int_0^\infty e^{-u}u^{r+s-1}\,du \int_0^1 dt\, t^{r-1}(1-t)^{s-1}$$
$$= \Gamma(r+s)B(r, s)$$

where $B(r, s)$ is the *beta function*

$$B(r, s) = \frac{\Gamma(r)\Gamma(s)}{\Gamma(r+s)} = \int_0^1 x^{r-1}(1-x)^{s-1}\,dx \tag{10-10}$$

This integral representation for $B(r, s)$ is obviously valid only for $\operatorname{Re} r > 0$, $\operatorname{Re} s > 0$; the first equality in (10-10) defines $B(r, s)$ for all r, s in the complex plane (see also Problem 10-22).

An interesting special case of this relation is

$$\Gamma(z)\Gamma(1-z) = \Gamma(1)B(z, 1-z)$$
$$= \int_0^1 x^{z-1}(1-x)^{-z}\,dx$$

Let $x = t/(1+t)$. Then

$$\Gamma(z)\Gamma(1-z) = \int_0^\infty \frac{t^{z-1}\,dt}{1+t}$$

This integral can be done by using the contour shown in Figure 10-5. The result is

$$\Gamma(z)\Gamma(1-z) = \frac{\pi}{\sin \pi z} \tag{10-11}$$

Our derivation has been valid only for $0 < \operatorname{Re} z < 1$, but both sides of (10-11)

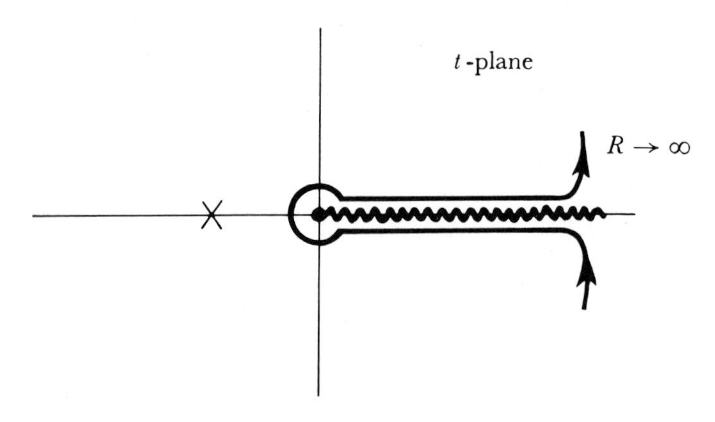

FIGURE 10-5 Contour for the integral $\int_0^\infty [t^{x-1}/(1+t)]\,dt$

are analytic functions of z (except at $z = 0, \pm 1, \pm 2, \ldots$), so we can extend the result into the entire plane.

Another integral that has been given a name is the so-called *exponential integral*

$$\mathrm{Ei}(x) = \int_{-\infty}^{x} \frac{e^t\,dt}{t} \tag{10-12}$$

This definition is conventionally supplemented by cutting the x-plane along the positive real axis. Thus $\mathrm{Ei}(x)$ is well defined for negative real x, but for positive real x we must distinguish between $\mathrm{Ei}(x + i\varepsilon)$ and $\mathrm{Ei}(x - i\varepsilon)$.

Two functions related to $\mathrm{Ei}(x)$ are the *sine* and *cosine integrals*:

$$\begin{aligned} \mathrm{Si}(x) &= \int_0^x \frac{\sin t}{t}\,dt \\ \mathrm{Ci}(x) &= \int_\infty^x \frac{\cos t}{t}\,dt \end{aligned} \tag{10-13}$$

The *error function* is defined as

$$\operatorname{erf} x = \frac{2}{\sqrt{\pi}} \int_0^x e^{-t^2}\,dt \tag{10-14}$$

The associated trigonometric integrals are known as *Fresnel integrals*:

$$C(x) = \int_0^x \cos\left(\frac{\pi t^2}{2}\right) dt \qquad S(x) = \int_0^x \sin\left(\frac{\pi t^2}{2}\right) dt \tag{10-15}$$

These integrals can be manipulated to obtain approximate expansions for the functions defined. Our use of the method of steepest descent to find an approximation to the gamma function illustrates only one possible such manipulation. Let us consider some others.

EXPANSIONS IN SERIES

One can often obtain a useful expression by expanding the integrand in some sort of series.

Example

$$\begin{aligned} \operatorname{erf} x &= \frac{2}{\sqrt{\pi}} \int_0^x e^{-t^2}\,dt \\ &= \frac{2}{\sqrt{\pi}} \int_0^x \left(1 - t^2 + \frac{t^4}{2!} - \frac{t^6}{3!} + - \cdots\right) dt \end{aligned}$$

$$= \frac{2}{\sqrt{\pi}}\left(x - \frac{x^3}{3} + \frac{x^5}{5\cdot 2!} - \frac{x^7}{7\cdot 3!} + - \cdots\right)$$

This series converges for all x, but is only useful for small x, ($x \lesssim 1$).

Integrations by parts are often useful.

Example

Suppose we want erf x for large x. As $x \to \infty$, erf $x \to 1$. Let us compute the difference from 1.

$$1 - \operatorname{erf} x = \frac{2}{\sqrt{\pi}} \int_x^\infty e^{-t^2}\, dt$$

We perform a sequence of integrations by parts.

$$\int_x^\infty e^{-t^2}\, dt = \frac{e^{-x^2}}{2x} - \int_x^\infty \frac{e^{-t^2}\, dt}{2t^2}$$

$$= \frac{e^{-x^2}}{2x} - \frac{e^{-x^2}}{4x^3} + \int_x^\infty \frac{3}{4}\frac{e^{-t^2}}{t^4}\, dt$$

etc. The result after n integrations by parts is ($n > 1$):

$$\operatorname{erf} x = 1 - \frac{2}{\pi} e^{-x^2}\left[\frac{1}{2x} - \frac{1}{2^2 x^3} + \frac{1\cdot 3}{2^3 x^5} - \frac{1\cdot 3\cdot 5}{2^4 x^7} + - \cdots + (-1)^{n-1}\frac{1\cdot 3\cdot 5\cdots(2n-3)}{2^n x^{2n-1}}\right]$$
$$+ (-1)^n \frac{1\cdot 3\cdot 5\cdots(2n-1)}{2^n}\frac{2}{\sqrt{\pi}}\int_x^\infty \frac{e^{-t^2}}{t^{2n}}\, dt \qquad (10\text{-}16)$$

The terms in the brackets do not form the beginning of a convergent infinite series. The series does not converge for any x, since the individual terms eventually increase as n increases. Nevertheless, this expression with a finite number of terms is very useful for large x.

The expression (10-16) is exact if we include the "remainder," that is, the last term, containing the integral. This remainder alternates in sign as n increases, which means that the error after n terms in the series is smaller in magnitude than the next term! Thus the accuracy of the approximate expression in the brackets is highest if we stop one term before the smallest.

The series in brackets in (10-16) is an example of an *asymptotic series* (provided it is continued indefinitely). In general

$$S(z) = C_0 + \frac{C_1}{z} + \frac{C_2}{z^2} + \cdots$$

is an asymptotic series expansion of $f(z)$ [written $f(z) \sim S(z)$, where $\sim$ reads "is asymptotically equal to"] provided, for any n, the error involved in terminating the series with the term $C_n z^{-n}$ goes to zero faster than z^{-n} as $|z|$ goes to ∞ for some range of arg z; that is,

$$\lim_{|z|\to\infty} z^n[f(z) - S_n(z)] = 0 \qquad (10\text{-}17)$$

for arg z in the given interval; $S_n(z)$ means $C_0 + C_1/z + \cdots + C_n/z^n$.

(A *convergent* series approaches $f(z)$ as $n \to \infty$ for given z, whereas an *asymptotic* series approaches $f(z)$ as $z \to \infty$ for given n).

From the definition, it is easy to show that asymptotic series may be added, multiplied, and integrated to obtain asymptotic series for the sum, product, and integral of the corresponding functions. Also, the asymptotic expansion of a given function $f(z)$ is unique, but the reverse is not true. An asymptotic series does not specify a function $f(z)$ uniquely.

As another example of the use of integrations by parts,

$$\begin{aligned}\mathrm{Ei}(-x) &= \int_{-\infty}^{-x} \frac{e^t}{t}\,dt = \int_{\infty}^{x} e^{-t}\frac{dt}{t} \qquad (x > 0)\\ &= -\frac{e^{-x}}{x} - \int_{\infty}^{x} \frac{e^{-t}}{t^2}\,dt\\ &= -\frac{e^{-x}}{x} + \frac{e^{-x}}{x^2} + 2\int_{\infty}^{x} \frac{e^{-t}}{t^3}\,dt\end{aligned}$$

Continuing in this way, we obtain

$$\begin{aligned}-\mathrm{Ei}(-x) = \frac{e^{-x}}{x}\left[1 - \frac{1}{x} + \frac{2!}{x^2} - \frac{3!}{x^3} + - \cdots + \frac{(-1)^n n!}{x^n}\right]&\\ + (-1)^n(n+1)!\int_{\infty}^{x} \frac{e^{-t}}{t^{n+2}}\,dt&\end{aligned} \qquad (10\text{-}18)$$

We can use this result in two ways:

1. As an asymptotic series for $\mathrm{Ei}(-x)$:

$$-\mathrm{Ei}(-x) \sim \frac{e^{-x}}{x}\left(1 - \frac{1}{x} + \frac{2!}{x^2} - \frac{3!}{x^3} + - \cdots\right)$$

2. As an exact expression for computing certain integrals, assuming we have available a table of $\mathrm{Ei}(-x)$. From (10-18)

$$\begin{aligned}\int_x^{\infty} \frac{e^{-t}\,dt}{t^n} = \frac{(-1)^n}{(n-1)!}\mathrm{Ei}(-x) + \frac{(-1)^n}{(n-1)!}\frac{e^{-x}}{x}&\\ \times\left[1 - \frac{1}{x} + \frac{2!}{x^2} - + \cdots + (-1)^n\frac{(n-2)!}{x^{n-2}}\right]&\end{aligned}$$

Finally, we can use the recursion relation (10-9) obtained by integration by parts, together with the approximation (8-18) obtained by the method of steepest descent, to obtain an asymptotic series for the gamma function. Set

$$\Gamma(z) \sim \sqrt{2\pi}\, z^{z-1/2} e^{-z}\left(1 + \frac{A}{z} + \frac{B}{z^2} + \cdots\right) \tag{10-19}$$

The constants $A, B, \ldots$ may be found from the recursion relation

$$\Gamma(z+1) = z\Gamma(z)$$

For

$$\begin{aligned}
\Gamma(z+1) &= \sqrt{2\pi}(z+1)^{z+1/2} e^{-(z+1)}\left[1 + \frac{A}{z+1} + \frac{B}{(z+1)^2} + \cdots\right] \\
&= \sqrt{2\pi}\exp\left[(z+\tfrac{1}{2})\log(z+1) - (z+1)\right] \\
&\quad \times \left[1 + \frac{A}{z+1} + \frac{B}{(z+1)^2} + \cdots\right] \\
&= \sqrt{2\pi}\exp\left[\left(z+\frac{1}{2}\right)\log z - z + \left(\frac{1}{12z^2} - \frac{1}{12z^3} + \frac{3}{40z^4} - + \cdots\right)\right] \\
&\quad \times \left[1 + \frac{A}{z+1} + \frac{B}{(z+1)^2} + \cdots\right] \\
&= \sqrt{2\pi}\, z^{z+1/2} e^{-z}\left(1 + \frac{1}{12z^2} - \frac{1}{12z^3} + \frac{113}{1440z^4} - + \cdots\right) \\
&\quad \times \left[1 + \frac{A}{z} + \frac{B-A}{z^2} + \frac{C-2B+A}{z^3} + \cdots\right]
\end{aligned}$$

On the other hand, (10-19) gives immediately

$$z\Gamma(z) = \sqrt{2\pi}\, z^{z+1/2} e^{-z}\left(1 + \frac{A}{z} + \frac{B}{z^2} + \cdots\right)$$

Equating corresponding terms in the two series we find

$$A = \tfrac{1}{12}, \qquad B = \tfrac{1}{288}, \ldots$$

LEGENDRE FUNCTIONS

Functions defined in other ways can often be expressed in terms of contour integrals. Consider the Legendre polynomials, given by Rodrigues' formula:

$$P_n(x) = \frac{1}{2^n n!}\left(\frac{d}{dx}\right)^n (x^2 - 1)^n$$

To find an integral representation, begin with Cauchy's formula,

$$f(z) = \frac{1}{2\pi i} \oint \frac{f(t)\, dt}{t - z}$$

Differentiating yields

$$\left(\frac{d}{dz}\right)^n f(z) = \frac{n!}{2\pi i} \oint \frac{f(t)\, dt}{(t - z)^{n+1}}$$

hence

$$P_n(z) = \frac{1}{2^n} \frac{1}{2\pi i} \oint \frac{(t^2 - 1)^n}{(t - z)^{n+1}}\, dt \tag{10-20}$$

where the contour encircles z once in a positive sense. This is called *Schläfli's integral representation* for the $P_n(z)$.

We can deduce another integral representation as follows. Let the contour in Schläfli's integral (10-20) be a circle about z with a radius $|\sqrt{z^2 - 1}|$. That is,

$$t = z + \sqrt{z^2 - 1}\, e^{i\varphi}$$

where φ goes from 0 to 2π. It then follows by straightforward algebra that

$$t^2 - 1 = 2(t - z)(z + \sqrt{z^2 - 1} \cos \varphi)$$

and

$$dt = i(t - z)\, d\varphi$$

Therefore, by substituting into (10-20), we find

$$P_n(z) = \frac{1}{\pi} \int_0^{\pi} (z + \sqrt{z^2 - 1} \cos \varphi)^n\, d\varphi \tag{10-21}$$

This is called *Laplace's integral representation* for the Legendre polynomials.

Laplace's integral representation can be used to derive a particularly useful function, called the *generating function* for the Legendre polynomials. Let us try to find a function $F(h, z)$ that has a power series expansion in h of the form:

$$\sum_{n=0}^{\infty} h^n P_n(z) = F(h, z)$$

Using Laplace's representation, we obtain

$$\begin{aligned} F(h, z) &= \frac{1}{\pi} \int_0^{\pi} \sum_{n=0}^{\infty} h^n (z + \sqrt{z^2 - 1} \cos \varphi)^n\, d\varphi \\ &= \frac{1}{\pi} \int_0^{\pi} \frac{1}{1 - hz - h\sqrt{z^2 - 1} \cos \varphi}\, d\varphi \end{aligned}$$

so that [by equation (8-3)],

$$F(h, z) = \frac{1}{\sqrt{1 - 2hz + h^2}} = \sum_{n=0}^{\infty} h^n P_n(z) \tag{10-22}$$

This $F(h, z)$ is the desired generating function.

The generating function can be used to derive recursion relations among the Legendre polynomials:

$$F(h, z) = \frac{1}{\sqrt{1 - 2hz + h^2}}$$

$$\frac{\partial F}{\partial h} = \frac{z - h}{1 - 2hz + h^2} F$$

$$(1 - 2hz + h^2) \frac{\partial F}{\partial h} = (z - h)F$$

Equating coefficients of h^n on both sides gives

$$(n + 1)P_{n+1}(z) - 2znP_n(z) + (n - 1)P_{n-1}(z) = zP_n(z) - P_{n-1}(z)$$

or

$$(n + 1)P_{n+1}(z) - (2n + 1)zP_n(z) + nP_{n-1}(z) = 0 \tag{10-23}$$

A second recursion relation is obtained by differentiating the generating function (10-22) with respect to z:

$$(1 - 2hz + h^2) \frac{\partial F}{\partial z} = hF$$

which gives

$$P_n'(z) - 2zP_{n-1}'(z) + P_{n-2}'(z) = P_{n-1}(z) \tag{10-24}$$

From the two recursion relations (10-23) and (10-24) we can derive many others; for example,

$$P_{n+1}'(z) - zP_n'(z) = (n + 1)P_n(z)$$

$$zP_n'(z) - P_{n-1}'(z) = nP_n(z)$$

$$P_{n+1}'(z) - P_{n-1}'(z) = (2n + 1)P_n(z)$$

$$(z^2 - 1)P_n'(z) = nzP_n(z) - nP_{n-1}(z)$$

etc.

For many purposes this method of obtaining recursion relations is more useful than the one outlined in Section 4-5 because the Legendre polynomials obtained by this method are already normalized [$P_n(1) = 1$].

Another use of the generating function is in evaluating $P_n(z)$ at various points. For example, at $z = 1$, (10-22) gives

$$\sum_{n=0}^{\infty} h^n P_n(1) = \frac{1}{1-h} = 1 + h + h^2 + h^3 + \cdots$$

Therefore,

$$P_n(1) = 1$$

At $z = 0$,

$$\sum_{n=0}^{\infty} h^n P_n(0) = (1 + h^2)^{-1/2} = 1 - \tfrac{1}{2}h^2 + (-\tfrac{1}{2})(-\tfrac{3}{2})\frac{h^4}{2!} + \cdots$$

Therefore,

$$P_n(0) = \begin{cases} 0 & \text{if } n \text{ is odd} \\ \dfrac{(n-1)!!(-1)^{n/2}}{2^{n/2}(n/2)!} & \text{if } n \text{ is even} \end{cases}$$

where we use the notation

$$n!!(n \text{ double factorial}) = n(n-2)(n-4)\cdots 5\cdot 3\cdot 1$$

In addition, the equation (10-22) gives a very useful representation for the *inverse distance* between two points in three-dimensional space. If the points are $\mathbf{x}$ and $\mathbf{x}'$ with $r' < r$, then

$$\frac{1}{|\mathbf{x} - \mathbf{x}'|} = \frac{1}{\sqrt{r^2 + r'^2 - 2rr' \cos\theta}} = \frac{1}{r}\frac{1}{\sqrt{1 - 2\dfrac{r'}{r}\cos\theta + \left(\dfrac{r'}{r}\right)^2}} \tag{10-25}$$

$$= \sum_{l=0}^{\infty} \frac{r'^l}{r^{l+1}} P_l(\cos\theta)$$

where θ is the angle between the vectors $\mathbf{x}$ and $\mathbf{x}'$. If $r < r'$, r and r' must be exchanged in formula (10-25); otherwise the series fails to converge. Sometimes both formulas are combined by writing

$$\frac{1}{|\mathbf{x} - \mathbf{x}'|} = \sum_{l=0}^{\infty} \frac{r_<^l}{r_>^{l+1}} P_l(\cos\theta) \tag{10-26}$$

Schläfli's formula (10-20) may be extended to noninteger n provided we can find an acceptable contour which takes into account the branch cuts of the integrand,

$$\frac{1}{(t-z)}\left[\frac{(t-1)(t+1)}{(t-z)}\right]^n$$

For $-1 \leq z \leq 1$, the contour C shown in Figure 10-6 encircles the point $t = z$ while remaining on a single sheet of the cut plane. Thus

$$\left(\frac{1}{2^\alpha}\right)\frac{1}{2\pi i}\int_C \frac{(t^2-1)^\alpha}{(t-z)^{\alpha+1}}\,dt = f_\alpha(z)$$

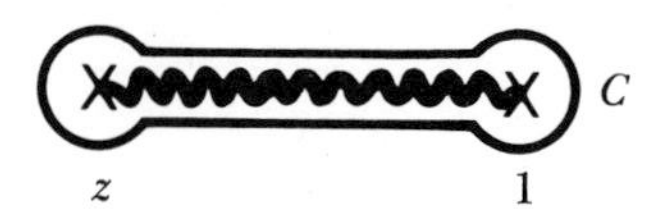

FIGURE 10-6 Contour for integral (10-27)

is a logical candidate for $P_\alpha(z)$. We must still check to see whether it obeys the Legendre equation. It is not difficult to show that

$$(1 - x^2)\frac{d^2f_\alpha(x)}{dx^2} - 2x\frac{df_\alpha(x)}{dx} + \alpha(\alpha + 1)f_\alpha(x)$$

$$= \left(\frac{1}{2^\alpha}\right)\frac{1}{2\pi i}(\alpha + 1)\int_C \frac{d}{dt}\left[\frac{(t^2 - 1)^{\alpha+1}}{(t - z)^{\alpha+2}}\right]dt$$

But this is zero because C is a closed contour (i.e., because the initial and final points are at the same t on the same sheet). Hence

$$P_\alpha(z) = \frac{1}{2^\alpha}\frac{1}{2\pi i}\int_C \frac{(t^2 - 1)^\alpha}{(t - z)^{\alpha+1}}\,dt \qquad (10\text{-}27)$$

This representation of the Legendre function allows us to derive an important analyticity property of the functions for noninteger α: $P_\alpha(z)$ has a cut in z from -1 to $-\infty$. To see this, let z be a negative number less than -1. Then the values $z + i\varepsilon$ and $z - i\varepsilon$ will clearly lead to contours summing different values of the integrand in the t-plane (see Figure 10-7), and thus to different

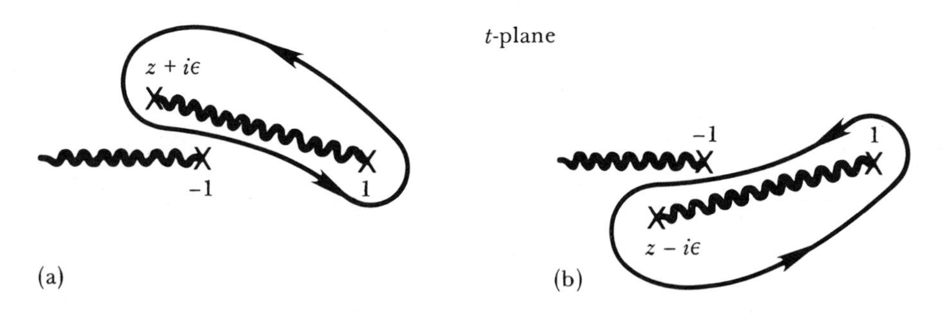

FIGURE 10-7 Two possible contours for integral (10-27) if $-\infty < z < -1$

values for $P_\alpha(z \pm i\varepsilon)$. Hence $P_\alpha(z)$ is multivalued for these z; to separate the sheets we should draw a branch cut in the z-plane (Figure 10-8). The branch cut disappears if α is an integer, since then the branch cuts in the t-plane in the defining integral (10-27) are no longer present.

Clearly the choice of cuts shown in Figure 10-6 is not unique. Another

FIGURE 10-8 Branch cut of $P_\alpha(z)$ for noninteger α

possible choice would be to connect -1 with z, and $+1$ with $+\infty$. This set of cuts in the t-plane would define a $P_\alpha(z)$ with a cut from $+1$ to $+\infty$ in the z-plane (show this!). Either set of cuts provides an acceptable definition for $P_\alpha(z)$; and in practical applications one definition may be preferable to another. Remember, however, that the different definitions produce functions with different properties; copying results from reference works without checking the defining contour can lead to disaster.

For $-1 < z < 1$, the integral (10-27) defines a well-behaved function. We should therefore be able to expand it in a series of Legendre functions of integral order $P_n(x)$ (see Chapter 4). This can be done most simply by a change of variables in (10-27). Define

$$u = \frac{1}{2}\frac{(t^2 - 1)}{(t - z)} \qquad t = u + \sqrt{u^2 - 2uz + 1} \tag{10-28}$$

Then

$$P_\alpha(z) = \frac{1}{2\pi i}\oint_{C'} \frac{u^\alpha\, du}{\sqrt{u^2 - 2uz + 1}} \tag{10-29}$$

where C' is the mapping into the u-plane of our original contour C. Using (10-28), we see that the circular integral around $t = z$ in Figure 10-6 reduces to a circular integral around $u = \infty$, the circular integral around $t = 1$ becomes an integral around $u = 0$, and the integral along the short branch cut in t becomes an integral along a branch cut in u from 0 to ∞ (this is the cut due to u^α). The only other singularity in the u-plane is a short branch cut, which is the mapping of the t-plane cut from -1 to $-\infty$. See Figure 10-9 where we have drawn a contour with these topological properties.

To evaluate (10-29), break up the integral into circular pieces and straight pieces.

$$P_\alpha(z) = \frac{1}{2\pi i}\int_{\substack{\text{little}\\ \text{circle}}} \frac{u^\alpha\, du}{\sqrt{u^2 - 2uz + 1}} + \frac{1}{2\pi i}\int_{\substack{\text{big}\\ \text{circle}}} \frac{u^\alpha\, du}{\sqrt{u^2 - 2uz + 1}} + $$
$$- \frac{1}{2\pi i}\int_{u=-R_1}^{-1} \frac{u^\alpha[1 - e^{2i\pi\alpha}]\, du}{\sqrt{u^2 - 2uz + 1}} - \frac{1}{2\pi i}\int_{u=-1}^{-R_2} \frac{u^\alpha[1 - e^{2i\pi\alpha}]\, du}{\sqrt{u^2 - 2uz + 1}}$$

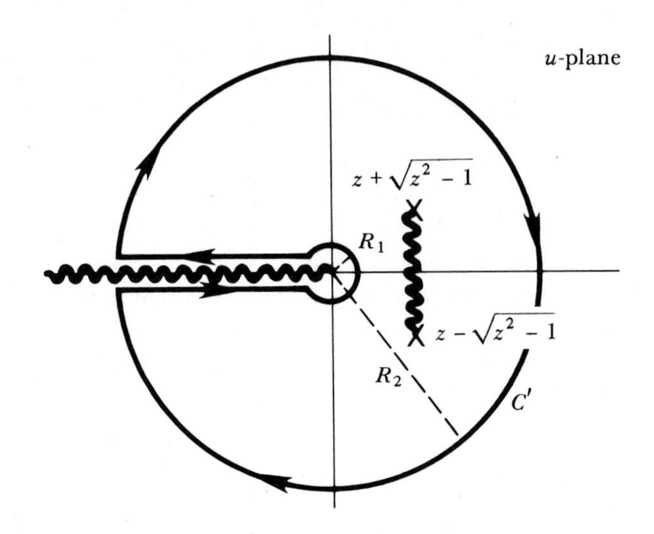

FIGURE 10-9 Contour for integral (10-29)

We now use (10-25) to expand the square root in the denominator of the integrand: for the inner circle and the corresponding piece of the straight line,

$$\frac{1}{\sqrt{u^2 - 2uz + 1}} = \sum_{l=0}^{\infty} u^l P_l(z)$$

whereas for the outer circle and line (see Problem 10-21)

$$\frac{1}{\sqrt{u^2 - 2uz + 1}} = -\sum_{l=0}^{\infty} \frac{1}{u^{l+1}} P_l(z)$$

Each piece of the integral is now straightforward. We leave the evaluation of these pieces to the student and quote the result:

$$P_\alpha(\cos\theta) = \frac{1}{\pi}\sum_{n=0}^{\infty} P_n(\cos\theta)(-1)^n \sin\pi\alpha\left[\frac{1}{\alpha - n} - \frac{1}{\alpha + n + 1}\right] \tag{10-30}$$

BESSEL FUNCTIONS

Similar integral representations can be found for Bessel functions. In this case it is most convenient to begin by finding the generating function and to use it to get the contour integrals. To find the generating function $F(z, h)$, we begin with the recursion relation

$$J_{m+1}(z) + J_{m-1}(z) = \frac{2m}{z} J_m(z)$$

Multiply by h^m and sum over all m from $-\infty$ to $+\infty$. The result is

$$\left(h + \frac{1}{h}\right)F(z,h) = \frac{2h}{z}\frac{\partial F(z,h)}{\partial h}$$

Integrating gives

$$F(z,h) = \varphi(z)\exp\left[\frac{z}{2}\left(h - \frac{1}{h}\right)\right]$$

where $\varphi(z)$ is some function of z, yet to be determined. Now adjust φ so that the coefficient of h^0 is $J_0(z)$. It is readily verified that this requires $\varphi(z) = 1$. Then, if we write

$$\exp\left[\frac{z}{2}\left(h - \frac{1}{h}\right)\right] = \sum_{n=-\infty}^{\infty} h^n U_n(z)$$

we know that:

1. $U_0(z) = J_0(z)$
2. $U_{n-1}(z) + U_{n+1}(z) = \frac{2n}{z} U_n(z)$
3. $U_n(-z) = (-1)^n U_n(z)$
4. $U_{-n}(z) = (-1)^n U_n(z)$

This is not quite enough to guarantee that $U_n(z) = J_n(z)$. Let us differentiate the generating function with respect to z:

$$\frac{1}{2}\left(h - \frac{1}{h}\right)\exp\left[\frac{z}{2}\left(h - \frac{1}{h}\right)\right] = \sum_n h^n U_n'(z)$$

$$\frac{1}{2}\left(h - \frac{1}{h}\right)\sum_n h^n U_n(z) = \sum_n h^n U_n'(z)$$

Equating coefficients of h^n on both sides of this equation gives

$$U_n'(z) = \tfrac{1}{2}[U_{n-1}(z) - U_{n+1}(z)]$$

Thus $U_1 = -U_0' = -J_0' = J_1$, and all the rest follows from the recursion relations.

As an example of the usefulness of this generating function, we have

$$\begin{aligned}\sum_n J_n(x+y)h^n &= \exp\left[\frac{x+y}{2}\left(h - \frac{1}{h}\right)\right] \\ &= \exp\left[\frac{x}{2}\left(h - \frac{1}{h}\right)\right]\exp\left[\frac{y}{2}\left(h - \frac{1}{h}\right)\right] \\ &= \sum_k h^k J_k(x) \sum_l h^l J_l(y)\end{aligned}$$

Therefore,

$$J_n(x+y) = \sum_k J_k(x) J_{n-k}(y) \tag{10-31}$$

From the generating function there follows immediately *Schläfli's integral representation*

$$J_n(z) = \frac{1}{2\pi i} \oint \frac{\exp\left[\frac{z}{2}\left(t - \frac{1}{t}\right)\right] dt}{t^{n+1}} \tag{10-32}$$

where the contour encloses the origin once in a positive sense. If we set $t = e^{i\theta}$, we get

$$\begin{aligned} J_n(z) &= \frac{1}{2\pi i} \int_0^{2\pi} \frac{e^{iz \sin \theta}}{e^{(n+1)i\theta}} ie^{i\theta}\, d\theta \\ &= \frac{1}{2\pi} \int_0^{2\pi} e^{i(z \sin \theta - n\theta)}\, d\theta \end{aligned}$$

Therefore,

$$J_n(z) = \frac{1}{\pi} \int_0^{\pi} \cos(n\theta - z \sin \theta)\, d\theta \tag{10-33}$$

This is called *Bessel's integral.*

We shall now discuss the case where n is no longer necessarily an integer. Consider the integral

$$f_n(z) = \frac{1}{2\pi i} \int_C \frac{\exp\left[\frac{z}{2}\left(t - \frac{1}{t}\right)\right]}{t^{n+1}} dt \tag{10-34}$$

Will any choice of C make $f_n(z)$ a solution of Bessel's equation? First, operate on $f_n(z)$ with the Bessel differential operator and obtain

$$\begin{aligned} &\left[z^2\left(\frac{d}{dz}\right)^2 + z\frac{d}{dz} + (z^2 - n^2)\right] f_n(z) \\ &\quad = \frac{1}{2\pi i} \int \frac{dt}{t^{n+1}} \exp\left[\frac{z}{2}\left(t - \frac{1}{t}\right)\right]\left[\frac{z^2}{4}\left(t - \frac{1}{t}\right)^2 + \frac{z}{2}\left(t - \frac{1}{t}\right) + z^2 - n^2\right] \\ &\quad = \frac{1}{2\pi i} \int dt \frac{d}{dt} \left\{ \frac{\exp\left[\frac{z}{2}\left(t - \frac{1}{t}\right)\right]}{t^n} \left[\frac{z}{2}\left(t + \frac{1}{t}\right) + n\right] \right\} \\ &\quad = \frac{1}{2\pi i} [F_n(z, t)] \end{aligned}$$

[where $F_n(z, t)$ is the function in braces, and we must take the difference of the values of this function at the two ends of the path of integration]. If we choose a path so that this difference is zero, then (10-34) will be a solution of Bessel's equation. If n is an integer, any closed path suffices; of course, the integral (10-34) is then zero if the path does not include the origin.

We shall suppose that n is not necessarily an integer and that z is real and positive. Then $F_n(z, t)$ vanishes at $t \to 0+$, and as $t \to -\infty$. Thus we may obtain two solutions defined by the contours of Figure 10-10. These are called Hankel

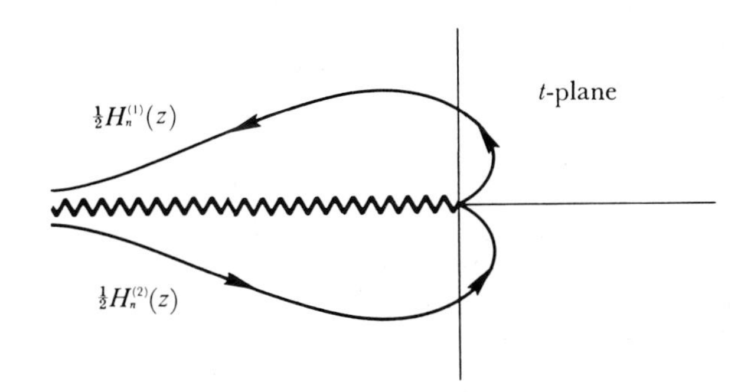

FIGURE 10-10 Contours such that the integral (10-34) is a solution of Bessel's equation. Because the integrand is, for n not an integer, not single-valued, we cut the t-plane along the negative real axis and define $t^{n+1} = \exp[(n+1)\ln t]$, with $\ln t$ real on the positive real axis.

functions of the first and second kind; they always provide us with two solutions of Bessel's equation.

These integral representations remain satisfactory for

$$|\arg z| < \frac{\pi}{2}$$

If we want a representation that is valid for $|\arg z - \alpha| < \pi/2$, we must rotate the contours in a manner that will be left as an exercise.

For noninteger m we may then define

$$J_m(z) = \tfrac{1}{2}[H_m^{(1)}(z) + H_m^{(2)}(z)] \tag{10-35}$$

when m becomes an integer, the sum of the two paths reduces to our original circle about the origin. Likewise the difference of the two Hankel functions is related to our second solution,

$$Y_m(x) = \frac{[\cos m\pi J_m(x) - J_{-m}(x)]}{\sin m\pi}$$

by

$$\begin{aligned} H_m^{(1)}(x) &= J_m(x) + iY_m(x) \\ H_m^{(2)}(x) &= J_m(x) - iY_m(x) \end{aligned} \tag{10-36}$$

By applying the method of steepest descent, we can find asymptotic forms of the Hankel functions as $z \to \infty$ from these contours. Begin by rewriting the integrand

$$\frac{1}{2\pi i}\int_C \frac{e^{z/2(h-1/h)}\,dh}{h^{m+1}} = \frac{1}{2\pi i}\int_C e^{z/2(h-1/h)-(m+1)\ln h}\,dh$$
$$= \frac{1}{2\pi i}\int_C e^{f(h)}\,dh$$

The points where $f'(h) = 0$ occur at

$$h = \frac{m+1}{z} \pm \frac{1}{z}\sqrt{(m+1)^2 - z^2}$$

For z much larger than $m + 1$, these are nearly at

$$h \approx \pm i$$

In the method of steepest descent, we deform the contour so it goes over the saddle point. Here we clearly will send the contour for $H_n^{(1)}(z)$ over the saddle near $h = +i$, and the contour for $H_n^{(2)}(z)$ over the other saddle. The values of $f''(z)$ at these saddle points are

$$f''(z) = \frac{-z}{(\pm i)^3} - (m+1) \approx z e^{\mp i\pi/2}$$

Hence for z large and positive,

$$\tfrac{1}{2}H_m^{(1)}(z) \approx \sqrt{\frac{2\pi}{z}}\,\frac{e^{i[z-(m+1)(\pi/2)]}}{2\pi i}\,e^{i[(\pm\pi/2)+(\pi/4)]}$$

$$\tfrac{1}{2}H_m^{(2)}(z) \approx \sqrt{\frac{2\pi}{z}}\,\frac{e^{-i[z-(m+1)(\pi/2)]}}{2\pi i}\,e^{i[(\pm\pi/2)-(\pi/4)]}$$

from (8-15). We must still determine the phase from the direction in which we go over the saddle point. A little thought shows that for $H_n^{(1)}(z)$, $\theta = 3\pi/4$ whereas for $H_n^{(2)}(z)$, $\theta = \pi/4$ (i.e., we choose $+\pi/2$ rather than $-\pi/2$ in both expressions). Hence

$$H_m^{(1)}(z) \simeq \sqrt{\frac{2}{\pi z}}\,e^{i(z-m\pi/2-\pi/4)}$$
$$H_m^{(2)}(z) \simeq \sqrt{\frac{2}{\pi z}}\,e^{-i(z-m\pi/2-\pi/4)} \tag{10-37}$$

These examples have illustrated certain types of questions that can be answered most easily from the contour integral representation. In general these representations are helpful in computing asymptotic behavior in all parameters (we might have looked for behavior for large n with z fixed), and in determining analytic properties for different values of the parameters.

PROBLEMS

10-5 Show that Ei $(x + i\varepsilon)$ is different from Ei $(x - i\varepsilon)$.

10-6 Evaluate $\int_0^\infty e^{-x^2} \operatorname{Ci}(ax)\, dx$

10-7 Evaluate $\int_0^\infty e^{-ax} \operatorname{erf} x\, dx$

10-8 Evaluate

(a) $\Gamma(\frac{1}{2})$ (d) $B(\frac{1}{2}, -\frac{1}{2})$

(b) $\Gamma(\frac{5}{2})$ (e) $\Gamma(\frac{1}{3})\Gamma(-\frac{1}{3})$

(c) $B(1, 3)$

10-9 Evaluate $\int_0^\infty e^{-sx}[-\operatorname{Ei}(-x)]\, dx$

10-10 Evaluate $\int_{-2}^{2} (4 - x^2)^{1/6}\, dx$ in terms of beta and gamma functions.

10-11 Consider the integral $F(z) = \int_c dt(-t)^{z-1} e^{-t}$, where the t-plane is cut along the positive real axis, $(-t)^{z-1}$ is defined to equal

$$\exp [(z - 1) \ln (-t)]$$

with $\ln(-t)$ real on the negative real t-axis, and the path of integration C comes in from $t = +\infty$ below the cut, goes around the origin, and returns to $t = +\infty$ above the cut. This integral defines the gamma function $\Gamma(z)$ throughout the complex plane [unlike the definition (10-8)]. More precisely, $f(z) = (\text{something}) \times \Gamma(z)$. Evaluate (something).

10-12 Evaluate $\oint \Gamma(z)e^{az}\, dz$ around the contour $|z| = \frac{5}{2}$ once in a positive sense.

10-13 Obtain two expansions of Si x, one useful for small x and one useful for large x.

10-14 (a) Consider the integral

$$I_n = \frac{1}{2\pi i} \int e^t\, t^{-(n+1)}\, dt$$

n is not necessarily an integer; the t-plane is cut along the negative real axis, with $\ln t$ defined to be real on the positive t-axis. The path of integration starts at $t = -\infty$ below the cut, circles the origin, and returns to $t = -\infty$ above the cut, as in Figure 10-10. Show that $I_n = 1/\Gamma(n + 1)$.

(b) Show that the integral (10-34), along the path of part (a), equals the Bessel function $J_n(z)$ as defined by the power series (3-69).

10-15 Evaluate $P_n'(1)$

(a) directly from Rodrigues' formula (3-63).

(b) from the generating function (10-22).

10-16 By using a generating function or in any other way, evaluate the sum

$$\sum_{n=0}^{\infty} \frac{x^{n+1}}{n+1} P_n(x)$$

where $P_n(x)$ are the Legendre polynomials.

10-17 Show that (3-67) implies

$$Q_n(z) = \frac{1}{2} \int_{-1}^{+1} \frac{P_n(t)\,dt}{z-t} \qquad (n = \text{integer})$$

by an application of Cauchy's formula.

10-18 Suppose the power series $\sum_{n=0}^{\infty} c_n z^n$ has a radius of convergence R. By using Laplace's integral representation for $P_n(z)$, find the region of the complex plane within which the series

$$\sum_{n=0}^{\infty} c_n P_n(z)$$

converges. (It is an ellipse.) (What happens if $R < 1$?)

10-19 The Hermite polynomials $H_n(x)$ may be defined by the generating function

$$e^{2hz-h^2} = \sum_{n=0}^{\infty} H_n(x) \frac{h^n}{n!}$$

(a) Find the recursion relation connecting H_{n-1}, H_n, and H_{n+1}.

(b) Evaluate

$$\int_{-\infty}^{\infty} e^{-x^2/2} H_n(x)\,dx$$

10-20 Consider functions $f_n(x)$ defined by

(a) $f_0(x) = \sum_{n=0}^{\infty} \frac{x^n}{(n!)^2}$

(b) $(n+1)f_{n+1} = xf_n - f_{n+2}$

(c) $f_n' = f_{n-1}$

Find a generating function $G(x, t)$ such that

$$G(x, t) = \sum_{n=-\infty}^{\infty} f_n(x) t^n$$

10-21 Contours for $H_n^{(1)}(z)$ and $H_n^{(2)}(z)$, suitable for $|\arg z| < \pi/2$, are sketched in the text. Draw a sketch of contours suitable for

$$-\frac{\pi}{4} < \arg z < \frac{3\pi}{4}$$

10-22 Consider the system of branch cuts shown in Figure 10-9. Show that the expansions

$$\frac{1}{\sqrt{u^2 - 2uz + 1}} = \sum_{l=0}^{\infty} u^l P_l(z)$$

and

$$\frac{1}{\sqrt{u^2 - 2uz + 1}} = -\sum_{l=0}^{\infty} \frac{1}{u^{l+1}} P_l(z)$$

have the proper signs to represent the behavior of the square root cut for real u. *Hint:* Consider the case $z = 0$.)

10-23 Use the "reflection property" of gamma functions, equation (10-11), to demonstrate that

$$B(r, s) = \frac{\Gamma(r)\Gamma(s)}{\Gamma(r+s)}$$

has only simple poles. (Let both r and s approach negative integers to see if a double pole develops.)

10-3 DERIVATION OF THE WKB CONNECTION FORMULAE

We shall conclude the chapter with another application of the method of steepest descent.

The WKB method provides approximate solutions of an equation of the form

$$\frac{d^2y}{dx^2} + f(x)y = 0 \tag{10-38}$$

provided $f(x)$ satisfies certain restrictions discussed below. Recall that any linear homogeneous second-order equation may be put in this form by the transformation (3-34). The one-dimensional Schrödinger equation is of this form and the method was developed for quantum mechanical applications by Wentzel, Kramers, and Brillouin, whence the name. The method had been given previously by Jeffreys [see Ref. (22), p. 522].

The solutions of equation (10-38) with $f(x)$ constant suggest the substitution

$$y = e^{i\phi(x)} \tag{10-39}$$

The equation becomes

$$-(\varphi')^2 + i\varphi'' + f = 0 \tag{10-40}$$

If we assume φ'' small, a first approximation is

$$\varphi' = \pm\sqrt{f} \qquad \varphi(x) = \pm\int\sqrt{f(x)}\,dx \tag{10-41}$$

The condition of validity (that φ'' be "small") is

$$\varphi'' \approx \frac{1}{2}\left|\frac{f'}{\sqrt{f}}\right| \ll |f| \tag{10-42}$$

From (10-38) and (10-39) we see that $1/\sqrt{f}$ is roughly one wavelength or exponential length of the solution y. Thus the condition of validity of our approximation is simply the reasonable one that the change in $f(x)$ in one wavelength should be small compared to $|f|$.

A second approximation is easily found by iteration. Set

$$\varphi'' \approx \pm\tfrac{1}{2}f^{-1/2}f'$$

in (10-40). Then

$$(\varphi')^2 \approx f \pm \frac{i}{2}\frac{f'}{\sqrt{f}}$$

$$\varphi' \approx \pm\sqrt{f} + \frac{i}{4}\frac{f'}{f}$$

$$\varphi(x) \approx \pm\int\sqrt{f(x)}\,dx + \frac{i}{4}\ln f$$

The two choices of sign give two approximate solutions that may be combined to give the general solution

$$y(x) \approx \frac{1}{[f(x)]^{1/4}}\left\{c_+ \exp\left[i\int\sqrt{f(x)}\,dx\right] + c_- \exp\left[-i\int\sqrt{f(x)}\,dx\right]\right\} \tag{10-43}$$

We have thus found an approximation to the general solution of the original equation (10-38) in any region where the condition of validity (10-42) holds. The method fails if $f(x)$ changes too rapidly or if $f(x)$ passes through zero. The latter is a serious difficulty since we often wish to join an oscillatory solution in a region where $f(x) > 0$ to an "exponential" one in a region where $f(x) < 0$.

We shall investigate this problem in some detail in order to derive the so-called *connection formulas* relating the constants c_+ and c_- of the WKB solutions on either side of a point where $f(x) = 0$.

Suppose $f(x)$ passes through zero at x_0 and is positive on the right as shown in Figure 10-11. Suppose further that $f(x)$ satisfies the condition of

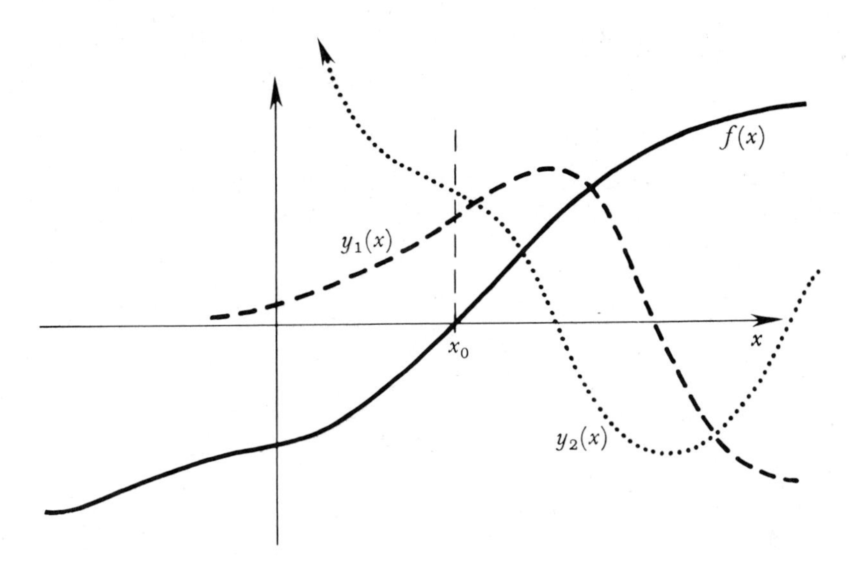

FIGURE 10-11 Graph of $f(x)$ and two exact solutions of the equation (10-38); one of them, $y_1(x)$, is the special solution that decreases to zero on the left.

validity (10-42) in regions both to the left and right of x_0 so that any specific solution $y(x)$ may be approximated in these regions by

$x \ll x_0, \qquad f(x) < 0:$

$$y(x) \approx \frac{a}{\sqrt[4]{-f(x)}} \exp\left[+\int_x^{x_0} \sqrt{-f(x)}\, dx\right] + \frac{b}{\sqrt[4]{-f(x)}} \exp\left[-\int_x^{x_0} \sqrt{-f(x)}\, dx\right] \tag{10-44}$$

$x \gg x_0, \qquad f(x) > 0:$

$$y(x) \approx \frac{c}{\sqrt[4]{f(x)}} \exp\left[+i\int_{x_0}^{x} \sqrt{f(x)}\, dx\right] + \frac{d}{\sqrt[4]{f(x)}} \exp\left[-i\int_{x_0}^{x} \sqrt{f(x)}\, dx\right] \tag{10-45}$$

where the symbols $\sqrt{\ }$ and $\sqrt[4]{\ }$ mean positive real roots throughout. If $f(x)$ is real, a solution that is real in the left region will also be real on the right. This "reality condition" states that if a and b are real, $d = c^*$.

Our problem is to "connect" the approximations on either side of x_0 so that they refer to the same exact solution; that is, to find c and d if we know a and b and vice versa. To make this connection, we need to use an approximate solution that is valid all along some path connecting the regions of x on either side of x_0, where the WKB approximations are valid. One procedure, used by Kramers, and by Jeffreys, is to use a solution valid on the real axis through x_0. Let us discuss this method in detail for a particular example.

Consider the differential equation

$$\frac{d^2y}{dx^2} + xy = 0 \tag{10-46}$$

This equation has a solution $y(x)$ that looks something like the sketch in Figure 10-12. This solution decreases exponentially as $x \to -\infty$ and therefore

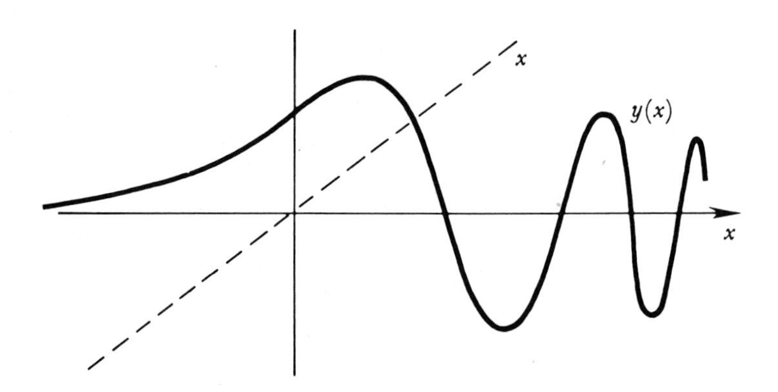

FIGURE 10-12 Sketch of the solution of (10-46), which has a decreasing exponential character for $x < 0$

we assume it has a Fourier transform. Let

$$g(\omega) = \int_{-\infty}^{\infty} y(x)e^{-i\omega x}\,dx$$

Then the Fourier transform of our differential equation (10-46) is

$$-\omega^2 g(\omega) + i\frac{dg}{d\omega} = 0$$

which gives

$$g(\omega) = Ae^{-i(\omega^3/3)} \qquad (A = \text{arbitrary constant})$$

Thus, inverting the transform,

$$y(x) = A\int_{-\infty}^{\infty} \frac{d\omega}{2\pi} \exp\left[i\left(\omega x - \frac{\omega^3}{3}\right)\right] \tag{10-47}$$

We shall drop the irrelevant constant $A/2\pi$. The solution (10-47) is good for all x.

Unfortunately, this integral is nontrivial. Partly for this reason it has a name, being known as an *Airy integral.* For the present application of finding the WKB connection formula, we only need the asymptotic forms for large $|x|$, which may be found by the saddle-point method of Section 8-2. The two cases $x \to \pm\infty$ must be considered separately:

1. Let $x \to +\infty$. The saddle points occur on the real axis, and we must go over both of them. Using the notation introduced in Section 8-2.

$$h(\omega) = i\left(\omega - \frac{\omega^3}{3x}\right) \qquad h'(\omega) = i\left(1 - \frac{\omega^2}{x}\right) \qquad h''(\omega) = -\frac{2i\omega}{x}$$

$$h'(\omega_0) = 0 \Rightarrow \omega_0 = \pm\sqrt{x}$$

$$h''(\omega_0) = \mp\frac{2i}{\sqrt{x}} \qquad |h''(\omega_0)| = \frac{2}{\sqrt{x}} \qquad \varphi = \mp\frac{\pi}{2}$$

$$\theta = \mp\frac{\pi}{4}$$

$$h(\omega_0) = \pm\tfrac{2}{3}i\sqrt{x}$$

Therefore,

$$\begin{aligned} y(x) &\sim \sum_{\pm}\sqrt{\frac{2\pi\sqrt{x}}{2x}}\exp\left[\pm\tfrac{2}{3}ix^{3/2}\right]\exp\left[\mp\frac{i\pi}{4}\right] \\ y(x) &\sim \frac{2\sqrt{\pi}}{x^{1/4}}\cos\left(\tfrac{2}{3}x^{3/2} - \frac{\pi}{4}\right) \end{aligned} \tag{10-48}$$

2. Let $x \to -\infty$. Now the large positive parameter is $-x$, and

$$h(\omega) = -i\left(\omega - \frac{\omega^3}{3x}\right) \qquad h'(\omega) = -i\left(1 - \frac{\omega^2}{x}\right) \qquad h''(\omega) = \frac{2i\omega}{x}$$

$$h'(\omega_0) = 0 \Rightarrow \omega_0 = \pm i\sqrt{-x}$$

The saddle points now occur on the imaginary axis.

$$h''(\omega_0) = \frac{2i}{x}(\pm i\sqrt{-x}) = \frac{\pm 2}{\sqrt{-x}}$$

$$|h''(\omega_0)| = \frac{2}{\sqrt{-x}} \qquad \varphi = \begin{pmatrix} 0 \\ \pi \end{pmatrix}$$

The topography of the surface $u = \operatorname{Re} h(\omega)$ is indicated in Figure 10-13. We go through the saddle point at $\omega_2 = -i\sqrt{-x}$, with $\theta = 0$. Then

$$\begin{aligned} y(x) &\sim \sqrt{\frac{2\pi\sqrt{-x}}{2(-x)}}\exp\left[-(-x)\tfrac{2}{3}\sqrt{-x}\right] \\ y(x) &\sim \frac{\sqrt{\pi}}{(-x)^{1/4}}\exp\left[-\tfrac{2}{3}(-x)^{3/2}\right] \end{aligned} \tag{10-49}$$

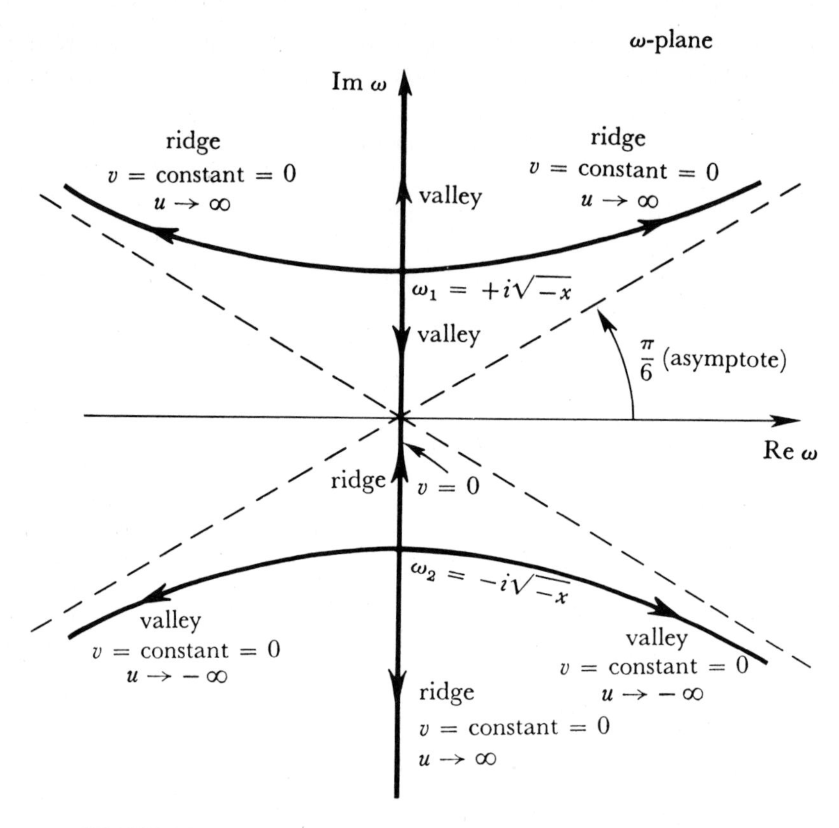

FIGURE 10-13 Topography of the surface $u = \text{Re}\, h(\omega)$ in the ω-plane, near the saddle points $\omega_{1,2} = \pm i\sqrt{-x}$, for the function $h(\omega) = -i[\omega + \omega^3/3(-x)]$. If $\omega = re^{i\theta}$, $u = \text{Re}\, h(\omega) = r \sin\theta + r^3/3(-x) \sin 3\theta$, and $v = \text{Im}\, h(\omega) = -r\cos\theta - r^3/3(-x)\cos 3\theta$.

We see that our two asymptotic formulas (10-48) and (10-49) are in fact the respective WKB solutions, for the case in which only an exponentially decreasing component is present for $x < 0$.

Generalizing these results to equation (10-38) for the case where $f(x)$ is approximately linear throughout the region under consideration, we have proved the *connection formula*

$$\frac{1}{\sqrt[4]{-f(x)}} \exp\left[-\int_x^{x_0} \sqrt{-f(x)}\, dx\right] \to 2\frac{1}{\sqrt[4]{-f(x)}} \cos\left[\int_{x_0}^{x} \sqrt{f(x)}\, dx - \frac{\pi}{4}\right] \tag{10-50}$$

If any of the exponentially increasing component were present, we would be unable to Fourier transform the equation and hence could not use this technique. In general, then, we cannot determine the relative amount of the two oscillating terms if there is any increasing exponential. We can, however, assume that an oscillating solution different from (10-50) requires some increasing exponential behavior to be present.

Suppose we write the general solution in the form given by equations (10-44) and (10-45). Since the equation is linear the coefficients must be linearly related:

$$c = Aa + Bb \qquad d = Ca + Db \tag{10-51}$$

From equation (10-50) we have found that $a = 0$ implies $c = be^{-i\pi/4}$, $d = be^{i\pi/4}$ so $B = e^{-i\pi/4}$, $D = e^{i\pi/4}$

In the $f(x) > 0$ region we may expect some oscillating behavior

$$\frac{1}{\sqrt[4]{f(x)}} \cos\left[\int_{x_0}^{x} \sqrt{f(x)}\, dx + \phi\right] \tag{10-52}$$

where ordinarily $\phi \neq -\pi/4$. We might expect to be able to determine from this the "large" (exponentially increasing) component in the $f(x) < 0$ region, even though information about the "small" (exponentially decreasing) component was lost. We therefore try inverting our equations (10-51) to obtain

$$a = \frac{Dc - Bd}{AD - BC}$$

$$b = \frac{-Cc + Ad}{AD - BC}$$

For the case of behavior in the $x \gg x_0$ region like equation (10-52) we have $c = (R/2)e^{i\phi}$, $d = (R/2)e^{-i\phi}$ for some (possibly complex) number R. Hence

$$a = \frac{2i \sin\left(\phi + \dfrac{\pi}{4}\right)\dfrac{R}{2}}{AD - BC}$$

The determinant of our transformation, $AD - BC$, can be computed by using the fact that the Wronskian $y_1 y_2' - y_2 y_1'$ of any two solutions of the equation $y'' + f(x)y = 0$ must be a constant K. This constant can be computed for $x \ll x_0$ and $x \gg x_0$, leading to the relation

$$K = e^{i\pi/2}(a_1 b_2 - a_2 b_1) = (c_1 d_2 - c_2 d_1) = (AD - BC)(a_1 b_2 - a_2 b_1) \tag{10-53}$$

(see Problem 10-24). We thus find that $AD - BC = i$, and the amount of increasing exponential is determined to be:

$$\sin\left(\phi + \frac{\pi}{4}\right) \frac{1}{\sqrt[4]{-f(x)}} \exp\left[\int_{x}^{x_0} \sqrt{-f(x)}\, dx\right]$$

$$\leftarrow \frac{1}{\sqrt[4]{f(x)}} \cos\left[\int_{x_0}^{x} \sqrt{f(x)}\, dx + \phi\right] \tag{10-54}$$

provided ϕ is not too near $-\pi/4$. The arrows cannot be reversed, as we have seen, since the phase ϕ cannot be determined if we only know how much large exponential is present.

It is very easy to remember the connection formulas (10-50) and (10-54) by drawing qualitative pictures like Figure 10-12. Consider first $y_1(x)$ with the decreasing exponential at the left of x_0. To the right of x_0, $y_1(x)$ looks like a cosine wave for which the phase is between $-\pi/2$ and 0 when $x = x_0$. We remember the phase is $-\pi/4$. Furthermore, the amplitude of the cosine wave on the right "looks" larger than that of the exponential on the left. We remember it is twice as large. These comments reproduce the result (10-50). A similar mnemonic for remembering (10-54) results from considering a solution $y_2(x)$ that is 90° out of phase with $y_1(x)$ in the right-hand region. By this same procedure, it is a straightforward matter to write down connection formulas similar to (10-50) and (10-54) for the case where the slope of $f(x)$ is negative at x_0 instead of positive.

We shall conclude by giving an example of the use of the WKB method in quantum mechanics. Consider the Schrödinger equation for a particle in a potential well

$$\frac{d^2\psi}{dx^2} + \frac{2m}{\hbar^2}[E - V(x)]\psi = 0$$

with $V(x)$ as shown in Figure 10-14.

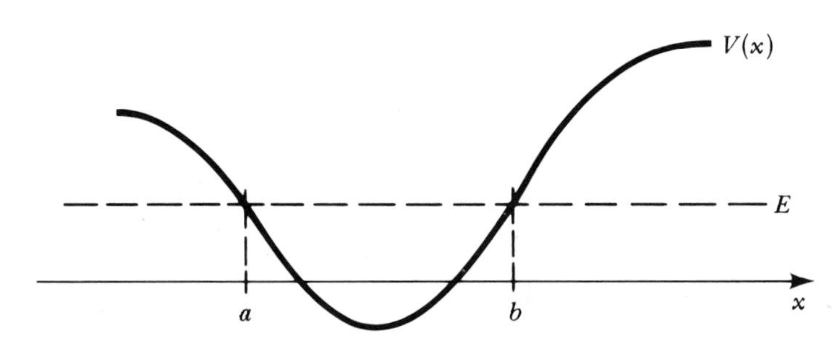

FIGURE 10-14 A typical one-dimensional potential well

$$f(x) = \frac{2m}{\hbar^2}[E - V(x)] \qquad \text{is } \begin{cases} \text{positive for } a < x < b \\ \text{negative for } x < a,\ x > b \end{cases}$$

If $\psi(x)$ is to be bounded for $x < a$, then in $a < x < b$ our connection formula (10-50) tells us

$$\psi(x) \approx \frac{A}{(E - V)^{1/4}} \cos\left[\int_a^x \sqrt{\frac{2m}{\hbar^2}}(E - V)\, dx - \frac{\pi}{4}\right]$$

If $\psi(x)$ is to be bounded for $x > b$, then in $a < x < b$ similarly

$$\psi(x) \simeq \frac{B}{(E-V)^{1/4}} \cos\left[\int_x^b \sqrt{\frac{2m}{\hbar^2}(E-V)}\,dx - \frac{\pi}{4}\right]$$

These two expressions must be the same, which gives the condition

$$\int_a^b \sqrt{2m(E-V)}\,dx = (n+\tfrac{1}{2})\pi\hbar$$

This result is very similar to the Bohr-Sommerfeld quantization condition of pre-1925 quantum mechanics.

Another common quantum mechanics application of the WKB method is transmission through a potential barrier (see Problem 10-31). These topics are discussed in many books on quantum theory, such as Schiff (29), Section 34; Merzbacher (26), Chapter VII; Landau and Lifshitz (24), Chapter VII. One feature we have not discussed that is important in applications is the estimation of errors in the approximation. This is covered in many of the quantum mechanics texts, as well as in Mathews and Walker (25), Chapter I.

PROBLEMS

10-24 (a) Consider the Wronskian $y_1 y_2' - y_2 y_1'$ of two solutions of the differential equation $y'' + f(x)y = 0$. By differentiating, prove that the Wronskian is a constant.

(b) Compute explicitly the Wronskian of our two independent solutions in the WKB approximation.

(c) Derive equation (10-53).

10-25 Consider the differential equation $y'' + xy = 0$.

(a) If $y \sim \dfrac{1}{x^{1/4}} \cos \tfrac{2}{3}x^{3/2}$ as $x \to \infty$, $y \sim\ ?$ as $x \to -\infty$.

(b) If $y \sim \dfrac{1}{(-x)^{1/4}} \exp\left[-\tfrac{2}{3}(-x)^{3/2}\right]$ as $x \to -\infty$, $y \sim\ ?$ as $x \to +\infty$.

(c) If $y \sim \dfrac{1}{(-x)^{1/4}} \exp\left[+\tfrac{2}{3}(-x)^{3/2}\right]$ as $x \to -\infty$, $y \sim\ ?$ as $x \to +\infty$.

Note: The answer to *one* of (a), (b), (c) is not defined. Be sure to indicate which one is not defined, as well as giving correct answers for the other two.

10-26 Use the WKB method to find approximate negative values of the constant E for which the equation

$$\frac{d^2y}{dx^2} + [E - V(x)]y = 0$$

has a solution that is finite for all x between $x = -\infty$ and $x = +\infty$, *inclusive.* $V(x)$ is the function shown in Figure 10-15.

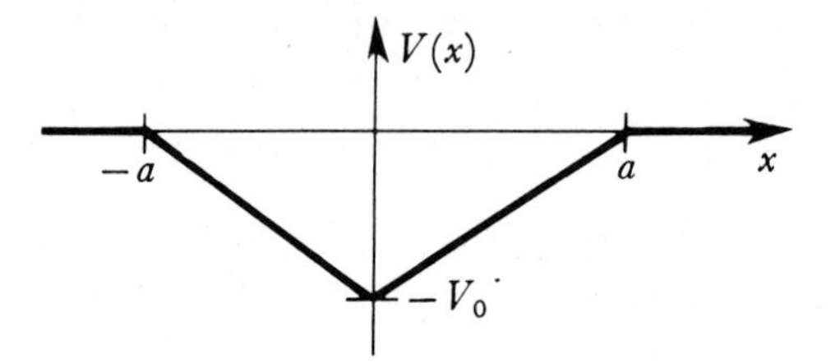

FIGURE 10-15 Potential well for Problem 10-26

10-27 Find a good approximation, for x large and positive, to the solution of the equation

$$y'' - \frac{3}{x}y' + \left(\frac{15}{4x^2} + x^{1/2}\right)y = 0$$

Hint: Remove first derivative term.

10-28 Obtain an approximate formula for the Bessel function $J_m(x)$ by the WKB method and give the limiting form of this expression for large $x(x \gg m)$. Do not worry about getting the constant in front correct. You may assume $m \gg \frac{1}{2}$.

10-29 Consider the solution of the differential equation

$$\frac{d^2y}{dx^2} + x^2y = 0$$

which has zero value and unit slope at $x = 10$.

(a) Give (approximately) the location of the next zero of $y(x)$ greater than 10.

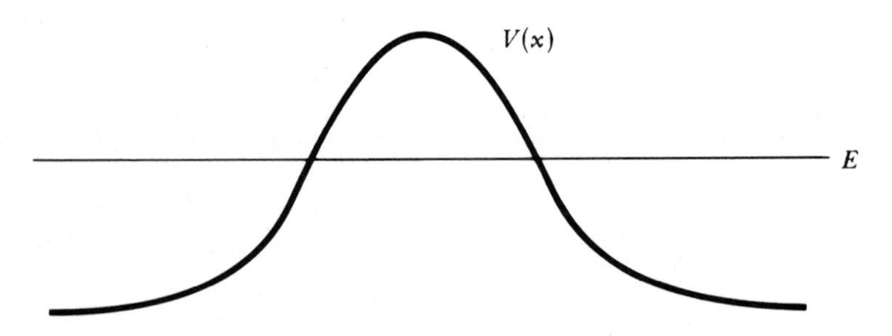

FIGURE 10-16 Potential barrier for Problem 10-31

(b) Give (approximately) the value of $y(x)$ at its first maximum for $x > 10$.

10-30 Consider a solution $y_1(x)$ of the differentiel equation

$$y'' + x^2 y = 0$$

such that $y_1(x)$ has a zero at $x = 5$. Give approximately the location of the 25th zero beyond the one at $x = 5$.

10-31 Consider a quantum mechanical wave of energy $E < V_{\max}$ incident from the left on the potential barrier shown in Figure 10-16. Use the WKB method to determine the transmitted and reflected waves.

SUMMARY

If $f(z)$ has no singularities in the upper half plane and $|f(z)| \to 0$ when $|z| \to \infty$, an application of Cauchy's integral theorem leads to

$$\operatorname{Re} F(x) = \frac{1}{\pi} P \int_{-\infty}^{\infty} \frac{\operatorname{Im} F(x')\, dx'}{x' - x}$$

$$\operatorname{Im} F(x) = -\frac{1}{\pi} P \int_{-\infty}^{\infty} \frac{\operatorname{Re} F(x')\, dx'}{x' - x}$$

dispersion relations for $F(x) = \lim_{\varepsilon \to 0} f(x + i\varepsilon)$. Evaluation of the integrals may be simplified in many cases by symmetries of the integrand. Subtracted dispersion relations may be used if $f(z)/z$ dies off properly at infinity.

Some important special functions are defined in terms of contour integrals. These include

$$\Gamma(z) = \int_0^{\infty} x^{z-1} e^{-x}\, dx \qquad \text{(Gamma function)}$$

$$B(r, s) = \int_0^1 x^{r-1}(1 - x)^{s-1}\, dx \qquad \text{(Beta function)}$$

$$\operatorname{Ei}(x) = \int_{-\infty}^{x} \frac{e^t\, dt}{t} \qquad \text{(Exponential integral)}$$

$$\operatorname{erf}(x) = \frac{2}{\sqrt{\pi}} \int_0^x e^{-t^2}\, dt \qquad \text{(Error function)}$$

A series $\sum_0^{\infty} a_n(z)$ is asymptotic to $S(z)$ if $\sum_0^m a_n(z)$ approaches the limit $S(z)$ faster than $1/z^m$ as $z \to \infty$ for a given m. Note that $\sum_0^{\infty} a_n(z)$ need not converge for a given z.

We used contour integral representations of Legendre functions to derive the following properties:

$$\frac{1}{\sqrt{1-2hz+h^2}} = \sum_{n=0}^{\infty} h^n P_n(z)$$

$$\frac{1}{|\mathbf{x}-\mathbf{x}'|} = \sum_{l=0}^{\infty} \frac{r_<^l}{r_>^{l+1}} P_l(\cos\theta)$$

$P_\alpha(z)$ has a branch cut from $z = -1$ to $z = -\infty$ if α is noninteger

$$P_\alpha(\cos\theta) = \frac{1}{\pi}\sum_{n=0}^{\infty} P_n(\cos\theta)(-1)^n \sin\pi\alpha \left[\frac{1}{\alpha-n} - \frac{1}{\alpha+n+1}\right]$$

for $-1 < \cos\theta < 1$

For Bessel functions we first derived the generating function from our previous work with the $J_n(x)$:

$$\exp\left[z\left(h-\frac{1}{h}\right)/2\right] = \sum_{n=-\infty}^{\infty} h^n J_n(z)$$

This was then used to obtain a contour integral representation for $J_n(z)$. We generalized this representation to find two contour integrals $H_n^{(1)}(x)$ and $H_n^{(2)}(x)$ that satisfied Bessel's equation. These were related to the J_n and Y_n discussed previously by

$$H_n^{(1)}(x) = J_n(x) + iY_n(x)$$

$$H_n^{(2)}(x) = J_n(x) - iY_n(x)$$

Application of the method of steepest descent to the contour integrals gave the asymptotic forms

$$H_n^{(1)}(z) \approx \sqrt{\frac{2}{\pi z}}\, e^{i(z-n\pi/2-\pi/4)}$$

$$H_n^{(2)}(z) \approx \sqrt{\frac{2}{\pi z}}\, e^{-i(z-n\pi/2-\pi/4)}$$

for large z.

The WKB connection formulas are

$$\frac{1}{\sqrt[4]{-f(x)}}\exp\left[-\int_x^{x_0}\sqrt{-f(x)}\,dx\right] \rightarrow \frac{2}{\sqrt[4]{f(x)}}\cos\left[\int_{x_0}^{x}\sqrt{f(x)}\,dx - \frac{\pi}{4}\right]$$

$$\sin\frac{\left(\phi+\dfrac{\pi}{4}\right)}{\sqrt[4]{-f(x)}}\exp\left[+\int_x^{x_0}\sqrt{-f(x)}\right] \leftarrow \frac{1}{\sqrt[4]{f(x)}}\cos\left[\int_{x_0}^{x}\sqrt{f(x)}\,dx + \phi\right]$$

relating solutions of the equation $y'' + fy = 0$ in the regions where $f(x)$ is negative (left-hand side) and positive (right-hand side). The arrows cannot be reversed.

CHAPTER 11 Integral Equations

This chapter is an introduction to the theory and methods of solution of integral equations. The discussion focusses on three aspects of solution of nonsingular integral equations. First, we point out a number of approaches that are "intuitively obvious" methods of attack: Section 11-2 explains the simplifications introduced by a degenerate (separable) kernel, and Section 11-3 demonstrates use of integral transforms, iterative solutions, and other devices. Second, we show how the theory of linear operators developed in connection with differential equations can be applied to linear integral equations (Schmidt-Hilbert theory). And third, in Section 11-5, we discuss the Wiener-Hopf technique, which is one of a number of more advanced transform techniques developed for solution of particular types of integral equations. This is a particularly beautiful example of the way in which singularities in a function of a complex variable may determine the function.

11-1 CLASSIFICATION

An integral equation is an equation in which an unknown function appears under an integral sign. We shall consider only a few simple types of such equations.

The general *linear* integral equation involving a single unknown function $f(x)$ may be written

$$\lambda \int_a^b K(x, y) f(y)\, dy + g(x) = h(x) f(x) \tag{11-1}$$

$h(x)$ and $g(x)$ are known functions of x, λ is a constant parameter, often playing the role of an eigenvalue, and $K(x, y)$ is called the *kernel* of the integral equation. If $h(x) = 0$, we have a *Fredholm equation of the first kind*; if $h(x) = 1$, we have a *Fredholm equation of the second kind.* In either case, if $g(x) = 0$, the equation is homogeneous.

Sometimes $K(x, y)$ is zero for $y > x$. In that case, the upper limit on the integral is x, and the equation is called a *Volterra equation.*

It is often convenient to write equation (11-1) in symbolic form:

$$\lambda Kf + g = hf$$

where K is the *operator* that means "multiply by the kernel $K(x, y)$ and integrate over y from a to b." Equations in this form may be easily compared with operator equations involving matrix or differential operators.

11-2 DEGENERATE KERNELS

If the kernel $K(x, y)$ is of the form

$$K(x, y) = \sum_{i=1}^{n} \varphi_i(x)\psi_i(y) \tag{11-2}$$

it is said to be *degenerate.* Integral equations with degenerate kernels may be solved by elementary techniques. Rather than giving a general discussion, we shall simply give an example.

Example

$$f(x) = x + \lambda \int_0^1 (xy^2 + x^2y)f(y)\, dy \tag{11-3}$$

Define

$$A = \int_0^1 y^2 f(y)\, dy \qquad B = \int_0^1 yf(y)\, dy \tag{11-4}$$

Then (11-3) becomes

$$f(x) = x + \lambda Ax + \lambda Bx^2 \tag{11-5}$$

Now substitute (11-5) back into the defining equations (11-4) for A and B:

$$\begin{aligned} A &= \tfrac{1}{4} + \tfrac{1}{4}\lambda A + \tfrac{1}{5}\lambda B \\ B &= \tfrac{1}{3} + \tfrac{1}{3}\lambda A + \tfrac{1}{4}\lambda B \end{aligned} \tag{11-6}$$

The solution of equations (11-6) is

$$A = \frac{60 + \lambda}{240 - 120\lambda - \lambda^2} \qquad B = \frac{80}{240 - 120\lambda - \lambda^2}$$

so that the solution of our original integral equation is

$$f(x) = \frac{(240 - 60\lambda)x + 80\lambda x^2}{240 - 120\lambda - \lambda^2} \tag{11-7}$$

Note that there are two values of λ for which our solution (11-7) becomes infinite. This sounds familiar; we call these the *eigenvalues* of the integral equation. The homogeneous equation has nontrivial solutions only if λ is one of these eigenvalues; these solutions are called eigenfunctions of the operator K.

Thus, if our kernel is degenerate, the problem of solving an integral equation is reduced to that of solving a system of algebraic equations, a much more familiar subject. If the degenerate kernel (11-2) contains N terms, we see that there will be N eigenvalues, not necessarily all different.

By observing that any reasonably behaved kernel can be written as an infinite series of degenerate kernels, Fredholm deduced a set of theorems that we shall state without proof. They should seem quite reasonable, however, from the student's previous experience with eigenvalues and algebraic equations. The theorems, as stated, refer to real kernels.

1. *Either* the inhomogeneous equation

$$f(x) = g(x) + \lambda \int_a^b K(x, y) f(y)\, dy \tag{11-8}$$

has a (unique) solution for *any* function $g(x)$ (that is, λ is *not* an eigenvalue), or the homogeneous equation

$$f(x) = \lambda \int_a^b K(x, y) f(y)\, dy$$

has at least one nontrivial solution (λ *is* an eigenvalue and the solution is an eigenfunction).

2. If λ is not an eigenvalue (first alternative), then λ is also not an eigenvalue of the "transposed" equation

$$f(x) = g(x) + \lambda \int_a^b K(y, x) f(y)\, dy$$

while if λ is an eigenvalue (second alternative), λ is also an eigenvalue of the transposed equation; that is,

$$f(x) = \lambda \int_a^b K(y, x) f(y)\, dy \tag{11-9}$$

has at least one nontrivial solution.

3. If λ is an eigenvalue, the inhomogeneous equation (11-8) has a solution if, and only if,

$$\int_a^b \varphi(x) g(x)\, dx = 0 \tag{11-10}$$

for every function $\varphi(x)$ that obeys the transposed homogeneous equation (11-9).

We shall not prove these theorems here. They are proved quite nicely in

Courant and Hilbert (10), Vol. 1. Statement 2, for example, is the analog of the fact that a matrix and its transpose have the same eigenvalues. The necessity of the condition (11-10) follows immediately if we multiply (11-8) by $\varphi(x)$ and integrate over x. We note in passing that the theorems strictly apply only to integral equations with *bounded* kernels and *finite* limits of integration, that is, to *nonsingular* integral equations. The theory of *singular* integral equations is a different matter.

PROBLEMS

11-1 Solve the integral equations

(a) $u(x) = e^x + \lambda \int_0^1 xtu(t)\, dt$

(b) $u(x) = \lambda \int_0^\pi u(t) \sin(x - t)\, dt$ (two eigenvalues; two solutions)

11-2 Solve

$$f(x) = x^2 + \int_0^1 xyf(y)\, dy$$

11-3 Solve the integral equations

(a) $u(x) = x + \lambda \int_0^\infty e^{-y} u(y)\, dy$

(b) $u(x) = \lambda \int_{-x}^{x} u(y) \cos y\, dy$

11-4 (a) Find the eigenfunction(s) and eigenvalue(s) of the integral equation

$$f(x) = \lambda \int_0^1 e^{x-y} f(y)\, dy$$

(b) Solve the integral equation

$$f(x) = e^x + \lambda \int_0^1 e^{x-y} f(y)\, dy$$

11-3 MISCELLANEOUS DEVICES

Volterra equations can often be turned into differential equations by differentiating.

Example

$$u(x) = x + \int_0^x xyu(y)\,dy = x + xf(x)$$

where

$$f(x) = \int_0^x yu(y)\,dy$$

Then

$$f'(x) = xu(x) = x[x + xf(x)]$$

Solving this differential equation gives

$$f(x) = -1 + Ce^{x^3/3}$$

so that

$$u(x) = Cxe^{x^3/3}$$

To find C, we substitute back into the original integral equation. The result is $C = 1$.

If the kernel is a function of $(x - y)$ only, a so-called *displacement kernel*, and if the limits are $-\infty$ to $+\infty$, we can use Fourier transforms. Consider the equation

$$f(x) = \varphi(x) + \lambda \int_{-\infty}^{\infty} K(x - y) f(y)\,dy \tag{11-11}$$

Take Fourier transforms (indicated by bars)

$$\int_{-\infty}^{\infty} dx\, f(x) e^{-ikx} = \bar{f}(k)$$

Then

$$\int_{-\infty}^{\infty} dx\, e^{-ikx} \int_{-\infty}^{\infty} K(x - y) f(y)\,dy = \bar{K}(k)\bar{f}(k)$$

by the convolution theorem (p. 209), and the transform of our integral equation (11-11) is

$$\bar{f}(k) = \bar{\varphi}(k) + \lambda \bar{K}(k) \bar{f}(k)$$

Therefore,

$$\bar{f}(k) = \frac{\bar{\varphi}(k)}{1 - \lambda \bar{K}(k)}$$

If we can invert this transform, we can solve the problem.

If the limits are 0 to x with a displacement kernel, and our functions vanish for $x < 0$, we use a Laplace transform, since it has the appropriate form of convolution integral for this case.

SERIES SOLUTIONS

A straightforward approach to solving the integral equation

$$f(x) = g(x) + \lambda \int_a^b K(x, y) f(y)\, dy \tag{11-12}$$

is iteration; we begin with the approximation

$$f(x) \approx g(x)$$

This is substituted into the original equation under the integral sign to obtain a second approximation, and the process is then iterated. The resulting series,

$$f(x) = g(x) + \lambda \int_a^b K(x, y) g(y)\, dy + \lambda^2 \int_a^b dy \int_a^b dy' K(x, y) K(y, y') g(y') + \cdots \tag{11-13}$$

is known as the *Neumann series*, or *Neumann solution*, of the integral equation (11-12). It will converge for sufficiently small λ, provided the kernel $K(x, y)$ is bounded.

Example

In the quantum mechanical theory of scattering by a potential $V(\mathbf{r})$ we seek a solution of the Schrödinger equation

$$\nabla^2\psi(\mathbf{r}) - \frac{2m}{\hbar^2} V(\mathbf{r})\psi(\mathbf{r}) + k^2\psi(\mathbf{r}) = 0$$

with boundary conditions that $\psi(\mathbf{r})e^{-iEt/\hbar}$ represent an incident plane wave, with wave vector $\mathbf{k}_0$, plus outgoing waves at $r \to \infty$. $k^2 = k_0^2 = 2mE/\hbar^2$.

The equation

$$\nabla^2\psi + k^2\psi = f(\mathbf{r})$$

with outgoing wave boundary conditions for the function $\psi(\mathbf{r})e^{-i\omega t}$ has the Green's function (see Chapter 12)

$$G(\mathbf{r}, \mathbf{r}') = -\frac{1}{4\pi} \frac{e^{ik|\mathbf{r}-\mathbf{r}'|}}{|\mathbf{r} - \mathbf{r}'|}$$

Thus we may transform the scattering problem into the integral equation

$$\psi(\mathbf{r}) = e^{i\mathbf{k}_0\cdot\mathbf{r}} - \frac{2m}{4\pi\hbar^2}\int d^3\mathbf{r}'\,\frac{e^{ik|\mathbf{r}-\mathbf{r}'|}}{|\mathbf{r}-\mathbf{r}'|}V(\mathbf{r}')\psi(\mathbf{r}')$$

where the term $e^{i\mathbf{k}_0\cdot\mathbf{r}}$ is the complementary function, adjusted to fit the boundary conditions. The solution may be written as the Neumann series (11-13). The first iteration gives already a very important result, known as the *Born approximation*,

$$\psi(\mathbf{r}) \approx e^{i\mathbf{k}_0\cdot\mathbf{r}} - \frac{m}{2\pi\hbar^2}\int d^3\mathbf{r}'\,\frac{e^{ik|\mathbf{r}-\mathbf{r}'|}}{|\mathbf{r}-\mathbf{r}'|}V(\mathbf{r}')e^{i\mathbf{k}_0\cdot\mathbf{r}'}$$

Returning to the integral equation (11-12), a more elegant, if more complicated, series solution was found by Fredholm, by subdividing the interval $a < x < b$, replacing the integral by a sum, solving the resulting algebraic equations, and then passing to the limit of infinitely many subdivisions. The result is that the solution of the integral equation (11-12) is

$$f(x) = g(x) + \lambda\int_a^b R(x, y, \lambda)g(y)\,dy \tag{11-14}$$

where $R(x, y, \lambda)$, the so-called *resolvent kernel*, is the ratio of two infinite series.

$$R(x, y, \lambda) = \frac{D(x, y, \lambda)}{D(\lambda)}$$

$$D(x, y, \lambda) = K(x, y) - \lambda\int dz\begin{vmatrix} K(x, y) & K(x, z) \\ K(z, y) & K(z, z)\end{vmatrix} + \frac{\lambda^2}{2!}\int dz\,dz'\begin{vmatrix} K(x, y) & K(x, z) & K(x, z') \\ K(z, y) & K(z, z) & K(z, z') \\ K(z', y) & K(z', z) & K(z', z')\end{vmatrix} - + \cdots \tag{11-15}$$

$$D(\lambda) = 1 - \lambda\int dz\,K(z, z) + \frac{\lambda^2}{2!}\int dz\,dz'\begin{vmatrix} K(z, z) & K(z, z') \\ K(z', z) & K(z', z')\end{vmatrix} - + \cdots \tag{11-16}$$

The importance of the Fredholm solution is that both power series (11-15) and (11-16) are guaranteed to converge (unlike the Neumann series, which often diverges). The eigenvalues can be found by looking for the zeros of the denominator function $D(\lambda)$.

PROBLEMS

11-5 Solve the following equations:

(a) $f(x) = x + \int_0^x f(y)\,dy$

(b) $u(x) = x^{\varepsilon} + \lambda \int_0^x (x - y)^{\varepsilon} u(y)\, dy$

for ε integer.

11-6 Solve the integral equation

$$f(x) = x + \lambda \int_0^1 y(x + y) f(y)\, dy$$

keeping terms through λ^2,
(a) by Fredholm's method.
(b) by Neumann's method.

11-7 Solve for $u(x)$:

(a) $u(x) = e^{-x^2} + \lambda \int_{-\infty}^{\infty} e^{-(x-y)^2} u(y)\, dy$

(Do not worry about expressing the answer in closed form.)

(b) $u(x) = \dfrac{1}{\cosh x} + \lambda \int_{-\infty}^{\infty} \dfrac{u(y)}{\cosh (x - y)}\, dy$

11-8 Solve the integral equation

$$f(x) = e^{-|x|} + \lambda \int_0^{\infty} f(y) \cos xy\, dy$$

Hint: Take the cosine transform of the integral equation.

11-9 Solve the integral equation

$$f(x) = e^{-|x|} + \lambda \int_{-\infty}^{\infty} e^{-|x-y|} f(y)\, dy$$

where $f(x)$ is to remain finite for $x \to \pm\infty$.

11-4 SCHMIDT-HILBERT THEORY

We shall now consider an approach quite different from the Neumann and Fredholm series. This approach is based on considering the eigenfunctions and eigenvalues of the homogeneous integral equation.

A kernel $K(x, y)$ is said to be *symmetric* if

$$K(x, y) = K(y, x)$$

and *Hermitian* if

$$K(x, y) = K^*(y, x)$$

The eigenvalues of a Hermitian kernel are real, and eigenfunctions belonging to different eigenvalues are orthogonal; two functions $f(x)$ and $g(x)$ are said to be orthogonal if

$$\int f^*(x)g(x)\,dx = 0$$

The proofs of these statements are practically identical with the corresponding proofs in Chapter 1, and will be left to the student to fill in. The identity is obvious if we write the integral equation in symbolic notation. The eigenvalue equation is then

$$\lambda K f = f$$

We shall restrict ourselves to Hermitian kernels, and begin by considering a degenerate one:

$$K(x, x') = \sum_{\alpha=1}^{N} f_\alpha(x) f_\alpha^*(x') \tag{11-17}$$

The corresponding homogeneous integral equation is

$$\begin{aligned} u(x) &= \lambda \int_a^b K(x, x')u(x')\,dx' \\ &= \lambda \sum_\alpha f_\alpha(x) \int f_\alpha^*(x')u(x')\,dx' \end{aligned}$$

Let us define

$$\int f_\alpha^*(x')u(x')\,dx' = c_\alpha \tag{11-18}$$

so that

$$u(x) = \lambda \sum_\alpha c_\alpha f_\alpha(x) \tag{11-19}$$

We see that any eigenfunction $u(x)$ corresponding to a finite eigenvalue λ may be expressed linearly in terms of the N functions $f_\alpha(x)$. Clearly there can be no more than N linearly independent eigenfunctions of this kind. [If the $f_\alpha(x)$ are not linearly independent, there may be fewer than N such eigenfunctions.] However, there may be an infinite number of eigenfunctions $g_n(x)$ corresponding to $\lambda = \infty$, $1/\lambda = 0$. Any function g_n is of this type if it is orthogonal to the kernel in the sense that $\int_a^b f_\alpha(x')g_n^*(x')\,dx' = 0$ for all the $f_\alpha(x)$. It is conventional to exclude these $g_n(x)$ from the list of eigenfunctions and to exclude $\lambda = \infty$ from the list of eigenvalues.

Substituting expression (11-19) into (11-18), and defining

$$A_{\alpha\beta} = \int f_\alpha^*(x) f_\beta(x)\,dx = A_{\beta\alpha}^*$$

we obtain

$$c_\alpha = \lambda \sum_\beta A_{\alpha\beta} c_\beta$$

Thus the eigenvalues λ are the reciprocals of the (nonzero) eigenvalues of the Hermitian matrix $A_{\alpha\beta}$. We know from Chapter 1 that the numbers $c_\alpha^{(i)}$, where the superscript i distinguishes the different eigenvectors of the matrix A, obey the orthogonality condition

$$c^{(i)} \cdot c^{(j)} = \sum_\alpha c_\alpha^{(i)*} c_\alpha^{(j)} = 0 \quad \text{if} \quad i \neq j$$

To see what normalization is convenient, let $u^{(i)}(x)$ denote the ith eigenfunction, and similarly for $u^{(j)}(x)$. Then

$$\begin{aligned}\int u^{(i)}(x)^* u^{(j)}(x)\, dx &= \lambda_i \lambda_j \sum_{\alpha\beta} c_\alpha^{(i)*} c_\beta^{(j)} \int f_\alpha(x)^* f_\beta(x)\, dx \\ &= \lambda_i \lambda_j \sum_{\alpha\beta} c_\alpha^{(i)*} c_\beta^{(j)} A_{\alpha\beta} \\ &= \lambda_i \sum_\alpha c_\alpha^{(i)*} c_\alpha^{(j)}\end{aligned}$$

Therefore, we choose

$$\sum_\alpha c_\alpha^{(i)*} c_\alpha^{(j)} = \frac{\delta_{ij}}{\lambda_i} \tag{11-20}$$

so that

$$\int u^{(i)}(x)^* u^{(j)}(x)\, dx = \delta_{ij}$$

that is, our eigenfunctions are orthonormal.

Note that according to (11-20) the numbers $\sqrt{\lambda_i}\, c_\alpha^{(i)}$ form a unitary matrix. Therefore

$$\sum_i \lambda_i c_\alpha^{(i)*} c_\beta^{(i)} = \delta_{\alpha\beta}$$

This is a useful orthogonality property. For example, it enables us to "solve" the relation

$$u^{(i)}(x) = \lambda_i \sum_\alpha c_\alpha^{(i)} f_\alpha(x)$$

for the $f_\alpha(x)$:

$$f_\alpha(x) = \sum_i c_\alpha^{(i)*} u^{(i)}(x)$$

This in turn enables us to express the kernel $K(x, x')$ given by (11-17) elegantly in terms of its eigenfunctions:

$$\begin{aligned} K(x, x') &= \sum_\alpha f_\alpha(x) f_\alpha^*(x') \\ &= \sum_\alpha \sum_{ij} c_\alpha^{(i)*} u^{(i)}(x) c_\alpha^{(j)} u^{(j)}(x')^* \\ &= \sum_i \frac{u^{(i)}(x) u^{(i)}(x')^*}{\lambda_i} \end{aligned} \tag{11-21}$$

The result, although shown here only for a degenerate kernel, is true in general for Hermitian kernels that are continuous and have only a finite number of eigenvalues of one sign. See mathematics books for proofs and conditions; Courant and Hilbert (10), Vol. 1, Chapter III, is particularly nice.

This approach can be extended to the inhomogeneous equation. We make use of a theorem that follows immediately from (11-21) but which also holds under more general conditions, as shown, for example, in Courant and Hilbert (10). Any function that can be represented "sourcewise" in terms of the kernel $K(x, y)$, that is, any function $\varphi(x)$ of the form

$$\varphi(x) = \int K(x, y) \psi(y)\, dy \qquad (\psi \text{ arbitrary and } K \text{ Hermitian})$$

can be expanded in a series of the eigenfunctions of $K(x, y)$:

$$\varphi(x) = \sum_i c_i u^{(i)}(x)$$

where

$$c_i = \int u^{(i)}(x)^* \varphi(x)\, dx = \frac{1}{\lambda_i} \int u^{(i)}(x)^* \psi(x)\, dx \tag{11-22}$$

In general, the eigenfunctions $u^{(i)}(x)$ do not form a complete set of functions. Not *any* function, but only sourcewise representable functions, can be expanded in terms of them. This is not an unfamiliar situation. For example, consider ordinary vectors in three dimensions, and an operator P that projects a vector $\mathbf{r}$ into the xy-plane. P has two independent eigenvectors corresponding to nonzero eigenvalues, $\mathbf{e}_x$ and $\mathbf{e}_y$, and any vector $\mathbf{r}'$ of the form $\mathbf{r}' = P\mathbf{r}$ can be expressed in terms of them. In order to expand a general vector, we must include the basis vector $\mathbf{e}_z$, which corresponds to a zero eigenvalue of P. Likewise, in order to expand a general function, we must include in our basis the $g_n(x)$ corresponding to zero eigenvalues of $1/\lambda$ for K.

Now we can solve the inhomogeneous equation

$$\varphi(x) = g(x) + \lambda \int K(x, y) \varphi(y)\, dy \tag{11-23}$$

We may expand

$$\varphi(x) - g(x) = \sum_i c_i u^{(i)}(x) \tag{11-24}$$

where, by (11-22),

$$c_i = \int dx\, u^{(i)}(x)^*[\varphi(x) - g(x)] = \lambda \frac{1}{\lambda_i} \int u^{(i)}(x)^* \varphi(x)\, dx \tag{11-25}$$

Thus if we define

$$d_i = \int dx\, u^{(i)}(x)^* \varphi(x) \qquad e_i = \int dx\, u^{(i)}(x)^* g(x)$$

(11-25) reads

$$c_i = d_i - e_i = \frac{\lambda}{\lambda_i} d_i$$

Therefore,

$$d_i = \frac{\lambda_i}{\lambda_i - \lambda} e_i$$

and

$$c_i = \frac{\lambda}{\lambda_i - \lambda} e_i$$

Substituting back into (11-24) gives for $\varphi(x)$

$$\varphi(x) = g(x) + \lambda \sum_i \frac{e_i u^{(i)}(x)}{\lambda_i - \lambda}$$

or

$$\varphi(x) = g(x) + \lambda \sum_i \frac{u^{(i)}(x)}{\lambda_i - \lambda} \int u^{(i)}(y)^* g(y)\, dy \tag{11-26}$$

If we compare this last result with the solution (11-14), we see that we have obtained a series representation for the resolvent kernel:

$$R(x, y, \lambda) = \sum_i \frac{u^{(i)}(x) u^{(i)}(y)^*}{\lambda_i - \lambda} \tag{11-27}$$

Symbolically, the original equation (11-23) was

$$\varphi = g + \lambda K \varphi \tag{11-28}$$

which has the formal solution

$$\varphi = (1 - \lambda K)^{-1} g$$

Note that the power series expansion of the operator in this expression gives just the Neumann series.

$$\varphi = g + \lambda K g + \lambda^2 K^2 g + \lambda^3 K^3 g + \cdots$$

Since, on the other hand, according to (11-26)

$$\varphi = (1 + \lambda R)g$$

we have the formal identity

$$(1 - \lambda K)^{-1} = 1 + \lambda R$$

It is interesting to compare our present results with those obtained for differential equations in Section (4-2). Suppose a differential operator L has eigenvalues k_i and eigenfunctions $u^{(i)}$ which satisfy given boundary conditions. The inhomogeneous differential equation has the form

$$L\varphi + f = k\varphi \tag{11-29}$$

where the inhomogeneous term f is a known function. The solution in terms of the Green's function is given, according to (4-42), by

$$\varphi(x) = \int G(x, x')f(x')\,dx' \tag{11-30}$$

where

$$G(x, x') = -\sum_i \frac{u^{(i)}(x)u^{(i)}(x')^*}{k_i - k} \tag{11-31}$$

Note that $G(x, x')$ is an integral kernel operator, and equation (11-30) is symbolically

$$\varphi = Gf$$

The differential equation (11-29) and the integral equation (11-28) have the same symbolic form if we set $k = 1/\lambda$ and $\lambda f = g$ and identify L with K. The solutions of these equations, according to (11-26) and (11-30), are

$$\varphi = (1 + \lambda R)g$$

$$\varphi = Gf = \frac{1}{\lambda}Gg$$

We may therefore expect that the operators $(1 + \lambda R)$ and $(1/\lambda)G$ can be written in the same form in terms of the corresponding eigenfunctions and eigenvalues. To verify this, write (11-31) with $k_i = 1/\lambda_i$ and $k = 1/\lambda$. Then

$$-\frac{1}{k_i - k} = \frac{\lambda\lambda_i}{\lambda_i - \lambda} = \lambda\left(1 + \frac{\lambda}{\lambda_i - \lambda}\right)$$

and (11-31) becomes

$$\begin{aligned} G(x, x') &= \lambda \sum_i u^{(i)}(x)u^{(i)}(x')^* + \lambda^2 \sum_i \frac{u^{(i)}(x)u^{(i)}(x')^*}{\lambda_i - \lambda} \\ &= \lambda\delta(x - x') + \lambda^2 \sum_i \frac{u^{(i)}(x)u^{(i)}(x')^*}{\lambda_i - \lambda} \end{aligned}$$

where we have used the completeness relation (2-7). The integral kernel $\delta(x - x')$ is equivalent to the identity operator, so we have

$$\frac{1}{\lambda} G = 1 + \lambda \sum_i \frac{u^{(i)}(x) u^{(i)}(x')^*}{\lambda_i - \lambda}$$

which has the same form as $1 + \lambda R$ with R given by the expansion (11-27).

PROBLEMS

11-10 Let $\varphi_\alpha(x)$ be a complete set of orthogonal functions, and expand the kernel $K(x, x')$ in the series

$$K(x, x') = \sum_{\alpha\beta} c_{\alpha\beta} \varphi_\alpha(x) \varphi_\beta^*(x')$$

(a) What is the condition that the coefficients $c_{\alpha\beta}$ must satisfy if $K(x, x')$ is to be a *Hermitian* kernel?

(b) Show that such a Hermitian kernel can always be written in the form (11-17), provided only a finite number of coefficients $c_{\alpha\beta}$ are nonzero.

11-5 WIENER-HOPF METHOD

A different integral transform method can be used to solve certain integral equations, said to be of the Wiener-Hopf type:

$$f(x) = \varphi(x) + \int_0^\infty K(x - y) f(y)\, dy \qquad (-\infty < x < \infty)$$

The displacement kernel and half-infinite range of integration are the significant features. Let $f(x) = f_+(x) + f_-(x)$ where $f_+(x) = 0$ for $x > 0$ and $f_-(x) = 0$ for $x < 0$. Then

$$f_+(x) + f_-(x) = \varphi(x) + \int_{-\infty}^{\infty} K(x - y) f_-(y)\, dy$$

Now take Fourier transforms:

$$\bar{f}_+(k) + \bar{f}_-(k) = \bar{\varphi}(k) + \bar{K}(k) \bar{f}_-(k)$$

which we shall write in the form

$$\bar{f}_-(k)[1 - \bar{K}(k)] + \bar{f}_+(k) = \bar{\varphi}(k) \tag{11-32}$$

Because $f_+(x) = 0$ for $x > 0$, $\bar{f}_+(k)$ has no singularities in the upper half k-plane. Likewise $\bar{f}_-(k)$ has no singularities in the lower half k-plane. Ordinarily, for some α and β,

$$|K(x)| \lesssim \begin{cases} e^{-\alpha x} & \text{as } x \to \infty \\ e^{\beta x} & \text{as } x \to -\infty \end{cases}$$

Therefore, $\bar{K}(k)$ is analytic in the strip $-\beta < \text{Im } k < \alpha$. We shall also assume that the function $\varphi(x)$ is at least this well-behaved at infinity, so $\bar{\varphi}(k)$ is also analytic in this strip.

Now factor $1 - \bar{K}(k)$ into two functions

$$1 - \bar{K}(k) = \frac{A(k)}{B(k)}$$

where $A(k)$ is analytic (and has no zeros) for $\text{Im } k < \alpha$, and $B(k)$ is analytic (and has no zeros) for $\text{Im } k > -\beta$.

Equation (11-32) now becomes

$$\bar{f}_-(k)A(k) + \bar{f}_+(k)B(k) = \bar{\varphi}(k)B(k) \tag{11-33}$$

where $\bar{f}_-(k)A(k)$ has no singularities for $\text{Im } k < 0$, and $\bar{f}_+(k)B(k)$ has no singularities for $\text{Im } k > 0$. The right-hand side, $\bar{\varphi}(k)B(k)$, may have singularities in both half planes. We separate these as follows:

$$\bar{\varphi}(k)B(k) = C(k) + D(k)$$

where $C(k)$ has no singularities in the upper half k-plane, and $D(k)$ has no singularities in the lower half k-plane.

Thus equation (11-33) becomes

$$\bar{f}_-(k)A(k) - D(k) = C(k) - \bar{f}_+(k)B(k) \tag{11-34}$$

where the left-hand side has no singularities in the lower half k-plane, and the right-hand side has no singularities in the upper half k-plane. Both sides must therefore equal an entire function. Furthermore, $\bar{f}_+$ and $\bar{f}_-$ go to zero at $\text{Im } k = \pm\infty$. [We used this implicitly in saying that $\bar{f}_+(k)$ had no singularities in the upper half plane because $f_+(x) = 0$ for $x > 0$.] The functions $A, \ldots, D$ can usually be shown to behave at worst like polynomials. Both sides of equation (11-34) are therefore equal to some polynomial. This fixes f_+ and f_- except possibly for some unknown constants that may be determined by physical reasoning or substitution in the original integral equation.

Example

$$f(x) = e^{-|x|} + \lambda \int_0^\infty e^{-|x-y|} f(y)\, dy \tag{11-35}$$

The Fourier transform of the function $e^{-|x|}$, which appears both as kernel and inhomogeneous term, is

$$\int_{-\infty}^{\infty} dx\, e^{-|x|}\, e^{-ikx} = \frac{2}{1 + k^2}$$

so that the Fourier transform of our integral equation (11-35) is

$$\bar{f}_+(k) + \bar{f}_-(k) = \frac{2}{1 + k^2} + \frac{2\lambda}{1 + k^2} \bar{f}_-(k)$$

or

$$\left(\frac{k^2 - \xi^2}{k^2 + 1}\right)\overline{f}_-(k) + \overline{f}_+(k) = \frac{2}{1 + k^2}$$

where $\xi^2 = 2\lambda - 1$. For definiteness, we shall assume ξ real ($\lambda > 1/2$).

Our next step is to represent the coefficient

$$\frac{k^2 - \xi^2}{k^2 + 1}$$

as a quotient $A(k)/B(k)$ where $A(k)$ is analytic for $\operatorname{Im} k < 0$, and $B(k)$ is analytic for $\operatorname{Im} k > -1$. Clearly only one such quotient is possible:

$$A(k) = \frac{k^2 - \xi^2}{k - i} \qquad B(k) = k + i$$

Substitution into equation (11-33) yields

$$\overline{f}_-(k)\,\frac{k^2 - \xi^2}{k - i} + \overline{f}_+(k)(k + i) = \frac{2}{k - i}$$

i.e.,

$$\overline{f}_-(k)\left(\frac{k^2 - \xi^2}{k - i}\right) - \left(\frac{2}{k - i}\right) = -\overline{f}_+(k)(k + i) \tag{11-36}$$

The left-hand side of (11-36) has no singularities in the lower half k-plane; the right-hand side has no singularities in the upper half k-plane. Each side must therefore be equal to a polynomial.

We now use arguments based on the asymptotic behavior of $\overline{f}_\pm$ to determine the order of the polynomial required. Let $P(k)$ be some polynomial in k such that

$$(k + i)\overline{f}_+(k) = P(k)$$

$$\overline{f}_+(k) = \frac{P(k)}{k + i}$$

If $\overline{f}_+(k)$ is to drop off as $|k| \to \infty$, $P(k)$ can only be a constant, which we will call iA. Hence

$$\overline{f}_+(k) = \frac{iA}{k + i}$$

$$f_+(x) = Ae^x \qquad x < 0$$

Likewise

$$\overline{f}_-(k)\left[\frac{k^2 - \xi^2}{k - i}\right] - \left[\frac{2}{k - i}\right] = -iA$$

$$\bar{f}_-(k) = \frac{2}{k^2 - \xi^2} - \frac{iA(k-i)}{k^2 - \xi^2}$$

$$f_-(x) = -\tfrac{2}{\xi} \sin \xi x + A\left(\cos \xi x + \frac{1}{\xi} \sin \xi x\right) \qquad x > 0$$

And

$$f(x) = \begin{cases} Ae^x & x < 0 \\ -\dfrac{2}{\xi} \sin \xi x + A\left(\cos \xi x + \dfrac{1}{\xi} \sin \xi x\right) & x > 0 \end{cases} \tag{11-37}$$

PROBLEMS

11-11 Solve integral equation (11-35)

$$f(x) = e^{-|x|} + \lambda \int_0^\infty e^{-|x-y|} f(y)\, dy \qquad (-\infty < x < \infty\,;\ \lambda > \tfrac{1}{2})$$

directly, without using Fourier transforms, and compare your solution with (11-37).

11-12 Find $u(x)$ if

$$u(x) = e^{-|x|} + \lambda \int_0^\infty |x - y|\, e^{-|x-y|} u(y)\, dy$$

SUMMARY

The following methods are often helpful in solving nonsingular integral equations:

1. Check to see whether the kernel is separable. In this case the problem can be solved by algebra. Often in practical cases it is useful to approximate a difficult kernel by a separable one, just to get some idea of the answer.

2. If the kernel is a function only of $(x - y)$, the convolution theorem may simplify evaluation of the Fourier or Laplace transform of the solution.

3. Iteration of the integral may be a useful approach if there is a small parameter multiplying it; otherwise a Fredholm solution may be necessary.

4. Once eigenvalues and eigenfunctions of a Hermitian kernel have been found, they can be used to construct a Green's function or resolvent kernel for the integral operator. Then any inhomogeneous equation can be solved by use of this Green's function.

We discussed the Wiener-Hopf technique for equations with displacement kernels and half-infinite ranges of integration. Many other integral transform techniques prove useful in particular situations.

CHAPTER 12 Partial Differential Equations

Partial differential equations (p.d.e.) describe many physical phenomena: propagation of waves, diffusion of heat, the behavior of quantum-mechanical wave functions. In this chapter we describe the prescription of boundary conditions for various types of p.d.e. (Section 12-2), and some methods for their solution (Sections 12-3 through 12-5). These methods of solution may all be viewed as straightforward application of our results in ordinary differential equations, perturbation theory, and transform techniques; hence this is a logical termination point for the course.

We lay particular stress on the use and determination of Green's functions for these equations. This is intended to extend the discussion of Chapters 2 and 4, and to expose the student to a number of real-life situations in which Green's functions are useful. This treatment focuses on volume Green's functions; the formal relationship between surface and volume Green's functions is left to more advanced texts.

Sections 12-3 and 12-4 are devoted to the separation of variables method. This method is designed to take advantage of our results for second-order ordinary differential equations, in particular the completeness of sets of Sturm-Liouville eigenfunctions. We can thus find eigenfunctions of partial differential operators, and use them to expand the general solution of the p.d.e. They can also be used as basis states for the perturbation theory developed in Chapter 2.

In Section 12-5, we demonstrate the use of Fourier and Laplace transforms in several variables to simplify p.d.e.'s. The treatment of poles on the real axis in the transform again is related to boundary conditions of the problem; we illustrate this in calculation of the retarded Green's function for the wave equation.

12-1 INTRODUCTION

Partial differential equations may be used to describe many physical phenomena. Some examples are the following:

1. Equation of the vibrating flexible string, or one-dimensional wave equation:

$$\frac{\partial^2 \psi}{\partial x^2} = \frac{1}{c^2} \frac{\partial^2 \psi}{\partial t^2}$$

where c is the speed of the waves. For the flexible string, $c^2 = T/\rho$ where T is the tension and ρ is the linear density.

2. Laplace's equation:

$$\nabla^2 \psi = \left(\frac{\partial^2 \psi}{\partial x^2} + \frac{\partial^2 \psi}{\partial y^2} + \frac{\partial^2 \psi}{\partial z^2} \right) = 0$$

3. Three-dimensional wave equation:

$$\nabla^2 \psi - \frac{1}{c^2} \frac{\partial \psi}{\partial t^2} = 0$$

4. Diffusion equation:

$$\nabla^2 \psi - \frac{1}{\kappa} \frac{\partial \psi}{\partial t} = 0$$

If ψ is temperature,

$$\kappa = \frac{K}{C\rho} = \frac{\text{thermal conductivity}}{(\text{specific heat}) \times (\text{density})}$$

5. Schrödinger equation:

$$-\frac{\hbar^2}{2m} \nabla^2 \psi + V(\mathbf{x})\psi = i\hbar \frac{\partial \psi}{\partial t}$$

or, if $\psi \propto e^{-iEt/\hbar}$,

$$\nabla^2 \psi + \frac{2m}{\hbar^2} [E - V(\mathbf{x})]\psi = 0$$

These are the equations with which we will primarily occupy ourselves. Note that they are all *linear, second-order* equations.

The equations above are all *homogeneous*, which means that, if ψ is a solution, so is any multiple of ψ. Many problems involve an inhomogeneous equation containing a term corresponding to applied "forces" or "sources." For example, if a force $f(x, t)$ per unit length is applied to a vibrating string, the equation is the inhomogeneous one:

$$\frac{\partial^2 \psi}{\partial x^2} - \frac{1}{c^2} \frac{\partial^2 \psi}{\partial t^2} = -\frac{1}{T} f(x, t)$$

A problem may be inhomogeneous because of the boundary conditions as well as because of the equation itself. The criterion for a homogeneous boundary-value problem is the one stated above; that is, if ψ is a solution of the equation *and* boundary conditions, then so is a multiple of ψ. An example of an inhomogeneous boundary condition is a vibrating string for which the end $x = 0$ is prescribed to move in a definite way $\psi(0, t) = g(t)$.

The general solution of an inhomogeneous problem is made up of any particular solution of the problem plus the general solution of the corresponding homogeneous problem, for which both the equation and boundary conditions are homogeneous. This composition of the solution has already been discussed for the case of ordinary differential equations in Chapter 3.

Because partial differential equations always involve more than one independent variable, the boundary assumes the form of a geometric shape in two or more dimensions. It will clearly be most simple to match specified conditions along the boundary if the solution has the same kind of symmetry as this boundary surface. For this reason, we always choose a coordinate system appropriate to the symmetry of the problem; when the solution is expressed in terms of these coordinates, it is easy to display its boundary values. Notice that four of the five examples given above involve the Laplacian operator, ∇^2. We will need the form of this operator in several coordinate systems; for purposes of reference, we list them here:

1. Rectangular coordinates

$$\nabla^2\psi = \frac{\partial^2\psi}{\partial x^2} + \frac{\partial^2\psi}{\partial y^2} + \frac{\partial^2\psi}{\partial z^2}$$

2. Spherical coordinates

$$\nabla^2\psi = \frac{1}{r^2}\frac{\partial}{\partial r}\left(r^2\frac{\partial\psi}{\partial r}\right) + \frac{1}{r^2\sin\theta}\frac{\partial}{\partial\theta}\left(\sin\theta\frac{\partial\psi}{\partial\theta}\right) + \frac{1}{r^2\sin^2\theta}\frac{\partial^2\psi}{\partial\varphi^2}$$

3. Cylindrical coordinates

$$\nabla^2\psi = \frac{1}{\rho}\frac{\partial}{\partial\rho}\left(\rho\frac{\partial\psi}{\partial\rho}\right) + \frac{1}{\rho^2}\frac{\partial^2\psi}{\partial\varphi^2} + \frac{\partial^2\psi}{\partial z^2}$$

12-2 BOUNDARY CONDITIONS

Before proceeding to methods for solving equations like those given above, we shall briefly discuss the general linear second-order partial differential equation. We shall make one restriction, however, and consider only two independent variables. This is done in order to simplify matters and to draw understandable pictures. Much of the reasoning can be immediately generalized to equations with more independent variables.

We have, then, a function $\psi(x, y)$ to be evaluated in some region of the

xy-plane. The partial differential equation will certainly need to be supplemented by boundary conditions of some sort; we will suppose these to involve ψ and/or some of its derivatives on the curve that encloses the region within which we are trying to solve the equation.

There are three common types of boundary conditions:

1. *Dirichlet* conditions: ψ is specified at each point of the boundary.

2. *Neumann* conditions: $(\nabla\psi)_n$, the normal component of the gradient of ψ, is specified at each point of the boundary.

3. *Cauchy* conditions: ψ and $(\nabla\psi)_n$ are specified at each point of some curve.

By analogy with ordinary second-order differential equations, we would expect that Cauchy conditions along a line would be the most natural set of boundary conditions. However, things are not quite so simple.

CHARACTERISTIC CURVES

For an ordinary (that is, one-dimensional) second-order differential equation for $\psi(x)$, the specification of ψ and ψ' at an ordinary point x_0 together with the differential equation itself is sufficient to determine the second and all higher derivatives at x_0 and thus ensure the existence of a solution in the form of a Taylor series near x_0.

We now investigate the corresponding question for second-order partial differential equations, namely, whether the specification of ψ and $(\nabla\psi)_n$ along a boundary curve together with the differential equation itself is sufficient to determine the second and higher derivatives of ψ on the boundary curve and thus ensure the existence of a solution in the form of a Taylor series near the curve,

$$\begin{aligned}\psi(x, y) = \psi(x_0, y_0) &+ \left.\frac{\partial\psi}{\partial x}\right|_{x=x_0}(x - x_0) + \left.\frac{\partial\psi}{\partial y}\right|_{y=y_0}(y - y_0) \\ &+ \frac{1}{2}\frac{\partial^2\psi}{\partial x^2}(x - x_0)^2 + \frac{1}{2}\frac{\partial^2\psi}{\partial x\,\partial y}(x - x_0)(y - y_0) \\ &+ \frac{1}{2}\frac{\partial^2\psi}{\partial y^2}(y - y_0)^2 + \cdots\end{aligned}$$

Let us suppose that our boundary curve is described parametrically by the equations

$$x = x(s) \qquad y = y(s)$$

where s is arc length along the boundary (see Figure 12-1). We shall suppose that we are given $\psi(s)$ and its normal derivative $N(s)$ along the boundary. The components of the unit normal $\hat{\mathbf{n}}$ are $-(dy/ds), (dx/ds)$ so that

$$N(s) = -\frac{\partial\psi}{\partial x}\frac{dy}{ds} + \frac{\partial\psi}{\partial y}\frac{dx}{ds}$$

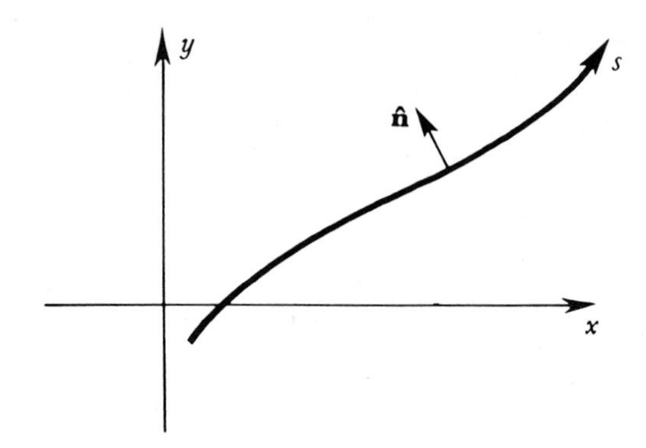

FIGURE 12-1 A boundary curve and unit vector $\hat{\mathbf{n}}$ normal to it

This equation and

$$\frac{d}{ds}\,\psi(s) = \frac{\partial \psi}{\partial x}\frac{dx}{ds} + \frac{\partial \psi}{\partial y}\frac{dy}{ds}$$

may be solved for the first partial derivatives of ψ:

$$\frac{\partial \psi}{\partial x} = -N(s)\frac{dy}{ds} + \left[\frac{d}{ds}\,\psi(s)\right]\frac{dx}{ds}$$

$$\frac{\partial \psi}{\partial y} = N(s)\frac{dx}{ds} + \left[\frac{d}{ds}\,\psi(s)\right]\frac{dy}{ds}$$

The trouble comes with the second derivatives. There are three:

$$\frac{\partial^2 \psi}{\partial x^2} \qquad \frac{\partial^2 \psi}{\partial x\,\partial y} \qquad \frac{\partial^2 \psi}{\partial y^2}$$

Two equations for these are found by differentiating the (known) first derivatives along the boundary

$$\frac{d}{ds}\frac{\partial \psi}{\partial x} = \frac{\partial^2 \psi}{\partial x^2}\frac{dx}{ds} + \frac{\partial^2 \psi}{\partial x\,\partial y}\frac{dy}{ds}$$

$$\frac{d}{ds}\frac{\partial \psi}{\partial y} = \frac{\partial^2 \psi}{\partial x\,\partial y}\frac{dx}{ds} + \frac{\partial^2 \psi}{\partial y^2}\frac{dy}{ds}$$

A third equation is provided by the original differential equation, which we shall write in the form

$$A\,\frac{\partial^2 \psi}{\partial x^2} + 2B\,\frac{\partial^2 \psi}{\partial x\,\partial y} + C\,\frac{\partial^2 \psi}{\partial y^2} = f\left(x, y, \frac{\partial \psi}{\partial x}, \frac{\partial \psi}{\partial y}\right)$$

where $f(x, y, \partial\psi/\partial x, \partial\psi/\partial y)$ is some known function. These three (inhomogeneous) equations can be solved for the second partial derivatives of ψ *unless* the determinant of the coefficients vanishes:

$$\begin{vmatrix} \frac{dx}{ds} & \frac{dy}{ds} & 0 \\ 0 & \frac{dx}{ds} & \frac{dy}{ds} \\ A & 2B & C \end{vmatrix} = 0$$

or

$$A\left(\frac{dy}{ds}\right)^2 - 2B\frac{dx}{ds}\frac{dy}{ds} + C\left(\frac{dx}{ds}\right)^2 = 0 \tag{12-1}$$

At each point in the xy-plane this equation determines two directions, the so-called *characteristic directions* at that point. Curves in the xy-plane whose tangents at each point lie along characteristic directions are called *characteristics* of the partial differential equation.

Thus the second derivatives are determined except in the case where the boundary curve is tangent to a characteristic somewhere. By further differentiations, a similar set of simultaneous equations for the third (and higher) derivatives may be found, and the condition for a solution involves a determinant that is exactly the same as the one above. Thus Cauchy boundary conditions will determine the solution if the boundary curve is nowhere tangent to a characteristic.

Returning to equation (12-1) for the characteristics, if the characteristics are to be *real* curves, we clearly must have $B^2 > AC$. Partial differential equations obeying this condition are called *hyperbolic* equations. If $B^2 = AC$, the equation is said to be *parabolic*; if $B^2 < AC$ the equation is *elliptic*. Of the examples in Section 12-1, numbers 1 and 3 are hyperbolic, number 2 is elliptic, and numbers 4 and 5 are parabolic. (5 is a little unusual, however, in that not all of its coefficients are real.)

HYPERBOLIC EQUATIONS

Let us discuss the choice of boundary conditions that is appropriate for each of the three types of equation, beginning with the hyperbolic equation. We have seen above that, generally speaking, Cauchy conditions along a curve that is not a characteristic are sufficient to specify the solution near the curve. A useful picture for visualizing the role of the characteristics and boundary conditions is obtained by thinking of the characteristics as curves along which partial information about the solution propagates. The meaning of this statement and the way in which it works are most easily understood with the aid of an elementary example.

Example

Consider the simplest hyperbolic equation, having $A = 1$, $B = 0$, $C =$ constant $= -1/c^2$. This is the one-dimensional wave equation

$$\frac{\partial^2\psi}{\partial x^2} - \frac{1}{c^2}\frac{\partial^2\psi}{\partial t^2} = 0 \tag{12-2}$$

for which the equation of characteristics (12-1) is

$$\left(\frac{dt}{ds}\right)^2 - \frac{1}{c^2}\left(\frac{dx}{ds}\right)^2 = 0$$

or

$$\left(\frac{dx}{dt}\right)^2 = c^2$$

Thus the characteristics are straight lines:

$$\begin{aligned} (x - ct) &= \xi = \text{constant} \\ (x + ct) &= \eta = \text{constant} \end{aligned} \tag{12-3}$$

These families of lines are shown in Figure 12-2.

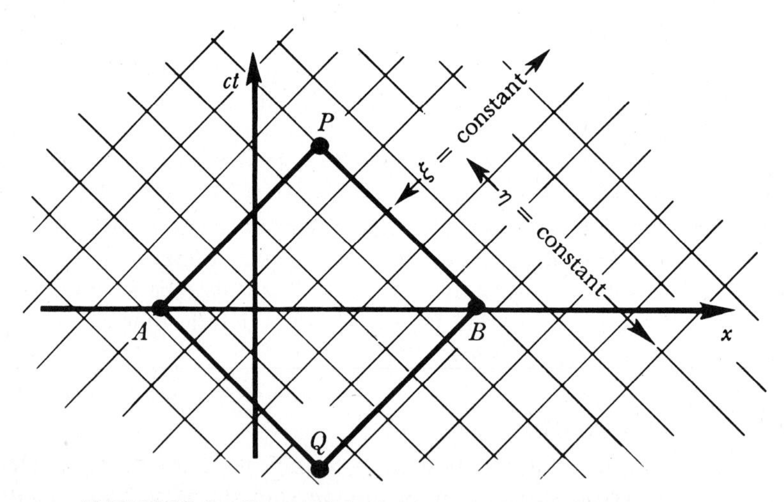

FIGURE 12-2 Characteristics for the one-dimensional wave equation

The characteristics form a "natural" set of coordinates for a hyperbolic equation. For example, if we transform equation (12-2) to the new coordinates ξ and η, defined by (12-3), we obtain the equation in so-called *normal form*:

$$\frac{\partial^2 \psi}{\partial \xi \, \partial \eta} = 0$$

The solution is immediate:

$$\psi = f(\xi) + g(\eta) = f(x - ct) + g(x + ct)$$

where f and g are arbitrary functions. Notice that f is constant on "wavefronts" $x = ct +$ constant that travel toward larger x as t increases, whereas g is constant on wavefronts that travel toward decreasing x. Any solution may therefore be expressed as the sum of two waves, one traveling to the right in x, and one traveling to the left; information propagates along with the waves.

Now, if we know $\psi(x)$ and its normal derivative $N(x) = 1/c(\partial\psi/\partial t)$ along the line segment AB of Figure 12-2, we can find the individual functions $f(\xi)$ and $g(\eta)$ all along this line segment, where they have the values $f(x)$ and $g(x)$. Specifically,

$$\psi(x, t = 0) = f(x) + g(x)$$

and

$$\frac{1}{c}\frac{\partial \psi}{\partial t}(x, t = 0) = -f'(x) + g'(x)$$

from which

$$f(x) = \frac{1}{2}\psi(x) - \frac{1}{2c}\int \frac{\partial \psi}{\partial t}\, dx$$

$$g(x) = \frac{1}{2}\psi(x) + \frac{1}{2c}\int \frac{\partial \psi}{\partial t}\, dx$$

The arbitrary constant associated with the integral is of no importance since it cancels in the sum $\psi = f + g$ everywhere.

The values of $f(x)$ along the line segment AB determine $f(\xi)$ along all the characteristics $\xi =$ constant that intersect AB. Similarly, the values of $g(x)$ determine $g(\eta)$ along all the curves $\eta =$ constant that intersect AB. Both $f(\xi)$ and $g(\eta)$, and thus $\psi(x, t)$ are determined within the common region traversed by both kinds of characteristics, which is the rectangle $AQBP$ in Figure 12-2.

The results obtained for the simple example above hold generally for hyperbolic equations. Suppose the net of characteristics has the form shown in Figure 12-3, where SS' is a boundary curve. Cauchy conditions along an arc AB of the boundary determine the solution within the "triangular"-shaped regions on each side, bounded by characteristics through A and B.

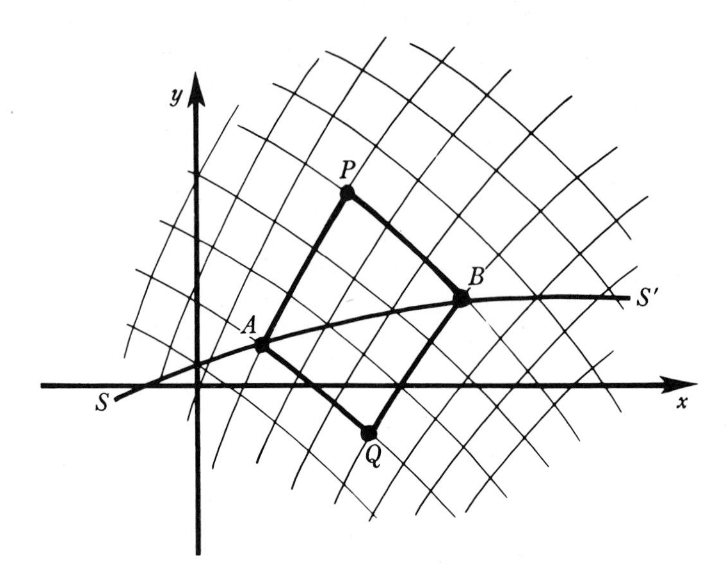

FIGURE 12-3 Characteristics for a hyperbolic equation, and a boundary curve SS' not tangent to a characteristic. Cauchy conditions along the arc AB of SS' fix the solution within the region $AQBP$.

The above picture enables us to discuss more complicated situations. For example, consider the boundary and net of characteristics shown in Figure 12-4. Cauchy conditions from A to B determine the behavior along *all* the vertical characteristics that intersect the boundary ABC and along those horizontal characteristics that intersect the arc AB. All that is left is the specification of

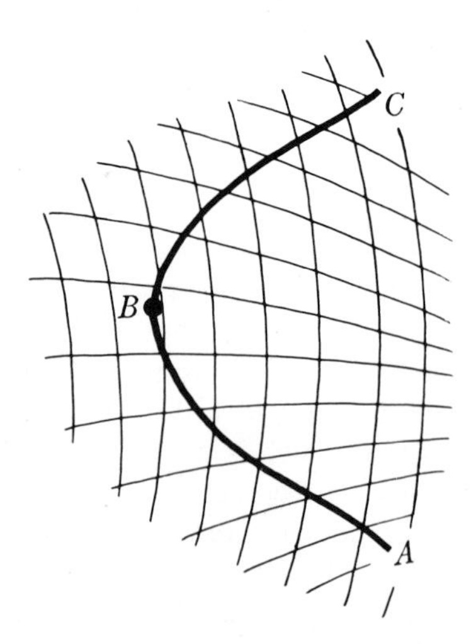

FIGURE 12-4 Net of characteristics with a boundary curve ABC that is tangent to a characteristic at point B

the behavior along the horizontal characteristics starting between B and C. Thus, Dirichlet or Neumann conditions along BC suffice; Cauchy conditions here are too much and *overdetermine* the solution.

Suppose we had a region bounded by characteristics, such as $APBQ$ in Figure 12-3. Dirichlet conditions along the sides AP and AQ then suffice to determine the solution inside. Hence the characteristics provide a "preferred" boundary in the sense that only one function, not two, must be specified.

ELLIPTIC EQUATIONS

Because elliptic equations have no real characteristic curves, one might expect that Cauchy conditions along an open boundary would suffice for any problem. Unfortunately, this is not a very useful procedure in practice. We do not ordinarily try to solve the elliptic equation $\nabla^2\varphi = 0$ in the neighborhood of some open curve; rather we seek a solution φ that holds within some particular region of the xy-plane. Such a region will have a closed boundary (although a portion of the boundary may be at infinity). This fact drastically influences the type of boundary conditions that are applicable.

If φ is a solution of $\nabla^2\varphi = 0$, it is the real part of some analytic function W and we may write

$$\varphi(z) = \operatorname{Re}\left[\frac{1}{2\pi i}\oint_{\text{boundary}} \frac{W(z')\,dz'}{z'-z}\right]$$

In particular, we may evaluate $\varphi(z)$ entirely in terms of the values $\varphi(z + re^{i\theta})$ on a circle of radius r about z:

$$\varphi(z) = \frac{1}{2\pi}\int_0^{2\pi} \varphi(z + re^{i\theta})\,d\theta$$

i.e.,

$$\varphi(0) = \frac{1}{2\pi}\int_0^{2\pi} \varphi(re^{i\theta})\,d\theta \tag{12-4}$$

This leads us to suspect that only Dirichlet conditions on a closed boundary are necessary. However, we must generalize the result to boundaries that are not circles and to points that are not in the center.

Application of the transformation

$$z = \frac{\rho - a}{1 - \dfrac{\rho a^*}{r^2}} \qquad \rho = \frac{z + a}{1 + \dfrac{z a^*}{r^2}} \qquad (|a|^2 < r^2)$$

maps the circle $|z| = r$ into the circle $|\rho| = r$; points interior to the circle in the z-plane are mapped into the interior of a circle in the ρ-plane, and the point $z = 0$ is mapped into $\rho = a$. The complex number z is an analytic function of ρ inside the circle; hence $W(z)$ is an analytic function of $\rho = \sigma + i\tau$ and φ will obey Laplace's equation with respect to these new variables,

$$\frac{\partial^2 \varphi}{\partial \sigma^2} + \frac{\partial^2 \varphi}{\partial \tau^2} = 0$$

If we write $u(\rho) = \varphi(z)$, the equivalent of equation (12-4) in the ρ-plane is then

$$u(0) = \frac{1}{2\pi} \int_0^{2\pi} u(re^{i\alpha})\, d\alpha$$

Rewriting this we find

$$\begin{aligned} \varphi(-a) &= \frac{1}{2\pi} \int_0^{2\pi} \varphi(re^{i\theta}) \frac{d\alpha}{d\theta}\, d\theta \\ &= \frac{1}{2\pi} \int_0^{2\pi} \varphi(re^{i\theta}) \left[\frac{r^2 - a^2}{|re^{i\theta} + a|^2} \right] d\theta \end{aligned}$$

Thus the potential at any point z within a circle of radius R can be expressed in terms of the values on the boundary by

$$\varphi(z) = \frac{1}{2\pi} \int_0^{2\pi} \varphi(Re^{i\theta}) \left[\frac{R^2 - |z|^2}{|Re^{i\theta} - z|^2} \right] d\theta$$

Likewise for any region of the xy-plane that can be mapped into a disk, we can express solutions $\varphi(x, y)$ of $\nabla^2 \varphi = 0$ in terms of the values on the boundary. It is a more difficult mathematical problem to prove the property for a general region of the xy-plane. The general proof may be found in Ahlfors (2) or other mathematics books under the name of the Dirichlet problem.

For some problems Neumann conditions are more appropriate. The important thing to remember about elliptic equations is that Dirichlet or Neumann conditions along a closed boundary provide a framework for most physical problems. (Occasionally one may use mixed boundary conditions, where a linear combination of φ and its normal derivative is specified around the boundary.)

We might inquire whether this type of condition is suitable for hyperbolic equations. The answer in general is no. We shall not go into the details, but the basic reason is that nonzero solutions of a hyperbolic equation can be found that vanish (or whose normal derivatives vanish) on suitable closed boundaries. The existence of these "normal modes" causes difficulties when one tries to impose Dirichlet or Neumann conditions on a closed boundary.

PARABOLIC EQUATIONS

Finally, we consider parabolic equations. The two-dimensional prototype is the one-dimensional diffusion problem

$$\frac{\partial^2 \psi}{\partial x^2} = \frac{\partial \psi}{\partial t} \tag{12-5}$$

It has as characteristics lines t = constant. Suppose we are given values $\psi(x, t_0)$ along a characteristic. The partial derivatives

$$\frac{\partial \psi(x, t_0)}{\partial x} \qquad \frac{\partial^2 \psi(x, t_0)}{\partial x^2} = g(x)$$

can then be computed, giving

$$\left[\frac{\partial \psi(x, t)}{\partial t}\right]_{t=t_0} = g(x)$$

Thus Dirichlet conditions along a characteristic give us the normal derivative automatically; this is why Cauchy conditions are improper for these characteristic lines.

Lines of t = constant are common boundaries for physical processes. Hence Dirichlet conditions along such lines are the most common boundary conditions for parabolic equations. One could alternatively use a Dirichlet condition along a boundary like that shown in Figure 12-5.

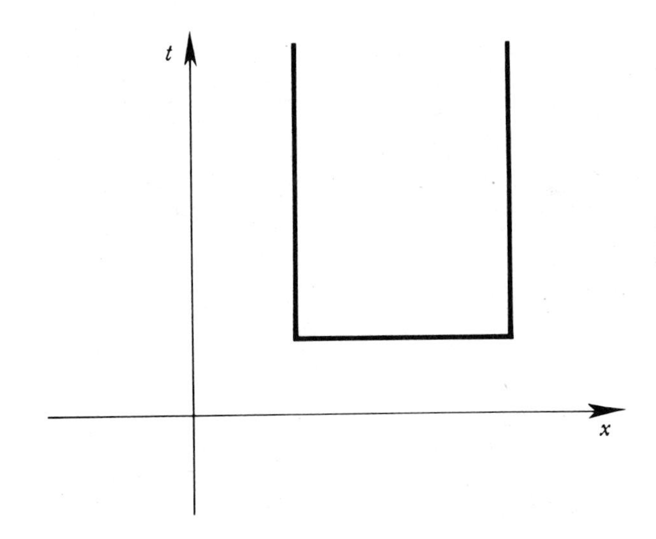

FIGURE 12-5 A suitable boundary for a simple diffusion problem

Notice that if $\psi(x, t)$ is a solution of equation (12-5), $\psi(x, -t)$ is not. That is, the solutions are not symmetric in time (unlike solutions to the wave equation, which can be made symmetric in time). We shall see, in fact, that the solutions are well-behaved only for times greater than the time t_0 labelling our boundary characteristic; for earlier times they develop singularities.

We summarize in Table 12-1 the types of boundary conditions appropriate for different types of equations.

TABLE 12-1

Types of Boundary Conditions Appropriate for the Three Classes of Equations

Equation	*Condition*	*Boundary*
Hyperbolic	Cauchy	Open
Elliptic	Dirichlet or Neumann	Closed
Parabolic	Dirichlet or Neumann	Open

12-3 SEPARATION OF VARIABLES I: BASIC METHODS

We now turn to the explicit solution of some partial differential equations. The methods most commonly used work by removing one or more partial derivative terms so that an equation with fewer variables is obtained. This may be repeated until an ordinary differential equation in one variable results. The first method we shall take up is known as the method of *separation of variables.* In this procedure one searches for "fundamental" solutions of a special form, in which dependence on each variable factors off from dependence on the other variables. The solution to a particular problem may then be expressed as a linear combination of these fundamental solutions.

To make this more concrete, suppose that for some reason we need a solution of the wave equation expressed in spherical polar coordinates. Let us look for a solution of

$$\nabla^2\psi - \frac{1}{c^2}\frac{\partial^2\psi}{\partial t^2} = 0 \tag{12-6}$$

of the form

$$\psi(\mathbf{x}, t) = X(\mathbf{x})T(t) \tag{12-7}$$

Substituting this trial solution into the partial differential equation (12-6) and dividing by XT gives

$$\frac{\nabla^2 X}{X} = \frac{1}{c^2}\frac{1}{T}\frac{d^2T}{dt^2}$$

The left side is a function of $\mathbf{x}$ only; the right side is a function of t only. They must therefore be constant, equal to (say) $-k^2$. ($-k^2$, or k, is called a *separation constant.*) Thus, the equation separates into two, the time-dependent one being

$$\frac{d^2T}{dt^2} + \omega^2 T = 0$$

where $\omega = kc$. The solution is

$$T = \begin{Bmatrix} \sin \omega t \\ \cos \omega t \end{Bmatrix} \quad \text{or} \quad T = e^{\pm i\omega t}$$

Now let us turn to the second equation, involving the space function $X(\mathbf{x})$. It is the so-called *Helmholtz equation*

$$\nabla^2 X + k^2 X = 0$$

or, in spherical coordinates $(r\theta\varphi)$,

$$\left[\frac{1}{r^2}\frac{\partial}{\partial r}\left(r^2 \frac{\partial}{\partial r}\right) + \frac{1}{r^2 \sin\theta}\frac{\partial}{\partial\theta}\left(\sin\theta \frac{\partial}{\partial\theta}\right) + \frac{1}{r^2 \sin^2\theta}\frac{\partial^2}{\partial\varphi^2}\right]X + k^2 X = 0 \qquad (12\text{-}8)$$

Let $X = R(r)\Theta(\theta)\Phi(\varphi)$. Substituting into (12-8) and dividing through by $R\Theta\Phi$ gives

$$\frac{1}{r^2 R}\frac{d}{dr}\left(r^2 \frac{dR}{dr}\right) + \frac{1}{r^2\Theta \sin\theta}\frac{d}{d\theta}\left(\sin\theta \frac{d\Theta}{d\theta}\right) + \frac{1}{r^2\Phi \sin^2\theta}\frac{d^2\Phi}{d\varphi^2} + k^2 = 0$$

If we were to multiply through by $r^2 \sin^2\theta$, the third term would depend only on φ, while the rest would depend only on r and θ. Therefore,

$$\frac{1}{\Phi}\frac{d^2\Phi}{d\varphi^2} = \text{constant} = -m^2$$

or

$$\frac{d^2\Phi}{d\varphi^2} + m^2\Phi = 0$$

with solutions

$$\Phi = \begin{pmatrix} \sin m\varphi \\ \cos m\varphi \end{pmatrix} \quad \text{or} \quad \Phi = e^{\pm im\varphi}$$

The r, θ equation becomes

$$\frac{1}{r^2 R}\frac{d}{dr}\left(r^2 \frac{dR}{dr}\right) + \frac{1}{r^2\Theta \sin\theta}\frac{d}{d\theta}\left(\sin\theta \frac{d\Theta}{d\theta}\right) - \frac{m^2}{r^2 \sin^2\theta} + k^2 = 0$$

If we multiply through by r^2, the first and fourth terms depend only on r, while the second and third depend only on θ. Thus

$$\frac{1}{\sin\theta}\frac{d}{d\theta}\left(\sin\theta \frac{d\Theta}{d\theta}\right) - \frac{m^2}{\sin^2\theta}\Theta = \text{constant} \times \Theta = -l(l+1)\Theta \qquad (12\text{-}9)$$

and

$$\frac{1}{r^2}\frac{d}{dr}\left(r^2 \frac{dR}{dr}\right) + \left[k^2 - \frac{l(l+1)}{r^2}\right]R = 0 \qquad (12\text{-}10)$$

If we set $\cos\theta = x$, the θ equation (12-9) becomes

$$(1-x^2)\frac{d^2\Theta}{dx^2} - 2x\frac{d\Theta}{dx} + \left[l(l+1) - \frac{m^2}{1-x^2}\right]\Theta = 0$$

This is just the associated Legendre equation (3-59). Its solutions are

$$\Theta = P_l^m(x), Q_l^m(x) \qquad \text{(associated Legendre functions)}$$

The radial equation (12-10), with the change of dependent variable $R = u/\sqrt{r}$, becomes

$$\frac{d^2u}{dr^2} + \frac{1}{r}\frac{du}{dr} + \left[k^2 - \frac{(l+\frac{1}{2})^2}{r^2}\right]u = 0$$

which is just Bessel's equation (3-68) with $x = kr$ and $m = l + 1/2$. Therefore,

$$R = \frac{J_{l+1/2}(kr)}{\sqrt{r}} \qquad \frac{Y_{l+1/2}(kr)}{\sqrt{r}}$$

It is conventional to define *spherical Bessel functions* by

$$j_l(x) = \sqrt{\frac{\pi}{2x}}\,J_{l+1/2}(x) \qquad n_l(x) = \sqrt{\frac{\pi}{2x}}\,Y_{l+1/2}(x) \tag{12-11}$$

$$h_l^{(1,2)}(x) = j_l(x) \pm i n_l(x) \qquad \text{(spherical Hankel functions)} \tag{12-12}$$

It can be shown that

$$j_l(x) = (-x)^l\left(\frac{1}{x}\frac{d}{dx}\right)^l \frac{\sin x}{x}$$

$$n_l(x) = (-x)^l\left(\frac{1}{x}\frac{d}{dx}\right)^l \left(-\frac{\cos x}{x}\right)$$

Abramowitz and Stegun (1), Chapter 10, gives a convenient summary of the recursion relations, asymptotic behavior, etc., of spherical Bessel functions.

If $k = 0$, so that $\partial\psi/\partial t = 0$ and we are really discussing Laplace's equation, the radial equation (12-10) becomes

$$R'' + \frac{2}{r}R' - \frac{l(l+1)}{r^2}R = 0$$

which has the solutions

$$R = \begin{Bmatrix} r^l \\ r^{-(l+1)} \end{Bmatrix}$$

Thus we have found the following "fundamental" solutions:

$$\nabla^2\psi - \frac{1}{c^2}\frac{\partial^2\psi}{\partial t^2} = 0 \Rightarrow \psi = \begin{Bmatrix} e^{+i\omega t} \\ e^{-i\omega t} \end{Bmatrix} \cdot \begin{Bmatrix} e^{+im\varphi} \\ e^{-im\varphi} \end{Bmatrix} \cdot \begin{Bmatrix} P_l^m(\cos\theta) \\ Q_l^m(\cos\theta) \end{Bmatrix} \cdot \begin{Bmatrix} j_l(kr) \\ n_l(kr) \end{Bmatrix} \tag{12-13}$$

$$\nabla^2\psi = 0 \Rightarrow \psi = \begin{Bmatrix} e^{+im\varphi} \\ e^{-im\varphi} \end{Bmatrix} \cdot \begin{Bmatrix} P_l^m(\cos\theta) \\ Q_l^m(\cos\theta) \end{Bmatrix} \cdot \begin{Bmatrix} r^l \\ r^{-(l+1)} \end{Bmatrix} \tag{12-14}$$

where each bracket represents a linear combination of the two functions inside. Any linear combination of such solutions is again a solution, because of the linearity of the original differential equation. Furthermore, because all the functions involved can be related to complete orthonormal sets, *any* solution to the original equation can be expressed as a linear combination of these. The separation of variables technique, then, leads us to write the general solution to (12-6) in the form

$$\psi = \sum_{\omega m l} [A_l j_l(kr) + B_l n_l(kr)][C_l^m P_l^m(\cos\theta) + D_l^m Q_l^m(\cos\theta)]^*$$
$$\times [E_m e^{im\varphi} + F_m e^{-im\varphi}][G_\omega e^{i\omega t} + H_\omega e^{-i\omega t}]$$

The constants $A, \ldots, H$ are determined by boundary conditions.

Example

To illustrate the usefulness of these solutions, we solve the following boundary-value problem. Consider the acoustic radiation from a split sphere antenna; that is, ψ obeys the wave equation (12-6), and at $r = a$,

$$\psi = \begin{cases} V_0 e^{-i\omega_0 t} & 0 < \theta < \dfrac{\pi}{2} \\ -V_0 e^{-i\omega_0 t} & \dfrac{\pi}{2} < \theta < \pi \end{cases}$$

To begin, we will clearly only use solutions with $\omega = \omega_0$, so that

$$k = k_0 = \frac{\omega_0}{c}$$

Also, since the boundary conditions are axially symmetric, we need only consider solutions with $m = 0$. Since everything must be well behaved at $\cos\theta = \pm 1$, we use only P_l with integral l. Thus we have reduced our trial solution to

$$\psi = e^{-i\omega_0 t} \sum_l P_l(\cos\theta)[A_l j_l(k_0 r) + B_l n_l(k_0 r)]$$

We now use the fact that at infinity only *outgoing* waves are present. For large x, using (10-37)

$$j_l(x) \sim \frac{1}{x}\cos\left[x - \frac{\pi}{2}(l+1)\right]$$

$$n_l(x) \sim \frac{1}{x}\sin\left[x - \frac{\pi}{2}(l+1)\right]$$

$$h_l^{(1,2)}(x) \sim \frac{1}{x} e^{\pm i[x-(l+1)\pi/2]}$$

Thus the radial functions we want are the $h_l^{(1)}(k_0 r)$, since then as $r \to \infty$

$$\psi \sim \frac{1}{r} e^{i(k_0 r - \omega_0 t)}$$

which is an outgoing wave. We have reduced our solution to

$$\psi = e^{-i\omega_0 t} \sum_l A_l P_l(\cos\theta) h_l^{(1)}(k_0 r) \tag{12-15}$$

We now impose the boundary condition at $r = a$:

$$\sum_l A_l P_l(\cos\theta) h_l^{(1)}(k_0 a) = \begin{cases} V_0 & 0 < \theta < \dfrac{\pi}{2} \\ -V_0 & \dfrac{\pi}{2} < \theta < \pi \end{cases}$$

To determine the A_l we multiply both sides of (12-15) by $P_m(\cos\theta) d\cos\theta$ and integrate from -1 to $+1$. Then, since

$$\int_0^\pi P_l(\cos\theta) P_m(\cos\theta) \sin\theta \, d\theta = \frac{2}{2l+1} \delta_{lm}$$

we obtain

$$\frac{2}{2l+1} A_l h_l^{(1)}(k_0 a) = V_0 \int_0^{\pi/2} P_l(\cos\theta) \sin\theta \, d\theta - V_0 \int_{\pi/2}^{\pi} P_l(\cos\theta) \sin\theta \, d\theta$$

Clearly only odd l contribute. If l is odd

$$A_l = \frac{(2l+1)V_0}{h_l^{(1)}(k_0 a)} \int_0^1 P_l(x) \, dx$$

But from (10-23) and (10-24)

$$\int P_l(x) \, dx = \frac{P_{l+1}(x) - P_{l-1}(x)}{2l+1}$$

Therefore,

$$A_l = \frac{V_0}{h_l^{(1)}(k_0 a)} [P_{l+1}(x) - P_{l-1}(x)]_0^1$$

Using the generating function we obtain (see Chapter 10)

$$P_l(1) = 1 \qquad P_l(0) = \begin{cases} 0 & (l \text{ odd}) \\ (-1)^{l/2} \dfrac{l!}{2^l \left[\left(\dfrac{l}{2}\right)!\right]^2} & (l \text{ even}) \end{cases}$$

The result is

$$A_l = (-1)^{(l-1)/2} \frac{V_0}{h_l^{(1)}(k_0 a)} \frac{(l+1)(2l+1)(l-1)!}{2^{l+1}\left[\left(\frac{l+1}{2}\right)!\right]^2} \qquad (l \text{ odd})$$

so that our final solution is

$$\psi(r, \theta, t) = V_0 e^{-i\omega_0 t} \sum_{l \text{ odd}} (-1)^{(l-1)/2} \frac{(l+1)(2l+1)(l-1)!}{2^{l+1}\left[\left(\frac{l+1}{2}\right)!\right]^2} \times \frac{h_l^{(1)}(k_0 r)}{h_l^{(1)}(k_0 a)} P_l(\cos\theta) .$$

Example

As a second exercise in separating variables, let us treat the vibration of a round drum head. The differential equation describing small oscillations is

$$\nabla^2 u = \frac{1}{c^2} \frac{\partial^2 u}{\partial t^2} \qquad (12\text{-}16)$$

Let us look for periodic solutions, which describe the *normal modes* of the drum. These have the form $u(\mathbf{x}, t) = u(\mathbf{x})e^{-i\omega t}$, and substituting into (12-16) we find the equation for $u(\mathbf{x})$:

$$\nabla^2 u + k^2 u = 0 \quad \text{where } k = \frac{\omega}{c} = \text{wave number} \qquad (12\text{-}17)$$

In two-dimensional polar coordinates

$$\nabla^2 = \frac{1}{r}\frac{\partial}{\partial r}\left(r \frac{\partial}{\partial r}\right) + \frac{1}{r^2}\frac{\partial^2}{\partial \theta^2}$$

so that (12-17) is

$$\frac{1}{r}\frac{\partial}{\partial r}\left(r \frac{\partial u}{\partial r}\right) + \frac{1}{r^2}\frac{\partial^2 u}{\partial \theta^2} + k^2 u = 0$$

Try a solution of the form

$$u(r, \theta) = R(r)\Theta(\theta)$$

Separating the variables gives

$$\begin{aligned} &\frac{d^2\Theta}{d\theta^2} + n^2\Theta = 0 \Rightarrow \Theta = e^{\pm in\theta} \\ &\frac{d^2R}{dr^2} + \frac{1}{r}\frac{dR}{dr} + \left(k^2 - \frac{n^2}{r^2}\right)R = 0 \end{aligned} \qquad (12\text{-}18)$$

The R equation is Bessel's equation, so that

$$R = \begin{Bmatrix} J_n(kr) \\ Y_n(kr) \end{Bmatrix} \tag{12-19}$$

Before proceeding with the solution of the drum-head problem, we shall give the generalization of this separation procedure for three-dimensional cylindrical coordinates. The Laplacian is

$$\nabla^2 = \frac{1}{\rho}\frac{\partial}{\partial\rho}\left(\rho\frac{\partial}{\partial\rho}\right) + \frac{1}{\rho^2}\frac{\partial^2}{d\varphi^2} + \frac{\partial^2}{\partial z^2}$$

Solutions of Laplace's equation

$$\nabla^2\psi = 0$$

are

$$\psi = \begin{Bmatrix} J_m(\alpha\rho) \\ Y_m(\alpha\rho) \end{Bmatrix} \begin{Bmatrix} e^{\alpha z} \\ e^{-\alpha z} \end{Bmatrix} \begin{Bmatrix} e^{im\varphi} \\ e^{-im\varphi} \end{Bmatrix} \tag{12-20}$$

while solutions of the Helmholtz equation

$$\nabla^2\psi + k^2\psi = 0$$

are given by

$$\psi = \begin{Bmatrix} J_m(\sqrt{k^2-\alpha^2}\,\rho) \\ Y_m(\sqrt{k^2-\alpha^2}\,\rho) \end{Bmatrix} \begin{Bmatrix} e^{i\alpha z} \\ e^{-i\alpha z} \end{Bmatrix} \begin{Bmatrix} e^{im\varphi} \\ e^{-im\varphi} \end{Bmatrix} \tag{12-21}$$

For $k = 0$, the solutions (12-21) reduce to solutions of the Laplace equation; for $\alpha = 0$ they reduce to our two-dimensional drum-head solutions (12-19); for $k = 0$ *and* $\alpha = 0$ we recover the familiar electrostatic potentials $\rho^{\pm m} e^{\pm im\varphi}$.

Example

Returning to the drum problem, the amplitude of oscillation must have the form

$$u = J_m(kr) \begin{Bmatrix} \cos m\theta \\ \sin m\theta \end{Bmatrix} e^{\pm i\omega t} \tag{12-22}$$

where the function $Y_m(kr)$ has been eliminated because it becomes infinite at $r \to 0$. The requirement that our solution be single valued means that

m must be an integer. If the drum head is clamped at its outer edge ($r = R$), we must have

$$u = 0 \qquad \text{at } r = R$$

Therefore,

$$J_m(kR) = 0$$

The zeroes of Bessel functions are tabulated in various places, for example, Abramowitz and Stegun (1). Some of the zeroes are (approximately) as follows:

$$\begin{aligned} J_0(x) &= 0: x \approx 2.40, 5.52, 8.65, \ldots \\ J_1(x) &= 0: x \approx 3.83, 7.02, 10.17, \ldots \\ J_2(x) &= 0: x \approx 5.14, 8.42, 11.62, \ldots \end{aligned} \tag{12-23}$$

Thus the lowest modes of our drum head are

$$k = \frac{2.40}{R} \qquad \omega = 2.40\frac{c}{R} \qquad u \propto J_0\left(2.40\frac{r}{R}\right)$$

$$k = \frac{3.83}{R} \qquad \omega = 3.83\frac{c}{R} \qquad u \propto J_1\left(3.83\frac{r}{R}\right)\begin{Bmatrix}\cos\theta \\ \sin\theta\end{Bmatrix}$$

$$k = \frac{5.14}{R} \qquad \omega = 5.14\frac{c}{R} \qquad u \propto J_2\left(5.14\frac{r}{R}\right)\begin{Bmatrix}\cos 2\theta \\ \sin 2\theta\end{Bmatrix}$$

$$k = \frac{5.52}{R} \qquad \omega = 5.52\frac{c}{R} \qquad u \propto J_0\left(5.52\frac{r}{R}\right)$$

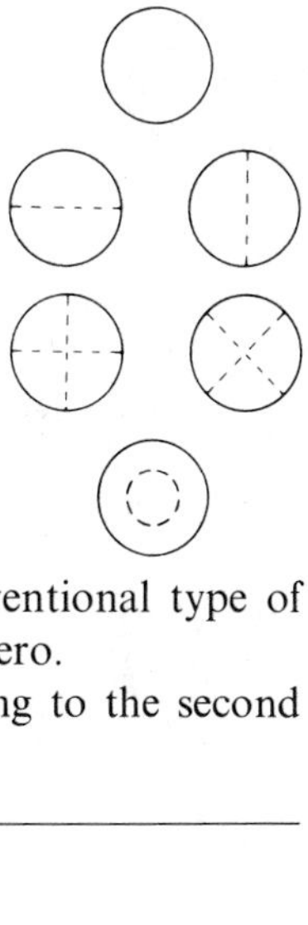

To the right of each mode we have drawn the conventional type of picture, indicating the *nodes* or places where u is always zero.

Note that there are two independent modes belonging to the second and third frequencies. This is an example of *degeneracy*.

Example

As a final example of the use of separation of variables in solving boundary-value problems, let us find the temperature within a cube of side L, initially at temperature $T = 0$, which at time $t = 0$ is immersed in a heat bath at temperature $T = T_0$. We must solve the equation

$$\nabla^2 T = \frac{1}{\kappa}\frac{\partial T}{\partial t} \tag{12-24}$$

Suppose

$$T \propto e^{-\lambda t}$$

Then

$$\nabla^2 T + \frac{\lambda}{\kappa} T = 0$$

$$\frac{\partial^2 T}{\partial x^2} + \frac{\partial^2 T}{\partial y^2} + \frac{\partial^2 T}{\partial z^2} = -\frac{\lambda}{\kappa} T$$

Separation of variables gives

$$T \quad \propto e^{i\alpha x}\, e^{i\beta y}\, e^{i\gamma z}$$

with

$$\alpha^2 + \beta^2 + \gamma^2 = \frac{\lambda}{\kappa}$$

Now one boundary condition is (choosing the origin at one corner of the cube)

$$T = T_0 \qquad \text{for } x = 0, L$$

$$T = T_0 \qquad \text{for } y = 0, L$$

$$T = T_0 \qquad \text{for } z = 0, L$$

Note that this is an inhomogeneous boundary condition, although a very simple one. A particular solution of (12-24) is $T = T_0$, and the complementary function must satisfy the corresponding homogeneous boundary conditions: $T = 0$ on the surface. Thus we write

$$T = T_0 + \sum_{lmn} C_{lmn} \sin\frac{l\pi x}{L} \sin\frac{m\pi y}{L} \sin\frac{n\pi z}{L} e^{-\lambda_{lmn} t}$$

where

$$\lambda_{lmn} = \kappa \frac{\pi^2}{L^2}(l^2 + m^2 + n^2)$$

To determine the constants C_{lmn}, we have the condition that $T = 0$ when $t = 0$. Therefore,

$$\sum_{lmn} C_{lmn} \sin\frac{l\pi x}{L} \sin\frac{m\pi y}{L} \sin\frac{n\pi z}{L} = -T_0$$

Multiply by $\sin(l'\pi x/L)\sin(m'\pi y/L)\sin(n'\pi z/L)$ and integrate over the whole cube. The result is

$$C_{lmn} = \begin{cases} -\dfrac{64 T_0}{\pi^3 lmn} & \text{if } l, m, n \text{ all odd} \\ 0 & \text{otherwise} \end{cases}$$

Thus, for $t > 0$

$$T = T_0\left\{1 - \frac{64}{\pi^3}\sum_{lmn\,\text{odd}}\frac{1}{lmn}\sin\frac{l\pi x}{L}\sin\frac{m\pi y}{L}\sin\frac{n\pi z}{L}\right.$$
$$\left.\times\exp\left[-(l^2+m^2+n^2)\kappa\frac{\pi^2 t}{L^2}\right]\right\} \qquad (12\text{-}25)$$

Our solution (12-25) is not very useful for small t. For $t \gg L^2/\kappa$, only the first term in the sum matters, and

$$T \approx T_0\left[1 - \frac{64}{\pi^3}\sin\frac{\pi x}{L}\sin\frac{\pi y}{L}\sin\frac{\pi z}{L}\exp\left(-\frac{3\kappa\pi^2}{L^2}t\right)\right]$$

SEPARATION INTO SUMMANDS

In all the examples above, the separation of variables has been achieved by looking for a solution in the form of a product of functions, each depending on fewer variables than are present in the original equation. For some problems, however, a separation of variables can be made by means of a solution having a different form; for example, a sum of functions of fewer variables.

Example

Consider an infinite heat conducting slab of thickness D with one surface ($x = D$) insulated. If, initially, the temperature T is zero and then heat is supplied (for example, by radiation) at a constant rate, Q calories per sec per cm^2 at the surface $x = 0$, find the temperature as a function of position and time within the slab. This situation is illustrated in Figure 12-6.

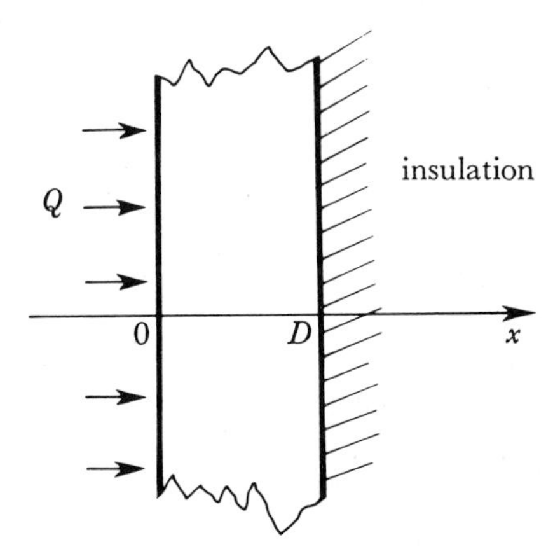

FIGURE 12-6 An infinite heat conducting slab of thickness D with the surface $x = D$ insulated and heat supplied at the surface $x = 0$ at a constant rate, Q calories per cm^2 per sec

We must solve the one-dimensional diffusion equation

$$\frac{\partial^2 T}{\partial x^2} - \frac{1}{\kappa}\frac{\partial T}{\partial t} = 0 \qquad \kappa = \frac{K}{C\rho}$$

with the given inhomogeneous boundary conditions. Physical considerations tell us the qualitative behavior of the temperature with time. After an initial transient period, we expect the temperature to rise linearly with time at any position. Thus we try a particular solution of the form

$$T_p = u(x) + \alpha t$$

which leads to a separation of variables, the x equation being

$$\frac{d^2 u}{dx^2} = \frac{\alpha}{\kappa}$$

with the solution

$$u(x) = \frac{1}{2}\frac{\alpha}{\kappa}x^2 + ax + b$$

b may be arbitrarily chosen for the particular solution, while the constants α and a are determined by the boundary conditions:

$$-Ku'(0) = Q \qquad u'(D) = 0$$

A solution satisfying these conditions is

$$u(x) = \frac{1}{2}\frac{Q}{KD}(x-D)^2 \qquad \alpha = \frac{Q\kappa}{KD} = \frac{Q}{C\rho D}$$

giving the particular solution

$$T_p = \frac{1}{2}\frac{Q}{KD}(x-D)^2 + \frac{Q}{C\rho D}t$$

We invite the student to finish the problem by finding the complementary function T_c and using it to satisfy the initial condition

$$T(x,0) = T_p + T_c = 0 \qquad \text{at } t = 0$$

Separation of variables by the method just illustrated finds its main application in classical mechanics in connection with the Hamilton-Jacobi differential equation. For these applications, we refer to Goldstein (15).

The separation of variables technique provides us with eigenfunctions of differential operators satisfying certain boundary conditions. These eigenfunctions may then be used for all the applications discussed in Chapter 2. In the next section we will illustrate such applications with examples.

PROBLEMS

12-1 Consider the motion of a stretched string, whose displacement $y(x,t)$ obeys the differential equation

$$\frac{1}{c^2}\frac{\partial^2 y}{\partial t^2} = \frac{\partial^2 y}{\partial x^2}$$

The boundary conditions are
(a) the ends of the string are fixed; $y(0,t) = y(L,t) = 0$
(b) the shape of the string is given at $t = 0$; $y(x,0) = f(x)$
(c) the shape of the string is given at another time T; $y(x,T) = g(x)$.
This is an example of the application of a Dirichlet boundary condition on a closed boundary (in the xt-plane) to a hyperbolic differential equation.

Does the problem have a solution? Discuss separately the cases:

(a) $\dfrac{cT}{L} = \text{integer}$

(b) $\dfrac{cT}{L} = \text{rational number}$

(c) $\dfrac{cT}{L} = \text{irrational number}$

12-2 A stretched string of length L has the end $x = 0$ fastened down, and the end $x = L$ moves in a prescribed manner:

$$y(L,t) = \cos \omega_0 t \qquad t > 0$$

Initially, the string is stretched in a straight line with unit displacement at the end $x = L$ and zero initial velocities everywhere. Find the displacement $y(x,t)$ as a function of x and t after $t = 0$.

12-3 Find the lowest frequency of a drum head in the shape of an isosceles right triangle with sides a, a, $a\sqrt{2}$.
Hint: Consider the modes of a square drum.

12-4 Find the three lowest eigenvalues of the Schrödinger equation for a particle confined in a cylindrical box of radius a and height h.

$$-\frac{\hbar^2}{2m}\nabla^2\psi = E\psi \qquad (\psi = 0 \text{ on walls}) \qquad a \approx h$$

12-5 Outside an infinitely long cylinder of radius a, a potential function $u(\mathbf{x},t)$ satisfies the wave equation

$$\nabla^2 u = \frac{1}{c^2}\frac{\partial^2 u}{\partial t^2}$$

The cylinder is split along its length, and on its surface

$$u = \begin{cases} u_0 e^{-i\omega_0 t} & (0 < \varphi < \pi) \\ -u_0 e^{-i\omega_0 t} & (\pi < \varphi < 2\pi) \end{cases}$$

Find u outside the cylinder if only outgoing waves are present at large distances.

12-6 A quantity u satisfies the wave equation

$$\nabla^2 u - \frac{1}{c^2}\frac{\partial^2 u}{\partial t^2} = 0$$

inside a hollow cylindrical pipe of radius a, and $u = 0$ on the walls of the pipe. If at the end $z = 0$, $u = u_0 e^{-i\omega_0 t}$, waves will be sent down the pipe with various spatial distributions (modes). Find the phase velocity of the fundamental mode as a function of the frequency ω_0 and interpret the result for small ω_0.

12-7 Find the lowest frequency of oscillation of acoustic waves in a hollow sphere of radius R. The boundary condition is $\partial\psi/\partial r = 0$ at $r = R$ and ψ obeys the differential equation

$$\nabla^2 \psi = \frac{1}{c^2}\frac{\partial^2 \psi}{\partial t^2}$$

12-8 A half cube of material (cut on a diagonal as shown in Figure 12-7) is

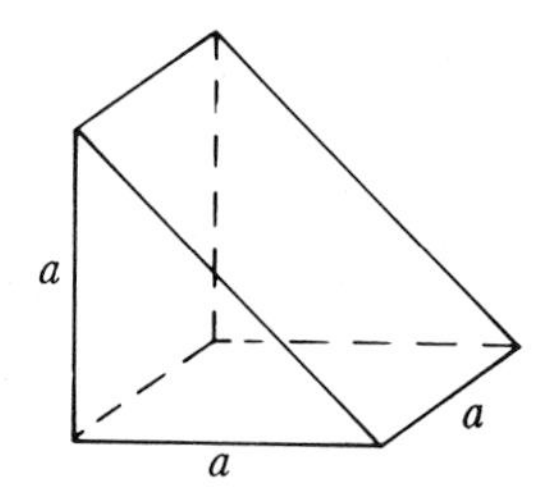

FIGURE 12-7

heated to temperature T_0 and then plunged into an oil bath that keeps the surface at zero temperature. Find the approximate temperature $T(\mathbf{x}, t)$ at late times when a single term is adequate. What happens if the two sides adjacent to the right angle are not of equal length?

12-9 A half cylinder of metal of length a and radius a is initially at temperature T_0. At time $t = 0$ the metal is immersed in a bath that maintains the surface at temperature $T = 0$. Find an approximate expression for $T(\rho, \theta, z, t)$ for large t.

12-10 The temperature in a homogeneous sphere of radius a obeys the differential equation

$$\nabla^2 T = \frac{1}{\kappa} \frac{\partial T}{\partial t}$$

By external means, the surface temperature of the sphere is forced to behave as shown in Figure 12-8. The alternations extend to $t = \pm\infty$. Find the temperature $T(t)$ at the center of the sphere.

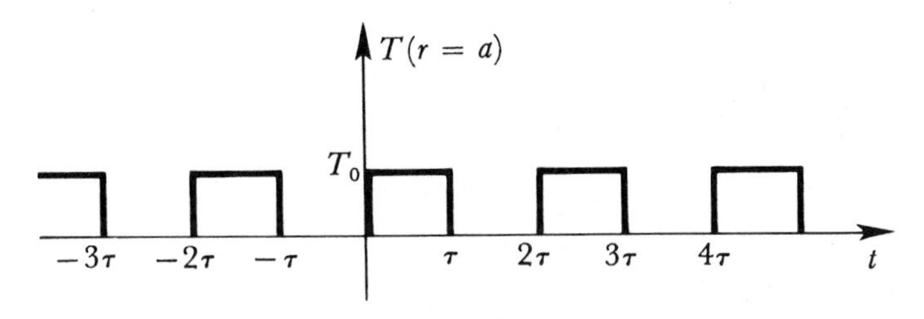

FIGURE 12-8

12-11 In a long rod of square cross section ($0 \leq x \leq s$; $0 \leq y \leq s$) the temperature obeys the differential equation

$$\nabla^2 T = \frac{1}{\kappa} \frac{\partial T}{\partial t}$$

At the end $z = 0$, the temperature is $T = T_0 \cos \omega t$ (independent of x and y).

(a) Find the temperature along the rod, if the sides are insulated.

(b) Find the temperature along the rod, if the sides are maintained at zero temperature.

In both cases, neglect transients.

12-12 The cube $|x| < L/2$, $|y| < L/2$, $|z| < L/2$ is immersed in a heat bath at temperature $T = 0$ and allowed to come to equilibrium. At time $t = 0$, a pulse of energy is released at the origin, so that $T = \delta(\mathbf{x})$. Find an expression for the temperature as a function of time at the point $(L/4, 0, 0)$.

12-13 A cylinder of length l has a semicircular cross section of radius a (see Figure 12-9). The cylinder has density ρ, specific heat C, and thermal conductivity K. It is initially at temperature T_0, and its surface is maintained at that temperature by immersion in an oil bath. At time $t = 0$, a pulse of energy is released at a point halfway between the ends of the cylinder, marked P in the figure below. If the total energy

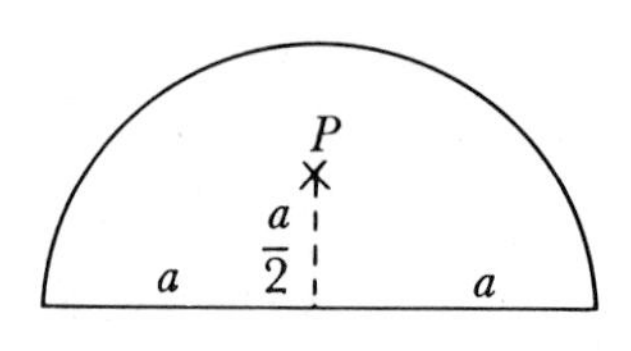

FIGURE 12-9

released is E, find the temperature distribution in the cylinder for large t.

12-14 Assume that the neutron density n inside U_{235} obeys the differential equation

$$\nabla^2 n + \lambda n = \frac{1}{\kappa}\frac{\partial n}{\partial t} \qquad (n = 0 \text{ on surface})$$

(a) Find the critical radius R_0 such that the neutron density inside a U_{235} sphere of radius R_0 or greater is unstable and increases exponentially with time.
(b) Suppose two hemispheres, each just barely stable, are brought together to form a sphere. This sphere is unstable, with

$$n \sim e^{t/\tau}$$

Find the "time-constant" τ of the resulting explosion.

12-15 Assume that in a block of uranium 235 the neutron density $n(\mathbf{x}, t)$ obeys the differential equation

$$\nabla^2 n + n = \frac{\partial n}{\partial t}$$

and that $n = 0$ at the surface. Find the minimum volume of material that can be made into a cylindrical piece that is supercritical.

12-16 Give a complete set of solutions of the equation

$$\nabla^4 \varphi = \nabla^2(\nabla^2 \varphi) = 0$$

in
(a) spherical coordinates (θ, φ functions orthogonal)
(b) rectangular coordinates (x, y functions orthogonal)

12-17 Consider a drum head in the shape of a sector of a circle of radius R and angle β.
(a) Which mode (that is, first, third, ninety-eighth, etc.) is the one illustrated in Figure 12-10? Give the answer for $\beta = \pi/2$, π, $3\pi/2$.
(b) Sketch on one graph the frequencies of the first half dozen modes as functions of β.

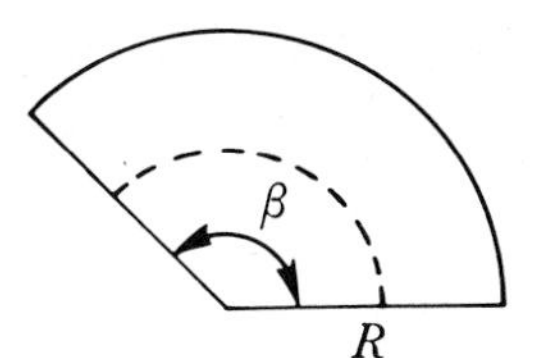

FIGURE 12-10

12-4 SEPARATION OF VARIABLES II: PERTURBATIONS AND GREEN'S FUNCTIONS

PERTURBATION THEORY

Perturbation theory calculations may be carried out by simple substitution in the formulae of Chapter 2.

Example

Consider a circular drum head of radius a with a point mass m attached a distance b from the center. Find the shift in the lowest frequency.

The basic equation is

$$T\nabla^2 u = \rho \frac{\partial^2 u}{\partial t^2}$$

If $u \sim e^{-i\omega t}$, then

$$\frac{T}{\rho}\nabla^2 u = -\omega^2 u$$

This is a standard eigenvalue problem with operator $(T/\rho)\nabla^2$ and eigenvalue $-\omega^2$. Let

$$\rho = \rho_0 + \Delta\rho$$

ρ_0 being a uniform density. Then the unperturbed operator is

$$\frac{T}{\rho_0}\nabla^2$$

and the perturbation Q is

$$Q = -\frac{T\Delta\rho}{\rho_0^2}\nabla^2$$

[One may object that the expansion $(\rho_0 + \Delta\rho)^{-1} \approx \rho_0^{-1} - \rho_0^{-2}\Delta\rho$, valid for small $\Delta\rho$, is not obviously appropriate when $\Delta\rho$ is a delta function. The procedure is nevertheless correct; the reader whom physical intuition does not convince may construct a rigorous proof.] The lowest unperturbed eigenvalue and eigenfunction are [see equation (12-23)]

$$\lambda_0^0 = -\left(\frac{2.40}{a}\right)^2 \frac{T}{\rho_0} \qquad u_0^0 = \frac{J_0\left(2.40\,\dfrac{r}{a}\right)}{\sqrt{\pi a^2}\, J_1(2.40)}$$

Note that we have normalized u_0^0 to agree with our initial assumption

$$\int_0^{2\pi} d\theta \int_0^a r\, dr\, (u_0^0)^2 = 1$$

Then

$$\lambda_0^{(1)} = \int_0^{2\pi} d\theta \int_0^a r\, dr\; u_0^0(Qu_0^0)$$

But

$$Qu_0^0 = -\frac{T\Delta\rho}{\rho_0^2}\nabla^2 u_0^0$$

$$= -\frac{\Delta\rho}{\rho_0}\lambda_0^0 u_0^0$$

Also, in the problem at hand, $\Delta\rho$ is m times a delta function at $r = b$. Thus

$$\lambda_0^{(1)} = -m\frac{\lambda_0^0}{\rho_0}[u_0^0(r = b)]^2$$

or

$$\frac{\lambda_0^{(1)}}{\lambda_0^0} = -\frac{m}{\rho_0}[u_0^0(r = b)]^2$$

$$= -\frac{m}{M}\frac{\left[J_0\left(2.40\,\dfrac{b}{a}\right)\right]^2}{[J_1(2.40)]^2}$$

where $M = \rho_0 \pi a^2$ is the total mass of the drum head. The frequency shift is given by

$$\frac{\delta\omega}{\omega} = \frac{1}{2}\frac{\lambda_0^{(1)}}{\lambda_0^0} = -\frac{m}{2M}\frac{\left[J_0\left(2.40\,\dfrac{b}{a}\right)\right]^2}{[J_1(2.40)]^2}$$

Example

For the circular drum head problem mentioned above, we found the second unperturbed eigenvalue

$$\lambda_2^0 = -\left(\frac{3.83}{a}\right)^2 \frac{T}{\rho_0}$$

to be degenerate. Find the correct zero-order eigenstates to use as a basis with perturbation Q.

Two normalized orthogonal eigenfunctions belonging to λ_2^0 are

$$u_\alpha^0 = \sqrt{\frac{2}{\pi a^2}} \frac{J_1\left(3.83 \frac{r}{a}\right) \cos \theta}{J_0(3.83)}$$

$$u_\beta^0 = \sqrt{\frac{2}{\pi a^2}} \frac{J_1\left(3.83 \frac{r}{a}\right) \sin \theta}{J_0(3.83)}$$

The perturbation Q is given by

$$Q = -\frac{T\Delta\rho}{\rho_0^2} \nabla^2$$

where $\Delta\rho$ is the density due to a point mass m, having polar coordinates (b, ψ). It is straightforward to verify that Q is represented by the matrix

$$\frac{2m}{\pi a^2 \rho_0} \frac{J_1\left(3.83 \frac{b}{a}\right)^2}{J_0(3.83)^2} \frac{T}{\rho_0} \left(\frac{3.83}{a}\right)^2 \begin{pmatrix} \cos^2 \psi & \sin \psi \cos \psi \\ \sin \psi \cos \psi & \sin^2 \psi \end{pmatrix}$$

The 2×2 matrix in brackets has eigenvalues and eigenvectors

$$\lambda_1 = 1 \qquad \lambda_2 = 0$$

$$a_1^0 = \begin{pmatrix} \cos \psi \\ \sin \psi \end{pmatrix} \qquad a_2^0 = \begin{pmatrix} -\sin \psi \\ \cos \psi \end{pmatrix}$$

Therefore, the correct zero-order eigenfunctions are

$$u_1^0 = (\cos \psi) u_\alpha^0 + (\sin \psi) u_\beta^0$$

$$u_2^0 = (-\sin \psi) u_\alpha^0 + (\cos \psi) u_\beta^0$$

and, in the system with these base vectors, Q becomes

$$\frac{2m}{\pi a^2 \rho_0} \frac{\left[J_1\left(3.83 \frac{b}{a}\right)\right]^2}{[J_0(3.83)]^2} \frac{T}{\rho_0} \left(\frac{3.83}{a}\right)^2 \begin{pmatrix} 1 & 0 \\ 0 & 0 \end{pmatrix} = \begin{pmatrix} Q_{11} & 0 \\ 0 & Q_{22} \end{pmatrix}$$

The degenerate frequency splits, one mode remaining at the unperturbed frequency, the other being lowered by

$$\omega^2 \to \omega_0^2 - Q_{11} = \left(\frac{3.83}{a}\right)^2 \frac{T}{\rho_0} - Q_{11}$$

Example

The Schrödinger equation for the hydrogen atom is

$$\left[-\frac{\hbar^2}{2m}\nabla^2 + V(\mathbf{x})\right]\psi(\mathbf{x}) = E\psi(\mathbf{x})$$

where m is the (reduced) mass of the electron and the energy E is an eigenvalue. $V(\mathbf{x})$ is the Coulomb potential $-e^2/r$. Let us consider the $2p$ state with magnetic quantum number $+1$

$$\psi_{2p}(x) = C(x + iy)e^{-r/2a_0} \qquad E_{2p} = -\frac{e^2}{8a_0}$$

a_0 is the Bohr radius, equal to $\hbar^2/(me^2)$; the constant C must equal $1/\sqrt{64\pi a_0^5}$ if $\psi_{2p}(\mathbf{x})$ is to be normalized. What is the shift in energy (ΔE) if the small perturbation $\Delta V(\mathbf{x}) = \lambda z^2/a_0^2$ is applied?

Our formula (2-27) yields immediately

$$\begin{aligned}\Delta E &= \int d^3x\, \psi_{2p}^*(\mathbf{x}) \frac{\lambda z^2}{a_0^2} \psi_{2p}(\mathbf{x}) \\ &= \frac{\lambda C^2}{a_0^2} \int_0^\infty r^2\, dr \int d\Omega (x^2 + y^2) z^2 e^{-r/a_0} \\ &= 6\lambda\end{aligned}$$

GREEN'S FUNCTIONS

Separation of variables can also be used to find Green's functions. The formal sum over eigenfunctions (4-42) may be used, but frequently other techniques provide the answer in a more convenient form.

Example

Consider the homogeneous equation

$$\nabla^2\psi = 0$$

in an infinite region. The boundary condition is that $\psi \to 0$ as $r \to \infty$. We can define a Green's function for this equation to be the solution of

$$\nabla^2\varphi(\mathbf{x}) = \delta(\mathbf{x} - \mathbf{x}') \tag{12-26}$$

with the same boundary conditions. Then any solution u of the inhomogeneous equation

$$\nabla^2 u = f(\mathbf{x})$$

with $u \to 0$ as $r \to \infty$ can be expressed in terms of φ by

$$u = \int d^3x' \, \varphi(\mathbf{x}', \mathbf{x}) f(\mathbf{x}')$$

(Show this by manipulation of the equations involved!)

Clearly, φ can depend only on the scalar quantity $r = |\mathbf{x} - \mathbf{x}'|$. Thus we take our origin of (spherical) coordinates to be at the point $\mathbf{x}'$. For all $\mathbf{x} \neq \mathbf{x}'$, $\nabla^2 \varphi(x) = 0$. Therefore,

$$\varphi \sim \begin{Bmatrix} r^l \\ r^{-(l+1)} \end{Bmatrix} P_l^m(\cos\theta) e^{\pm im\varphi}$$

Because of the spherical symmetry and the boundary condition at infinity, the only possibility is

$$\varphi = \frac{A}{r}$$

To find A, integrate the differential equation (12-26) over the volume of a sphere of radius R surrounding the origin. This gives

$$1 = \int d\mathbf{S} \cdot \nabla\varphi$$

$$= 4\pi r^2 \left(\frac{-A}{r^2} \right) \Rightarrow A = -\frac{1}{4\pi}$$

The Green's function (that is, potential of a point charge or Coulomb potential) is therefore

$$G(\mathbf{x}, \mathbf{x}') = \frac{-1}{4\pi |\mathbf{x} - \mathbf{x}'|}$$

(This means we must supplement (12-14) with the statement that $1/r$ is not a solution of $\nabla^2 \psi = 0$ right at the origin.)

Example

Find the Green's function G for the Helmholtz equation

$$\nabla^2 u + k^2 u = 0 \tag{12-27}$$

in a circular drum. G is the solution of

$$\nabla^2 G + k^2 G = \delta(\mathbf{x} - \mathbf{x}') \tag{12-28}$$

and according to equation (4-42) can be expressed as

$$G = \sum_{k_n} \frac{\psi_n(\mathbf{x}) \psi_n^*(\mathbf{x}')}{k^2 - k_n^2}$$

where the functions $\psi_n(\mathbf{x})$ are normalized eigenfunctions of ∇^2 with eigenvalues $-k_n^2$. We found in equation (12-22) that these take the form

$$\psi_n = \frac{1}{N_n} J_m(k_p^m r) \cos m\varphi$$

or

$$\psi_n = \frac{1}{N_n} J_m(k_p^m r) \sin m\varphi$$

where $k_p^m R$ is the pth zero of $J_m(x)$, $k_n^2 = (k_p^m)^2$, and

$$N_n^2 = \frac{R^2\pi}{2} [J_{m+1}(k_p^m R)]^2$$

Thus

$$G = \sum_{mp} \frac{2J_m(k_p^m r) J_m(k_p^m r') \cos m(\varphi - \varphi')}{R^2\pi[J_{m+1}(k_p^m R)]^2[k^2 - (k_p^m)^2]}$$

This form of the function may, however, not be particularly useful in all cases. Let us explore some other methods that give the answer in a different form.

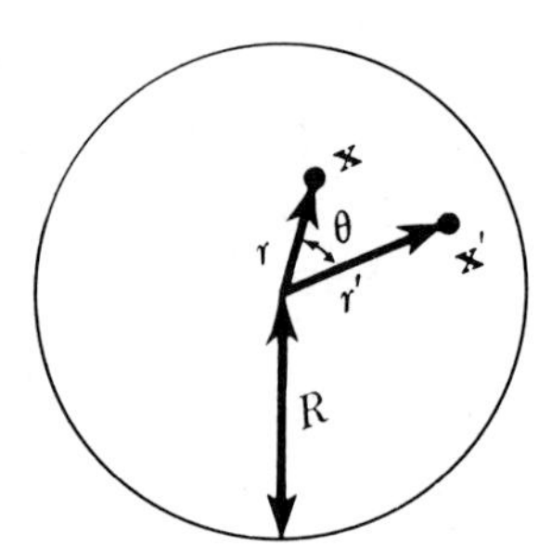

FIGURE 12-11 Coordinates for the circular drum Green's function

It is clear from physical considerations that $G(\mathbf{x}, \mathbf{x}')$ can depend only on r, r', and θ, the angle between $\mathbf{x}$ and $\mathbf{x}'$. Choose $\mathbf{x}'$ on the axis of polar coordinates, as shown in Figure 12-11. Now G is the solution of

$$\nabla^2 G + k^2 G = \delta(\mathbf{x} - \mathbf{x}')$$

where $\delta(\mathbf{x} - \mathbf{x}')$ is the two-dimensional delta function; $\int \delta(\mathbf{x} - \mathbf{x}')\, d^2x = 1$ for any area of integration which includes $\mathbf{x}'$.

For $\mathbf{x} \neq \mathbf{x}'$,

$$\nabla^2 G + k^2 G = 0$$

The solution of this equation satisfying the boundary conditions may be written in the form

$$G = \begin{cases} \sum_m A_m J_m(kr) \cos m\theta & (r < r') \\ \sum_m B_m[J_m(kr) Y_m(kR) - Y_m(kr) J_m(kR)] \cos m\theta & (r > r') \end{cases} \tag{12-29}$$

Note that we choose the solution for $r > r'$ so that it vanishes automatically at $r = R$; we have also used the physically obvious fact that G is an *even* function of θ.

We determine the constants A_m and B_m by fitting our solutions (12-29) together along the circle $r = r'$. G is continuous but its derivative (that is, its *gradient*) has a discontinuity at the point $\mathbf{x} = \mathbf{x}'$. To find the singularity, integrate the differential equation (12-28) for G over an infinitesimal area that includes the point $\mathbf{x} = \mathbf{x}'$. We obtain

$$\int \nabla^2 G \, d^2x = \int (\nabla G)_n \, dl = 1 \tag{12-30}$$

where we have used the two-dimensional analog of Gauss' theorem:

$$\int_V \nabla \cdot \mathbf{u} \, d^3x = \int_\Sigma \mathbf{u} \cdot d\mathbf{S}$$

Σ being the surface which encloses the volume V.

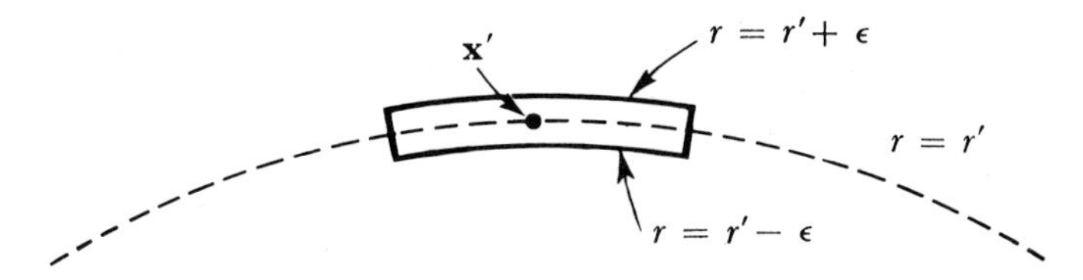

FIGURE 12-12 The area for application of Gauss's theorem

For an area enclosing the point $\mathbf{x}'$, we shall choose one shaped as shown in Figure 12-12. Neglecting the ends, (12-30) gives us

$$\int_{r'+\varepsilon} \frac{\partial G}{\partial r} dl - \int_{r'-\varepsilon} \frac{\partial G}{\partial r} dl = 1$$

where l measures arc length. Since $dl = r' \, d\theta$, this gives

$$\int d\theta \left(\left.\frac{\partial G}{\partial r}\right|_{r'+\varepsilon} - \left.\frac{\partial G}{\partial r}\right|_{r'-\varepsilon} \right) = \frac{1}{r'}$$

(provided the range of integration includes the point $\mathbf{x}'$; otherwise we get zero). Therefore,

$$\left(\left.\frac{\partial G}{\partial r}\right|_{r'+\varepsilon} - \left.\frac{\partial G}{\partial r}\right|_{r'-\varepsilon} \right) = \frac{1}{r'} \delta(\theta)$$

Let

$$\left.\frac{\partial G}{\partial r}\right|_{r'+\varepsilon} - \left.\frac{\partial G}{\partial r}\right|_{r'-\varepsilon} = \sum_m C_m \cos m\theta$$

We multiply both sides by $\cos m'\theta$, in the usual way, and integrate over θ from $-\pi$ to π. The result is

$$\frac{1}{r'} = \pi C_{m'} \varepsilon_{m'} \qquad \text{where } \varepsilon_{m'} = \begin{cases} 2 & \text{if } m' = 0 \\ 1 & \text{if } m' > 0 \end{cases} \tag{12-31}$$

Thus

$$\left.\frac{\partial G}{\partial r}\right|_{r'+\varepsilon} - \left.\frac{\partial G}{\partial r}\right|_{r'-\varepsilon} = \frac{1}{\pi r'} \sum_m \frac{1}{\varepsilon_m} \cos m\theta \tag{12-32}$$

This is the condition on the discontinuity in the gradient of G that we were after.

Now we can write down two simultaneous equations for the A_m and B_m of (12-29). From the continuity of G at $r = r'$, we obtain

$$A_m J_m(kr') = B_m[J_m(kr') Y_m(kR) - Y_m(kr') J_m(kR)]$$

For the condition (12-32) on the discontinuity in $\partial G/\partial r$, we obtain

$$B_m[J_m'(kr') Y_m(kR) - Y_m'(kr') J_m(kR)] - A_m J_m'(kr') = \frac{1}{\pi \varepsilon_m k r'}$$

The solution is

$$A_m = \frac{J_m(kR) Y_m(kr') - J_m(kr') Y_m(kR)}{2\varepsilon_m J_m(kR)}$$

$$B_m = \frac{-J_m(kr')}{2\varepsilon_m J_m(kR)}$$

where we have used the relation

$$J_m(x) Y_m'(x) - J_m'(x) Y_m(x) = \frac{2}{\pi x}$$

Note that $G(\mathbf{x}, \mathbf{x}') = G(\mathbf{x}', \mathbf{x})$, as before.

Let us now return to the original Green's function equation (12-28) and solve it by another method.

$$\nabla^2 G + k^2 G = \delta(\mathbf{x} - \mathbf{x}')$$

where $\delta(\mathbf{x} - \mathbf{x}')$ is the two-dimensional delta function located at the "source" point $\mathbf{x}'$.

There are two difficulties in finding the solution: one is to insure the proper singularity at the source point $\mathbf{x}'$ and the other is to satisfy the boundary conditions. It is sometimes convenient to separate the two difficulties by finding a solution of the form

$$G = u(\mathbf{x}, \mathbf{x}') + v(\mathbf{x}, \mathbf{x}')$$

where u has the singularity at $\mathbf{x}'$ [that is, obeys equation (12-28)] without necessarily satisfying the boundary conditions, while v is "smooth" at $\mathbf{x}'$ [that is, obeys the homogeneous equation (12-27)] but is adjusted to make $G(\mathbf{x}, \mathbf{x}')$ satisfy the boundary conditions. The function $u(\mathbf{x}, \mathbf{x}')$ is sometimes called a *fundamental solution* of (12-28).

The singularity required at $\mathbf{x}'$ may be found by integrating equation (12-28) over a small area element centered at $\mathbf{x}'$ and finding u in the form $u(\rho)$ where $\rho = |\mathbf{x} - \mathbf{x}'|$. From Gauss' theorem

$$\int_0^\rho \nabla^2 G \cdot 2\pi\rho \, d\rho = 2\pi\rho \frac{\partial G}{\partial \rho}$$

so that

$$2\pi\rho \frac{\partial G}{\partial \rho} + k^2 \int_0^\rho G \cdot 2\pi\rho \, d\rho = 1$$

and

$$G(\rho) \to \frac{1}{2\pi} \ln \rho + \text{constant} \qquad \text{as } \rho \to 0$$

$u(\mathbf{x}, \mathbf{x}')$ should then have this behavior near $\rho = 0$. The singular solution of equation (12-18), $Y_0(k\rho)$, has the behavior for small ρ:

$$Y_0(k\rho) \to \frac{2}{\pi} \ln \rho + \text{constant}$$

Thus

$$G = \tfrac{1}{4} Y_0(k\rho) + v(\mathbf{x}, \mathbf{x}')$$

We now write

$$v = \sum_n A_n J_n(kr) \cos n\theta$$

with A_n chosen so that G satisfies the boundary conditions. That is,

$$G(r = R) = 0 = \frac{1}{4} Y_0(k\rho_R) + \sum_n A_n J_n(kR) \cos n\theta$$

$$A_n = -\frac{1}{4\pi J_n(kR)\varepsilon_n} \int_0^{2\pi} Y_0(k\rho_R) \cos n\theta \, d\theta$$

where

$$\rho_R^2 = R^2 + r'^2 - 2Rr' \cos \theta$$

[See Figure 12-13; ε_n is defined in (12-31).] The result is

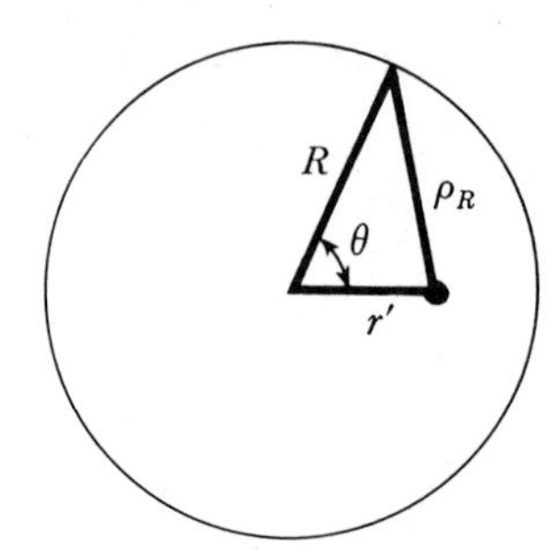

FIGURE 12-13 The coordinate ρ_R for the drum head

$$G(\mathbf{x}, \mathbf{x}') = \frac{1}{4} Y_0(k\rho) - \sum_{n=0}^{\infty} \frac{J_n(kr) \cos n\theta}{2\pi J_n(kR)\, \varepsilon_n} \int_0^{\pi} Y_0(k\rho_R) \cos n\theta \, d\theta$$

This form of the Green's function is convenient for some purposes; it is easy to visualize for low frequencies ω and source positions $\mathbf{x}'$ not too near the edge.

In summary, we recall the three forms found for a Green's function:

1. The formal sum over eigenfunctions (4-42).

2. A solution of the homogeneous equation and boundary conditions on either side of a "surface" containing the source point; the two solutions being matched on this surface in such a way as to produce the source-point singularity.

3. The form $G(\mathbf{x}, \mathbf{x}') = u(\mathbf{x}, \mathbf{x}') + v(\mathbf{x}, \mathbf{x}')$, where u is the fundamental solution, and v takes care of the boundary conditions.

PROBLEMS

Solve Problems 12-18 through 12-22 by perturbation theory.

12-18 Find the three lowest frequencies of oscillation of a string of length L, tension T, and mass per unit length ρ, if a mass m is fastened to the string a distance $L/4$ from one end.

$$\frac{\partial^2 y}{\partial x^2} - \frac{\rho}{T} \frac{\partial^2 y}{\partial t^2} = 0 \qquad (m \ll \rho L)$$

12-19 A square drum head of side a has surface density ρ_0 except in the corner $0 < x < a/2$, $0 < y < a/2$, where the density is 10% greater (see Figure 12-14). Find first-order approximate values for the three lowest frequencies of oscillation.

$$T \nabla^2 u = \rho \frac{\partial^2 u}{\partial t^2}$$

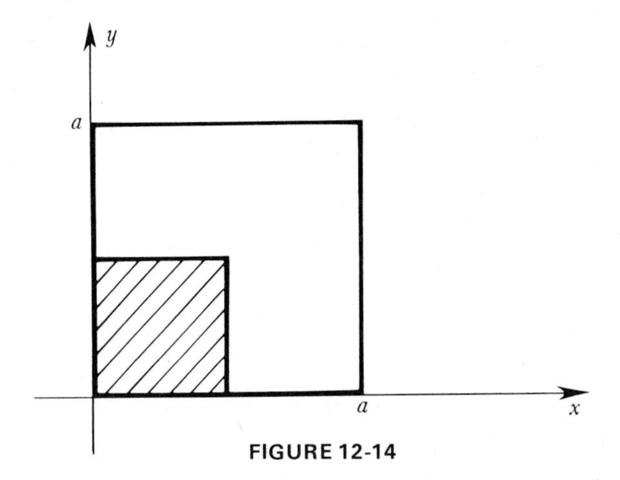

FIGURE 12-14

12-20 A square drum head with tension T, surface density ρ, and side a is perturbed by fastening two small masses, each of mass m, to the drum head as shown in Figure 12-15. Find the three lowest frequencies of the drum, neglecting terms of order $(m/\rho a^2)^2$.

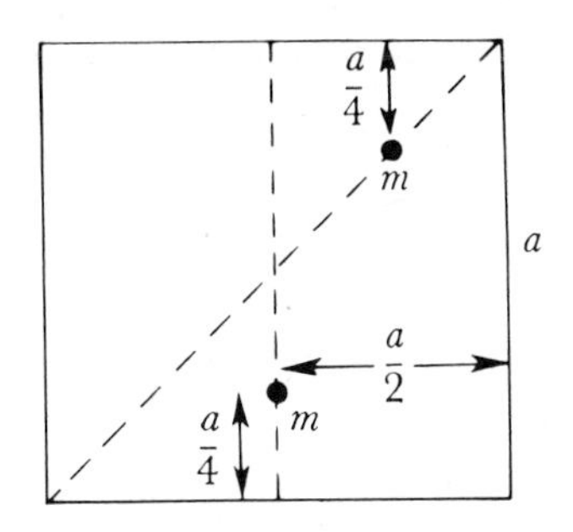

FIGURE 12-15

12-21 Find $\lambda(\alpha)$, the lowest eigenvalue of

$$\nabla^2\varphi + \lambda(1 + \alpha r^2)\varphi = 0 \qquad (0 \le r \le R)$$

$$(\varphi = 0 \text{ on surface of sphere } r = R)$$

for small α.

12-22 (a) Use perturbation theory to obtain an approximate value for the lowest eigenvalue of the Schrödinger equation

$$\left[-\frac{\hbar^2}{2m}\nabla^2 + V(\mathbf{x})\right]\psi(\mathbf{x}) = E\psi(\mathbf{x})$$

if $V(\mathbf{x}) = \alpha r^2 + \beta z^2$ (β small).

(b) Verify your answer by separating the differential equation in rectangular coordinates and solving exactly.

In Problems 12-23 through 12-26, use separation of variables to find the Green's functions.

12-23 Find the Green's function of the Helmholtz equation in the cube $0 \leq x, y, z \leq L$ by solving the equation

$$\nabla^2 u + k^2 u = \delta(\mathbf{x} - \mathbf{x}') \qquad (u = 0 \text{ on surface of cube})$$

12-24 Find the Green's function $G(\mathbf{x}, \mathbf{x}')$ for the Helmholtz equation

$$\nabla^2 u + k^2 u = 0$$

inside the sphere $r = a$, with the boundary condition $u(r = a) = 0$. Find G by solving the equation, not just as a formal sum over eigenfunctions. Note that G can depend only on r, r', and θ, the angle between $\mathbf{x}$ and $\mathbf{x}'$.

12-25 Green's functions can be written for *homogeneous* equations with *inhomogeneous* boundary conditions. To illustrate this, consider the Helmholtz equation

$$\nabla^2 u + k^2 u = 0$$

inside the circle $r = R$, with the boundary condition $u = f(\theta)$ at $r = R$, $f(\theta)$ being some given function. The solution can be written

$$u(\mathbf{x}) = \int_0^{2\pi} G(\mathbf{x}, \theta') f(\theta')\, d\theta'$$

Find $G(\mathbf{x}, \theta)$.

12-26 Consider the solution $u(\mathbf{x})$ of the two-dimensional inhomogeneous Helmholtz equation

$$\nabla^2 u(\mathbf{x}) + k^2 u(\mathbf{x}) = f(\mathbf{x})$$

inside the circle $r = R$, the boundary condition being $u(\mathbf{x}) = g(\theta)$ at $r = R$. $f(\mathbf{x})$ and $g(\theta)$ are given functions. By manipulating the differential equations obeyed by $u(\mathbf{x})$ and $G(\mathbf{x}, \mathbf{x}')$, where the latter is the Green's function (12-28), obtain an explicit expression for $u(\mathbf{x})$ in terms of $f(\mathbf{x})$, $g(\theta)$, and $G(\mathbf{x}, \mathbf{x}')$.

Discuss and verify the relationship between $G(\mathbf{x}, \mathbf{x}')$ and the solution of Problem 12-25.

12-5 INTEGRAL TRANSFORM METHODS

We shall next work a few problems to show how various integral transforms can be used to solve boundary-value problems. The essential feature is the transformation of the equation to one containing derivatives with respect to a smaller number of variables. Also, the boundary conditions may sometimes be inserted in a more automatic way.

Example: Laplace transform

Consider the temperature distribution in the semi-infinite region $x > 0$ when initially $T = 0$ and the plane $x = 0$ is maintained at $T = T_0$. The diffusion equation becomes

$$\frac{\partial^2 T}{\partial x^2} = \frac{1}{\kappa}\frac{dT}{\partial t}$$

The boundary conditions are

1. $T(x, 0) = 0$
2. $T(0, t) = T_0$

The fact that we are only interested in $t > 0$ suggests the use of the Laplace transform with respect to t. (We are also interested only in $x > 0$, but a Laplace transform with respect to x is not useful. Why?) Let

$$F(x, s) = \int_0^{\infty} e^{-st}\, T(x, t)\, dt$$

The Laplace transform of the differential equation is

$$\frac{\partial^2 F}{\partial x^2} = \frac{sF}{\kappa} \tag{12-33}$$

since $T = 0$ for $t = 0$.

The second boundary condition becomes

$$F(0, s) = \frac{T_0}{s} \tag{12-34}$$

Thus, solving (12-33) and making use of (12-34), we obtain

$$F(x, s) = \frac{T_0}{s} \exp\left(-\sqrt{\frac{s}{\kappa}}\, x\right)$$

as we certainly reject the solution $\sim \exp\left(+\sqrt{s/\kappa}\, x\right)$.

Now we must invert this transform to obtain our solution.

$$T(x, t) = \frac{1}{2\pi i}\int_{c-i\infty}^{c+i\infty} \frac{T_0}{s} \exp\left(-\sqrt{\frac{s}{\kappa}}\, x\right) e^{st}\, ds \tag{12-35}$$

The integrand is multivalued. We shall cut the s-plane along the negative real axis and consider the contour shown in Figure 12-16. Since the total integral on the closed contour of Figure 12-16 is zero, the contribution

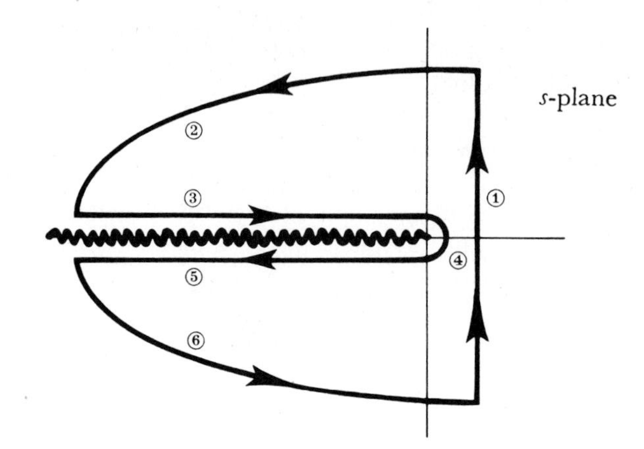

FIGURE 12-16 The s-plane with branch cut and contour used for evaluating the Laplace inversion integral (12-35)

from ① is given by

$$① = -② - ③ - ④ - ⑤ - ⑥$$

② and ⑥ give no contribution.

$$④ = -2\pi i \frac{1}{2\pi i} T_0 = -T_0$$

$$③ = \frac{1}{2\pi i}\int_{-\infty}^{0} \frac{T_0}{s} \exp\left(-ix\sqrt{\frac{1}{\kappa}}\sqrt{-s}\right) e^{st}\, ds$$

$$= \frac{-1}{2\pi i}\int_{0}^{\infty} \frac{T_0}{s} \exp\left(-ix\sqrt{\frac{1}{\kappa}}\sqrt{s}\right) e^{-st}\, ds$$

Similarly,

$$⑤ = \frac{1}{2\pi i}\int_{0}^{\infty} \frac{T_0}{s} \exp\left(ix\sqrt{\frac{1}{\kappa}}\sqrt{s}\right) e^{-st}\, ds$$

Therefore,

$$③ + ⑤ = \frac{1}{\pi}\int_{0}^{\infty} \frac{T_0}{s} e^{-st} \sin\left(x\sqrt{\frac{s}{\kappa}}\right) ds$$

Let $s = z^2$. Then

$$③ + ⑤ = \frac{2T_0}{\pi}\int_{0}^{\infty} \frac{dz}{z} e^{-tz^2} \sin\left(x\sqrt{\frac{1}{\kappa}}z\right)$$

and

$$T(x, t) = T_0\left[1 - \frac{2}{\pi}\int_{0}^{\infty} \frac{dz}{z} e^{-tz^2} \sin\left(x\sqrt{\frac{1}{\kappa}}z\right)\right]$$

It will be left to the student to do this integral and obtain

$$T(x,t) = T_0\left[1 - \operatorname{erf}\left(\frac{x}{2}\sqrt{\frac{1}{\kappa t}}\right)\right] \tag{12-36}$$

This evaluation of the integral (12-35) is straightforward and instructive, but a simpler method exists. Namely, if one differentiates both sides of (12-35) with respect to x, and changes the variable of integration to $u = \sqrt{s}$, one finds by a trivial contour deformation that

$$\frac{\partial T}{\partial x} = \frac{-T_0}{\sqrt{\pi \kappa t}} e^{-x^2/4\kappa t}$$

Note that we have obtained an answer in closed form, rather than an infinite series such as the separation of variables usually gives. This feature results, however, from the infinite range of the variable x rather than from the transform method. To solve this problem by the separation of variables technique, we would consider a slab of finite thickness D and in the end let $D \to \infty$. In this limiting process, the series solution would be transformed into an integral whose evaluation would give the result (12-36). It is instructive for the student to carry out this procedure.

Example: Fourier transform

As a second example of the use of integral transform methods in solving partial differential equations, consider another diffusion problem. Find the temperature distribution $T(x,t)$ in an infinite solid if we are given $T(x,0) = f(x)$. Nothing depends on y or z, so that

$$\frac{\partial^2 T}{\partial x^2} = \frac{1}{\kappa}\frac{\partial T}{\partial t}$$

For variety, we shall Fourier transform the x variable.

$$T(x,t) = \int_{-\infty}^{\infty} \frac{dk}{2\pi} F(k,t) e^{ikx}$$

$$F(k,t) = \int_{-\infty}^{\infty} T(x,t) e^{-ikx}\, dx$$

Thus

$$-k^2 F(k,t) = \frac{1}{\kappa}\frac{\partial F(k,t)}{\partial t}$$

with the solution

$$F(k,t) = \varphi(k) e^{-k^2 \kappa t}$$

The initial condition gives

$$F(k,0) = \int_{-\infty}^{\infty} T(x,0)e^{-ikx}\,dx$$
$$= \int_{-\infty}^{\infty} f(x)e^{-ikx}\,dx$$

Therefore,

$$\varphi(k) = \int_{-\infty}^{\infty} f(x)e^{-ikx}\,dx$$

and

$$F(k,t) = \int_{-\infty}^{\infty} f(x)\,e^{-ikx}\,e^{-k^2\kappa t}\,dx$$

Inverting the Fourier transform, we obtain

$$T(x,t) = \int_{-\infty}^{\infty} \frac{dk}{2\pi}\,e^{ikx}\int_{-\infty}^{\infty} dx'\,f(x')e^{-ikx'}\,e^{-k^2\kappa t}$$
$$= \int_{-\infty}^{\infty} dx'\,f(x')\int_{-\infty}^{\infty} \frac{dk}{2\pi}\,e^{ik(x-x')}\,e^{-k^2\kappa t}$$

The k integral is easy and gives

$$\sqrt{\frac{1}{4\pi\kappa t}}\,e^{-(x-x')^2/4\kappa t}$$

Therefore,

$$T(x,t) = \int_{-\infty}^{\infty} dx'\,f(x')\sqrt{\frac{1}{4\pi\kappa t}}\,e^{-(x-x')^2/4\kappa t}$$

GREEN'S FUNCTIONS

The function

$$G(x,t;x') = \sqrt{\frac{1}{4\pi\kappa t}}\,e^{-(x-x')^2/4\kappa t}$$

is the Green's function for the above problem. Suppose the initial heat source $f(x)$ (that is, the initial temperature distribution) consists of the *plane source* $\delta(x)$. Then

$$T(x,t) = \sqrt{\frac{1}{4\pi\kappa t}}\,e^{-x^2/4\kappa t} = G(x,t;0) \qquad \text{(12-37)}$$

This is a Gaussian dependence on x, with a width increasing as $\sqrt{t}$, as shown in Figure 12-17.

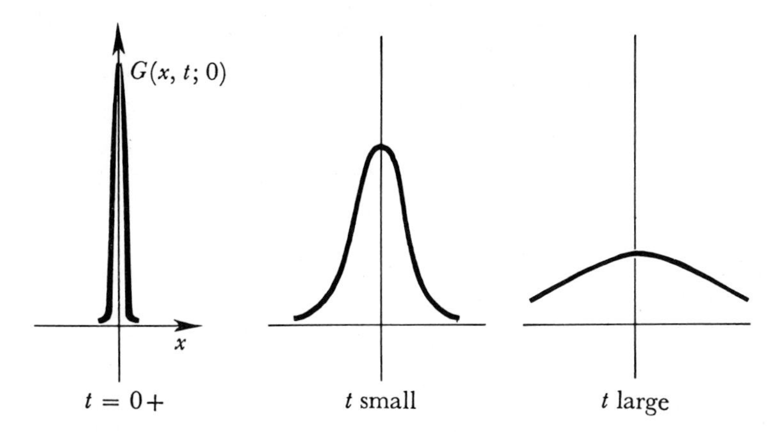

FIGURE 12-17 The Green's function $G(x, t; 0)$ for the one-dimensional heat problem with given initial temperature distribution

What is the distribution resulting from a *point source* $\delta(\mathbf{x})$ at time $t = 0$? We can get this from our preceding result as follows. Let $f(x, t)$ be the response

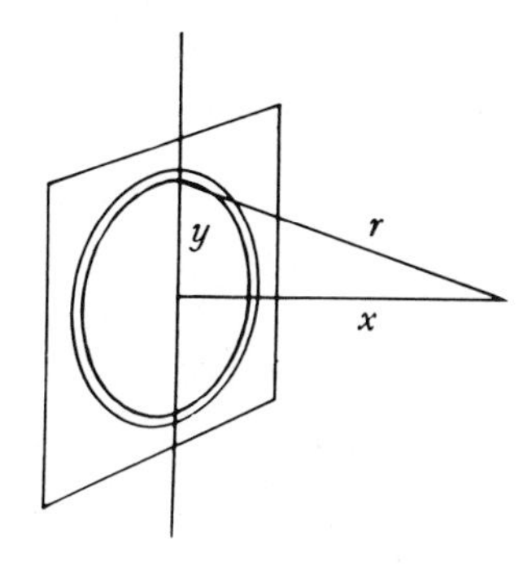

FIGURE 12-18 Relationships between variables x and r appropriate for a plane source and a point source, respectively

to a plane source $\delta(x)$ at $t = 0$. Let $g(r, t)$ be the response to a point source $\delta(\mathbf{x})$ at $t = 0$. A plane source is just the sum of many point sources. Then from Figure 12-18 we see

$$f(x, t) = \int_0^\infty 2\pi y \, dy \, g(r, t)$$

$$= \int_x^\infty 2\pi r \, dr \, g(r, t)$$

Therefore,

$$\frac{\partial f}{\partial x} = -2\pi x g(x, t)$$

$$g(r, t) = \frac{-1}{2\pi r}\left[\frac{\partial f(x, t)}{\partial x}\right]_{x=r}$$

Using our previously found response $f(x, t)$ for a plane source (12-37), we immediately obtain

$$g(r, t) = \left(\frac{1}{4\pi\kappa t}\right)^{3/2} e^{-r^2/4\kappa t} \tag{12-38}$$

as the response to a point initial distribution $T(\mathbf{x}, 0) = \delta(\mathbf{x})$. Notice that $g(r, t) = f(x, t)\cdot f(y, t)\cdot f(z, t)$. Why is this? This point source distribution is another example of a Green's function. As usual, clever use of Green's functions can greatly simplify solution of some problems.

For example, consider the response of the semi-infinite solid $x > 0$ to a point initial temperature distribution at $x = a$, $y = z = 0$, if the entire solid was initially ($t = 0$) at temperature $T = 0$ (except at the above point) and the boundary $x = 0$ is maintained at $T = 0$. The answer is obtained by superimposing a source function about $x = a$, $y = z = 0$:

$$\left(\frac{1}{4\pi\kappa t}\right)^{3/2} \exp\left[-\frac{(x-a)^2 + y^2 + z^2}{4\kappa t}\right]$$

and a negative source function about $x = -a$, $y = z = 0$:

$$-\left(\frac{1}{4\pi\kappa t}\right)^{3/2} \exp\left[-\frac{(x+a)^2 + y^2 + z^2}{4\kappa t}\right]$$

This second "fictitious" source is called an *image*. This *method of images* may be familiar from electrostatics, where it is often used.

As a second example, consider the problem of a cube initially at zero temperature, whose sides are maintained at zero temperature, with $T(\mathbf{x}, t = 0) = \delta(\mathbf{x})$, $\mathbf{x} = 0$ being the center of the cube. We can solve for the temperature distribution at later times either by separating variables or by superimposing an infinite set of images. The first solution is useful for large t; the second is useful for small t.

In Section 9-1 we used transform techniques to compute the Green's function for the one-dimensional Helmholtz equation. Poles appeared on the real axis of $\tilde{G}(\omega)$, and we found that different physical boundary conditions required different treatment of these poles. We are now in a position to find the Green's function for the entire wave equation, including the time dependence. This calculation shows a similar relationship between placement of poles on the real axis and boundary conditions.

Consider the wave equation in an infinite region

$$\nabla^2\varphi(\mathbf{x}, t) - \frac{1}{c^2}\frac{\partial^2\varphi(\mathbf{x}, t)}{\partial t^2} = 0 \tag{12-39}$$

To find the Green's function, we wish to solve the equation

$$\nabla^2\varphi - \frac{1}{c^2}\frac{\partial^2\varphi}{\partial t^2} = \delta(\mathbf{x}-\mathbf{x}')\delta(t-t') \tag{12-40}$$

First, it is clear that the solution depends only on $\mathbf{x}-\mathbf{x}'$ and $t-t'$; in other words, the wave equation (12-39) possesses *translational invariance* in $\mathbf{x}$ and t. Therefore we can, without loss of generality, set $\mathbf{x}'=0$, $t'=0$. Now introduce the Fourier transform Φ of φ:

$$\varphi(\mathbf{x},t) = \int \frac{d^3k}{(2\pi)^3}\int \frac{d\omega}{2\pi}\,\Phi(\mathbf{k},\omega)e^{i(\mathbf{k}\cdot\mathbf{x}-\omega t)}$$

$$\Phi(\mathbf{k},\omega) = \int d^3x \int dt\ \varphi(\mathbf{x},t)e^{-i(\mathbf{k}\cdot\mathbf{x}-\omega t)}$$

The Fourier transform of our differential equation (12-40) is

$$\left(-k^2+\frac{\omega^2}{c^2}\right)\Phi = 1$$

so that

$$\Phi(\mathbf{k},\omega) = \frac{c^2}{\omega^2-k^2c^2}$$

and

$$\varphi(\mathbf{x},t) = \int \frac{d^3k}{(2\pi)^3}\int \frac{d\omega}{2\pi}\,c^2\,\frac{e^{i(\mathbf{k}\cdot\mathbf{x}-\omega t)}}{\omega^2-k^2c^2}$$

If we choose the axis of polar coordinates in $\mathbf{k}$-space along the vector $\mathbf{x}$, the angular integrations are straightforward, and we obtain

$$\varphi(\mathbf{x},t) = \frac{1}{(2\pi)^3}\frac{c^2}{ir}\int_{-\infty}^{\infty} k\,dk \int_{-\infty}^{\infty} d\omega\,\frac{e^{-i(kr-\omega t)}}{\omega^2-k^2c^2} \qquad (r=|\mathbf{x}|) \tag{12-41}$$

We shall now do the ω integral. A problem arises, in that there are two poles on the path of integration, as shown in Figure 12-19. What path of integration do we follow? The uncertainty is directly related to our failure to specify the boundary conditions for our original problem. To see this, note that the only difference between going over, under, or through each pole is that φ picks up a different amount of

$$\frac{e^{ik(r\pm ct)}}{r}$$

But these functions are the (spherically symmetric) solutions of the homogeneous equation (12-39).

Now our differential equation (12-40) involves a disturbance localized at $t=0$, $\mathbf{x}=0$. A very reasonable boundary condition is that $\varphi=0$ for $t<0$;

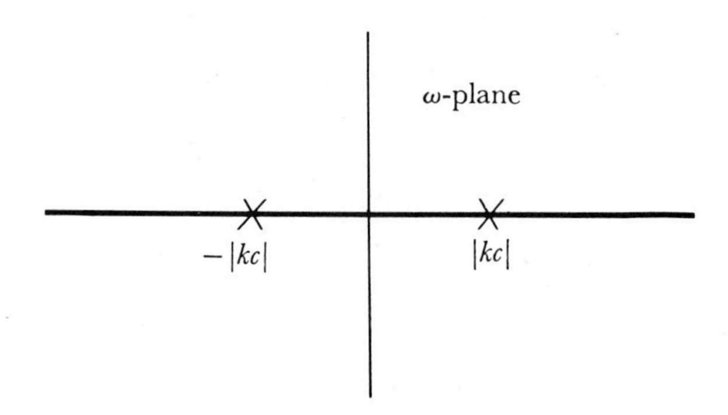

FIGURE 12-19 The ω-plane showing the location of the poles of the integrand in (12-41)

that is, nothing happens *before* the disturbance. In order to satisfy this condition, we make our contour go *over* both poles in Figure 12-19. Then, when $t < 0$ and we complete the contour by an *upper* semicircle, we enclose no poles.

When $t > 0$, we must complete the contour by a *lower* semicircle, and

$$\begin{aligned}
\varphi(\mathbf{x}, t) &= \frac{1}{(2\pi)^3}\frac{c^2}{ir}\int_{-\infty}^{\infty} k\,dk\,e^{ikr}\int_{-\infty}^{\infty} d\omega\,\frac{e^{-i\omega t}}{\omega^2 - k^2c^2} \\
&= \frac{1}{(2\pi)^3}\frac{c}{2}\frac{1}{ir}\int_{-\infty}^{\infty} dk\,e^{ikr}\int_{-\infty}^{\infty} d\omega\,e^{-i\omega t}\left(\frac{1}{\omega - kc} - \frac{1}{\omega + kc}\right) \\
&= -\frac{c}{8\pi^2 r}\int_{-\infty}^{\infty} dk\,e^{ikr}(e^{-ikct} - e^{ikct}) \\
&= -\frac{c}{4\pi r}[\delta(r - ct) - \delta(r + ct)]
\end{aligned}$$

The second delta function doesn't contribute because r and t are both positive. Therefore, the desired Green's function is

$$G(\mathbf{x} - \mathbf{x}', t - t') = \begin{cases} 0 & \text{if } t < t' \\ -\dfrac{c}{4\pi|\mathbf{x} - \mathbf{x}'|}\delta[|\mathbf{x} - \mathbf{x}'| - c(t - t')] & \text{if } t > t' \end{cases} \tag{12-42}$$

Let us briefly illustrate the use of this Green's function. The solution of

$$\nabla^2\varphi - \frac{1}{c^2}\frac{\partial^2\varphi}{\partial t^2} = f(\mathbf{x}, t)$$

is

$$\varphi(\mathbf{x}, t) = -\frac{c}{4\pi}\int d^3x'\,dt'\,f(\mathbf{x}', t')\frac{\delta[|\mathbf{x} - \mathbf{x}'| - c(t - t')]}{|\mathbf{x} - \mathbf{x}'|}$$

The t' integration may be done because of the δ-function, with the result

$$\varphi(\mathbf{x}, t) = -\frac{1}{4\pi}\int d^3x' \frac{f\left(\mathbf{x}', t - \frac{1}{c}|\mathbf{x} - \mathbf{x}'|\right)}{|\mathbf{x} - \mathbf{x}'|}$$

This is called a *retarded potential* because the source term in the integrand is evaluated at the previous time $t - (1/c)|\mathbf{x} - \mathbf{x}'|$, $(1/c)|\mathbf{x} - \mathbf{x}'|$ being the time for its influence to travel a distance $|\mathbf{x} - \mathbf{x}'|$ at the speed c. For a specific example, let $f(\mathbf{x}', t') = \delta[\mathbf{x}' - \xi(t')]$. This describes a point source moving along the path $\xi(t)$. Then

$$\varphi(\mathbf{x}, t) = -\frac{1}{4\pi}\int d^3x' \int dt' \frac{\delta[\mathbf{x}' - \xi(t')]\delta\left(t - t' - \frac{1}{c}|\mathbf{x} - \mathbf{x}'|\right)}{|\mathbf{x} - \mathbf{x}'|} \tag{12-43}$$

Clearly, the only contribution to the integral comes when $\mathbf{x}'$ and t' obey the equations

$$\mathbf{x}' = \xi(t')$$

$$t - t' = \frac{1}{c}|\mathbf{x} - \mathbf{x}'|$$

The question is: What contribution arises from this point?

Let us consider more generally the evaluation of

$$\int dx\, dy\, dz\, dt\; \delta[f_1(xyzt)]\delta[f_2(xyzt)]\delta[f_3(xyzt)]\delta[f_4(xyzt)]$$

This integral is done by changing the variables of integration from $xyzt$ to $f_1 f_2 f_3 f_4$. The integral becomes

$$\int df_1\, df_2\, df_3\, df_4 \left|\frac{\partial(xyzt)}{\partial(f_1 f_2 f_3 f_4)}\right| \delta(f_1)\delta(f_2)\delta(f_3)\delta(f_4) = \left|\frac{\partial(xyzt)}{\partial(f_1 f_2 f_3 f_4)}\right|$$

evaluated at $f_1 = f_2 = f_3 = f_4 = 0$, where this last expression involves the Jacobian

$$\begin{vmatrix} \frac{\partial x}{\partial f_1} & \frac{\partial x}{\partial f_2} & \frac{\partial x}{\partial f_3} & \frac{\partial x}{\partial f_4} \\ \frac{\partial y}{\partial f_1} & \frac{\partial y}{\partial f_2} & \frac{\partial y}{\partial f_3} & \frac{\partial y}{\partial f_4} \\ \frac{\partial z}{\partial f_1} & \frac{\partial z}{\partial f_2} & \frac{\partial z}{\partial f_3} & \frac{\partial z}{\partial f_4} \\ \frac{\partial t}{\partial f_1} & \frac{\partial t}{\partial f_2} & \frac{\partial t}{\partial f_3} & \frac{\partial t}{\partial f_4} \end{vmatrix} = \frac{\partial(xyzt)}{\partial(f_1 f_2 f_3 f_4)}$$

A useful theorem from calculus tells us that

$$\frac{\partial(xyzt)}{\partial(f_1 f_2 f_3 f_4)} = \left[\frac{\partial(f_1 f_2 f_3 f_4)}{\partial(xyzt)}\right]^{-1}$$

Now what is this Jacobian in our particular case?

$$f_1 = x' - \xi_x(t') \qquad f_3 = z' - \xi_z(t')$$

$$f_2 = y' - \xi_y(t') \qquad f_4 = t - t' - \frac{1}{c}|\mathbf{x} - \mathbf{x}'|$$

The evaluation of the determinant is straightforward and gives

$$\frac{\partial(f_1 f_2 f_3 f_4)}{\partial(x'y'z't')} = -1 - \frac{1}{c}\frac{\boldsymbol{\xi}\cdot(\mathbf{x}' - \mathbf{x})}{|\mathbf{x}' - \mathbf{x}|}$$

Therefore,

$$\varphi(\mathbf{x}, t) = -\frac{1}{4\pi}\frac{1}{|\mathbf{x} - \mathbf{x}'| + \frac{1}{c}\dot{\boldsymbol{\xi}}(t')\cdot(\mathbf{x}' - \mathbf{x})}$$

where $\mathbf{x}' = \boldsymbol{\xi}(t')$ and $t' = t - (1/c)|\mathbf{x} - \mathbf{x}'|$. In electrodynamics, this is referred to as the *Lienard-Wiechert potential.*

Actually, the integral (12-43) may be done more simply by doing first the d^3x' integral, and then the dt' integral. This is left as an exercise for the student.

PROBLEMS

12-27 Use Fourier transforms to find the motion of an infinitely long stretched string with initial displacement $\varphi(x)$ and initial velocity zero.

$$\frac{\partial^2 y}{\partial x^2} = \frac{1}{c^2}\frac{\partial^2 y}{\partial t^2}$$

12-28 An infinite stretched membrane has a real density ρ and tension T. Initially it has a given displacement $f(\mathbf{r})$ and zero initial velocity everywhere. Find the subsequent motion.

12-29 A straight wire of radius a is immersed in an infinite volume of liquid. Initially the wire and the liquid have temperature $T = 0$. At time $t = 0$, the wire is suddenly raised to temperature T_0 and maintained at that temperature. Find $F(r, s)$, the Laplace transform of the resulting temperature distribution $T(r, t)$ in the liquid.

12-30 A sphere of radius R is at temperature $T = 0$ throughout. At time $t = 0$, it is immersed in a liquid bath at temperature T_0. Find the subsequent temperature distribution $T(r, t)$ inside the sphere. [Let $\kappa =$ thermal conductivity/(density $\times$ specific heat).]

12-31 Show that the (retarded) Green's function for the Helmholtz equation

$$\nabla^2 u(\mathbf{x}) + k^2 u(\mathbf{x}) = 0$$

with the boundary condition that $u(\mathbf{x})e^{-i\omega t}$ represents outgoing waves only at infinity is

$$G(\mathbf{x}, \mathbf{x}') = -\frac{e^{ikr}}{4\pi r}$$

where $r = |\mathbf{x} - \mathbf{x}'|$. What is the connection between this result and the Green's function (12-42)?

12-32 The temperature in an infinite cylindrical rod of radius a satisfies the conditions

(a) $\nabla^2 T = \alpha^{-1}\dfrac{\partial T}{\partial t} \qquad \left(\alpha = \dfrac{K}{c\rho}\right)$

(b) $T = 0$ at $t = 0$
(c) $T = T_0 \cos \varphi$ at $\rho = a$
Find $T(\rho, \varphi, t)$ for $t > 0$.

12-33 In an absorbing medium, the neutron density obeys the differential equation

$$\kappa \nabla^2 n - \frac{n}{T} = \frac{\partial n}{\partial t}$$

where T is a constant. At $t = 0$, a burst of neutrons is produced on the yz-plane of an infinite medium, so that

$$n(\mathbf{x}, 0) = \delta(x) \qquad [\text{not } \delta(\mathbf{x})]$$

Find the neutron density, and the total number of neutrons (per unit area), for later times.

12-34 A semi-infinite body $x < 0$ has thermal conductivity K_1, density ρ_1, and specific heat C_1. It is initially at temperature T_0. At time $t = 0$, it is placed in thermal contact with the semi-infinite body $x > 0$, which has parameters K_2, ρ_2, C_2, and is initially at temperature $T = 0$. Find $T_2(x, t)$, the temperature in the second body

$$\frac{\partial^2 T}{\partial x^2} = \frac{C\rho}{K}\frac{\partial T}{\partial t}$$

SUMMARY

The solution, ψ, of a second-order linear p.d.e. is determined completely in the neighborhood of some *open* curve C by specification of values of ψ, and $\partial\psi/\partial n$ along the curve, provided the curve is not anywhere tangent to a characteristic for the equation. If the curve is tangent to a characteristic, less information need be specified. *Closed* curves are commonly used as boundaries for elliptic equations. In this case, specification of either ψ or $\partial\psi/\partial n$ along the boundary is adequate.

When the derivatives in a p.d.e. can be broken into two pieces in the form (differentiation with respect to y) + (differentiation with respect to x) then a trial solution of the form $X(x)Y(y)$ leads to independent equations for X and Y. In cases of physical interest, these ordinary differential equations have well-known solutions. The general solution of the p.d.e. may then be expanded in terms of these factored solutions.

One can apply Fourier or Laplace transforms separately in each variable of a p.d.e. Often the transformed problem is much easier to solve.

The solution of the inhomogeneous equation $\mathcal{O}u = f(x, y, \ldots)$, where $\mathcal{O}$ is a partial differential operator, may be expressed in terms of f and a Green's function for the particular operator and boundary conditions specified. These Green's functions can be found by various techniques using separation of variables or integral transforms.

References

1. Abramowitz, M., and Stegun, I., *Handbook of Mathematical Functions*, New York: Dover Publications (1965).
   Contains a very readable and complete summary of properties of special functions, as well as tables of values.

2. Ahlfors, L. V., *Complex Analysis*, New York: McGraw-Hill (1966).
   A mathematician's point of view.

3. Apostol, T. M., *Mathematical Analysis*, Reading, Mass.: Addison-Wesley (1974).
   Good for clearing up fine points we have skimmed over. See in particular the discussion of Fourier series (Chapter 15) and infinite series (Chapters 12 and 13).

4. Baker, George A. Jr., *Essentials of Padé Approximants*, New York: Academic Press (1975).

5. Burkhill, J. C., *The Theory of Ordinary Differential Equations*, New York: Longman (1975).
   Small but remarkably complete.

6. Carrier, G. F., Krook, M., and Pearson, C. E., *Functions of a Complex Variable*, New York: McGraw-Hill (1966).
   A good graduate level text for applications.

7. Churchill, Ruel V., Brown, J. W., and Verhey, R. F., *Complex Variables and Applications*, New York: McGraw-Hill (1974).
   One of the best introductory books for engineers and scientists. Clear and complete.

8. Copson, E. T., *Theory of Functions of a Complex Variable*, New York: Oxford Univ. Press (1962).
   The classic text on this subject. Unfortunately, many of the most useful theorems are given as exercises.

9. Courant, E., Livingston, M. S., and Snyder, H., *Phys. Rev.* 88: 1190 (1952).

10. Courant, R., and Hilbert, D., *Methods of Mathematical Physics*, New York: Wiley Interscience (1962).
    Slightly more abstract than Morse and Feshbach. Contains a great deal of information that can't be found elsewhere.

11. Dennery, P., and Krzywicki, A., *Mathematics for Physicists*, New York: Harper and Row (1967).
A somewhat more rigorous book at about the same level as this one.

12. Dirac, P. A. M., *The Principles of Quantum Mechanics*, Clarendon Press (1958).
The original exposition of the mathematical basis of quantum mechanics. This book has been known to produce a feeling of wild exhilaration in some students. Try it!

13. Erdélyi, A. (Ed.), *Bateman Manuscript Project*, New York: McGraw-Hill (1954).
These five volumes are a gold mine of formulae for special functions and integral transforms.

14. Forsyth, A. R., *A Treatise on Differential Equations*, New York: Macmillan (1951).
Excellent, even if it is slightly old fashioned.

15. Goldstein, H., *Classical Mechanics*, Reading, Mass.: Addison-Wesley (1950).
The Lagrangian and Hamiltonian formulations of mechanics discussed here are important mathematical methods which we have omitted in the text.

16. Halmos, P. R., *Finite-Dimensional Vector Spaces*, New York: Van Nostrand (1958).
This book provides a particularly clear mathematical treatment of matrices, vector spaces, and linear algebra.

17. Hilgevoord, J., *Dispersion Relations and Causal Description*, Amsterdam: North-Holland Pub. (1960).

18. Hille, E., *Analytic Function Theory*, New York: Chelsea Publishing Co. (1973).

19. Ince, E. L., *Integration of Ordinary Differential Equations*, New York: Wiley-Interscience (1956).

20. Jackson, J. D., *Classical Electrodynamics*, New York: Wiley (1975).
Lots of applications of the theory of partial differential equations.

21. Jahnke, E., Emde, F., and Losch, F., *Tables of Higher Functions*, New York: McGraw-Hill (1960).
The drawings that supplement the tables can be immensely helpful in visualizing functional behavior.

22. Jeffreys, H., and Jeffreys, B. S., *Methods of Mathematical Physics*, New York: Cambridge Univ. Press (1956).
Chapter 17 contains a detailed discussion of asymptotic expansions and saddle point methods.

23. Kronig, R. de L., *J. Opt. Soc. Amer. Rev. Sci. Inst.* 12:547 (1926).

24. Landau, L. D., and Lifshitz, E. M., *Quantum Mechanics: Non-Relativistic Theory*, Elmsford, N.Y.: Pergamon Press (1977).
Chapter VI of this book gives a readable discussion of applications of perturbation theory in quantum mechanics.

25. Mathews, J., and Walker, R., *Mathematical Methods of Physics*, Reading, Mass.: Benjamin (1970).
Contains a more advanced treatment of the applications discussed in this text, as well as coverage of a number of additional topics.

26. Merzbacher, E., *Quantum Mechanics*, New York: Wiley (1970).
Good treatment of Schrödinger equation problems, and of the WKB method.

27. Morse, P. M., and Feshbach, H., *Methods of Theoretical Physics* (2 Vols), New York: McGraw-Hill (1953).
Chapters 6 and 7 of this book give one of the most complete treatments of Green's functions (and related applications of eigenfunctions) to be found.
Because it is so complete, the casual reader will find it difficult. However, the presentation is both beautiful and clear, if one has ample time to spend. Chapter 5 contains a great deal of material on integral representations of special functions, and Chapter 4 has a discussion of saddle point methods.

28. Nussenzweig, H. M., *Causality and Dispersion Relations*, Vol. 95 Mathematics in Science and Engineering. New York: Academic Press (1972).

29. Schiff, L., *Quantum Mechanics*, New York: McGraw-Hill (1968).

30. Seshu, S., and Balabanian, N., *Linear Network Analysis*, New York: Wiley (1959).

31. Sommerfeld, A., *Partial Differential Equations in Physics*, New York: Academic Press (1967).

32. Volkovyskii, L. I., Lunts, G. L., and Aramovich, I. G., *A Collection of Problems on Complex Analysis*, Elmsford, N.Y.: Pergamon Press (1965).

33. Watson, G. N., *Proc. Roy. Soc. London* 95: 83 (1918).

34. Whittaker, E. T., and Watson, G. N., *A Course of Modern Analysis*, New York: Cambridge Univ. Press (1958).
Very good for mathematical properties of functions, infinite series (Chapters II and III), and asymptotic expansions (Chapter VIII).

35. Wyld, H. W., *Mathematical Methods for Physics*, Reading, Mass.: Benjamin (1976).
Further development of most of the topics of our text can be found here.

Index